DATE DUE

Chapman & Hall/CRC Mathematical and Computational Biology Series

Pattern Discovery in Bioinformatics

Theory & Algorithms

CHAPMAN & HALL/CRC
Mathematical and Computational Biology Series

Aims and scope:
This series aims to capture new developments and summarize what is known over the whole spectrum of mathematical and computational biology and medicine. It seeks to encourage the integration of mathematical, statistical and computational methods into biology by publishing a broad range of textbooks, reference works and handbooks. The titles included in the series are meant to appeal to students, researchers and professionals in the mathematical, statistical and computational sciences, fundamental biology and bioengineering, as well as interdisciplinary researchers involved in the field. The inclusion of concrete examples and applications, and programming techniques and examples, is highly encouraged.

Series Editors
Alison M. Etheridge
Department of Statistics
University of Oxford

Louis J. Gross
Department of Ecology and Evolutionary Biology
University of Tennessee

Suzanne Lenhart
Department of Mathematics
University of Tennessee

Philip K. Maini
Mathematical Institute
University of Oxford

Shoba Ranganathan
Research Institute of Biotechnology
Macquarie University

Hershel M. Safer
Weizmann Institute of Science
Bioinformatics & Bio Computing

Eberhard O. Voit
The Wallace H. Couter Department of Biomedical Engineering
Georgia Tech and Emory University

Proposals for the series should be submitted to one of the series editors above or directly to:
CRC Press, Taylor & Francis Group
24-25 Blades Court
Deodar Road
London SW15 2NU
UK

Published Titles

Cancer Modelling and Simulation
Luigi Preziosi

Computational Biology: A Statistical Mechanics Perspective
Ralf Blossey

Computational Neuroscience: A Comprehensive Approach
Jianfeng Feng

Data Analysis Tools for DNA Microarrays
Sorin Draghici

Differential Equations and Mathematical Biology
D.S. Jones and B.D. Sleeman

Exactly Solvable Models of Biological Invasion
Sergei V. Petrovskii and Bai-Lian Li

Introduction to Bioinformatics
Anna Tramontano

An Introduction to Systems Biology: Design Principles of Biological Circuits
Uri Alon

Knowledge Discovery in Proteomics
Igor Jurisica and Dennis Wigle

Modeling and Simulation of Capsules and Biological Cells
C. Pozrikidis

Niche Modeling: Predictions from Statistical Distributions
David Stockwell

Normal Mode Analysis: Theory and Applications to Biological and Chemical Systems
Qiang Cui and Ivet Bahar

Pattern Discovery in Bioinformatics: Theory & Algorithms
Laxmi Parida

Stochastic Modelling for Systems Biology
Darren J. Wilkinson

The Ten Most Wanted Solutions in Protein Bioinformatics
Anna Tramontano

Chapman & Hall/CRC Mathematical and Computational Biology Series

Pattern Discovery in Bioinformatics

Theory & Algorithms

Laxmi Parida

Chapman & Hall/CRC
Taylor & Francis Group

Boca Raton   London   New York

Chapman & Hall/CRC is an imprint of the
Taylor & Francis Group, an informa business

Chapman & Hall/CRC
Taylor & Francis Group
6000 Broken Sound Parkway NW, Suite 300
Boca Raton, FL 33487-2742

© 2008 by Taylor & Francis Group, LLC
Chapman & Hall/CRC is an imprint of Taylor & Francis Group, an Informa business

No claim to original U.S. Government works
Printed in the United States of America on acid-free paper
10 9 8 7 6 5 4 3 2 1

International Standard Book Number-13: 978-1-58488-549-8 (Hardcover)

Library of Congress Cataloging-in-Publication Data

Parida, Laxmi.
    Pattern discovery in bioinformatics / Laxmi Parida.
        p. ; cm. -- (Chapman & Hall/CRC mathematical and computational biology
    series)
    Includes bibliographical references and index.
    ISBN-13: 978-1-58488-549-8 (alk. paper)
    ISBN-10: 1-58488-549-1 (alk. paper)
    1. Bioinformatics. 2. Pattern recognition systems. I. Title. II. Series: Chapman
and Hall/CRC mathematical & computational biology series.
    [DNLM: 1. Computational Biology--methods. 2. Pattern Recognition,
Automated. QU 26.5 P231p 2008]
    QH324.2.P373 2008
    572.80285--dc22   2007014582

Visit the Taylor & Francis Web site at
http://www.taylorandfrancis.com

and the CRC Press Web site at
http://www.crcpress.com

Dedicated to Ma and Bapa

Contents

Acknowledgments

I owe the completion of this book to the patience and understanding of Tuhina at home and of friends and colleagues outside of home. I am particularly grateful for Tuhina's subtle, quiet cheer-leading without which this effort may have seemed like a thankless chore.

Behind every woman is an army of men. My sincere thanks to Alberto Apsotolico, Saugata Basu, Jaume Bertranpetit, Andrea Califano, Matteo Comin, David Gilbert, Danny Hermelin, Enam Karim, Gadi Landau, Naren Ramakrishnan, Ajay Royyuru, David Sankoff, Frank Suits, Maciej Trybilo, Steve Oshry, Samaresh Parida, Rohit Parikh, Mike Waterman and Oren Weimann for their, sometimes unwitting, complicity in this endeavor.

Chapman & Hall/CRC Mathematical and Computational Biology Series

Pattern Discovery in Bioinformatics

Theory & Algorithms

Chapter 1

Introduction

Le hasard favorise l'esprit preparé. [1]
- attributed to Louis Pasteur

1.1 Ubiquity of Patterns

Major scientific discoveries have been made quite by accident: however a closer look reveals that the scientist was intrigued by a specific *pattern* in the observations. Then some diligent persuasion led to an important discovery. A classic example is that of the English doctor from Gloucestershire, England, by the name of Edward Jenner. His primary observation was that milkmaids were immune to smallpox even though other family members would be infected with the disease. The milkmaids were routinely exposed to cowpox and subsequently Jenner's successful experiment of inducing immunity to smallpox in a little boy by first infecting him with cowpox led to the world's first smallpox vaccination. A sharp observation in 1796 ultimately led to the eradication of smallpox on this planet in the late 1970s.

A more recent story (1997) that has caught the attention of scientists and media alike is the story of a group of Nairobi women, immune to AIDS. While researchers pondered the possibility of these women acquiring immunity from the environment (like the case of cows for smallpox), a chance conversation of the attending doctor Dr. Joshua Kimani with the immune patients revealed that about half of them were close relatives. This sent the doctors scrambling to look for genetic similarity and the discovery of the presence of 'killer' T-cells in the immune system of these women. This has led researchers in the path of exploring a vaccine for AIDS.

Stories abound in our scientific history to suggest that these chance observations are key starting points in the process of major discoveries.

---

[1] Chance favors the prepared mind.

1

1.2   Motivations from Biology

The biology community is inundated with a large amount of data, such as the genome sequences of organisms, microarray data, interactions data such as gene-protein interactions, protein-protein interactions, and so on. This volume is rapidly increasing and the process of understanding the data is lagging behind the process of acquiring it. The sheer enormity of this calls for a systematic approach to understanding using (quantitative) computational methods. An inevitable first step towards making sense of the data is to study the regularities and hypothesize that this reveals vital information towards a greater understanding of the underlying biology that produced this data.

In this compilation we explore various modes of regularities in the data: string patterns, patterned clusters, permutation patterns, topological patterns, partial order patterns, boolean expression patterns and so on. Each class captures a different form of regularity in the data enabling us to provide possible answers to a wide range of questions.

1.3   The Need for Rigor

Unmistakeably, the nature of the subject of biology has changed in the last decades: the transition has been from 'what' to 'how'.

Just as a computer scientist or a mathematician or a physicist needs to have a fair understanding of biology to pose meaningful questions or provide useful answers, so does a biologist need to have an understanding of the computational methods to employ them correctly and provide the correct answers to the difficult questions.

While the easy availability of search engines makes access to exotic as well as simple-minded systems very easy, it is unclear that this is always a step forward. The burden is on the user to understand how the methods work, what problems they solve and how correct are the offered answers. This book aims at clarifying some of the mist that may accompany such encounters.

One of the features of the treatment in the book is that in each case, we attack the problem in a model-less manner. Why is this any good? As application domains change the underlying models change. Often our existing understanding of the domain is so inadequate that coming up with an appropriate model is difficult and sometimes even misleading. So much so that there is little consensus amongst researchers about the correct model. This has prompted many to resort to a model-less approach. Often the correct model can be used to refine the existing system or appropriately pre- or post-process the data. The model-less approach is not to be misconstrued as neglect of the domain specifications, but a tacit acknowledgment that each domain deserves much closer attention and elaborate treatment that goes well beyond the scope of this book. Also, this approach compels us to take a hard look at the problem domain, often giving rise to elegant and clean definitions with a sound mathematical foundation as well as efficient algorithms.

1.4 Who is a Reader of this Book?

This book is intended for biologists as well as for computer scientists and mathematicians. In fact it is aimed at anyone who wants to understand the workings and implications of pattern discovery. More precisely, to appreciate the contents of this book, it is sufficient to be familiar with the prerequisites of a regular bioinformatics course.

Often some readers are turned off by the use of terms such as 'theorem' or 'lemma', however the book does make use of these. Let me spend a few words justifying the use of these words and at the same time encouraging the community to embrace this vocabulary. Loosely speaking, a theorem is a statement accepted or proposed as a demonstrable truth. Once proven, the validity of the statement is unquestioned. It provides a means for organizing one's thoughts in a reusable manner. This mechanism that the mathematical sciences has to offer is so compelling that it will be a mistake not to assimilate it in this collection of logical thoughts.

Clearly, it is easier to have a theorem in mathematics than in the physical sciences. Most of the theorems in this book are to be simply viewed as *concise factual statements*. And, the *proof* is merely a justification for the claims. Lemmas, though traditionally used as supporting statements for theorems, is used here for simpler claims. All the proofs in this book require the logical thinking at the level of a college freshman, albeit a motivated one.

The proofs of the theorems and lemmas are given for a curious and suspicious reader. However, no continuity in thought is lost by skipping the proofs.

If I could replace a theorem with an example, I did. If I could replace an exposition with an exercise, I did. *An illustrative example is worth a thousand words and an instructive exercise is worth a thousand paragraphs.* I have made heavy use of these two tools throughout the book in an attempt to convey the underlying ideas.

The body of the chapter and the exercise problems accompanying it have a spousal relationship: one is incomplete without the other. These problems are not meant to test the reader but provide supplemental points for thought. Each exercise is designed carefully and simply requires 'connecting the dots' on part of the reader, while the body of the chapters explains the 'dots'. The challenging exercises (and sections) are marked with **.

1.4.1   About this book

Consider the scenario of getting a sculptor and a painter together to produce a work of art. Usually, the sculptor will not paint and the painter will not sculpt, however the results (if any) of their synergy could be incredibly spectacular!

Interdisciplinary areas such as bioinformatics must deal with such issues where each expert uses a different language. Establishing a common vocabulary across disciplines is ideal but has not been very realistic. Even sub-disciplines such as 'systems', 'computational logic', 'artificial intelligence' and so on, within the umbrella discipline of computer science, have known to have developed their very own dialects. Many seasoned researchers may also have witnessed in their lifetimes the rediscovery of the same theorems in different contexts or disciplines.

Sometimes, the problem is compounded by the fact that the sculptor dabbles in paint and the painter in clay. I believe I am cognizant of the agony and the ecstasy of cross-disciplinary areas. Yet I write these chapters.

Roadmap of the book.   The book is organized in three parts. Part I provides the prerequisites for the remainder of the book. Chapters 2 and 3 are designed as journeys, with a hypothetical but skeptical biologist, that takes the reader through the corridors of algorithms and statistics. We follow a story-line, and the ideas from these areas are presented on a need-to-know basis for a leery reader.

Chapter 4 discusses the connotation of *patterns* used in this book. The nuance of repetitiveness that is associated with patterns in this book is not universal and we reorient the reader to this view through this chapter.

Part II of the book focuses on patterns on linear (string) data. Chapter 5 discusses possible statistical models for such data. Patterns on strings are conceptually the simplest and the ramifications of these are discussed in Chapters 6 and 7.

String patterns are simple, yet complex! This is explored in the following two chapters. Chapter 8 discusses different (probabilistic) motif learning methods where the pattern or motif is a consequence of local multiple alignment. Chapter 9 focuses on methods, primarily combinatorial, where the pattern or motif is viewed as the (inexact) consensus of multiple occurrences.

Part III of the book deals with patterns that have more sophisticated specifications. The complexity is in the characterizations but not necessarily in the implications. A string pattern on DNA may have as strong a repercussion as any other.

Permutation patterns are mathematically elegant, algorithmically interesting, statistically challenging and biologically meaningful as well! A well-rounded topic as this is always a delight and is discussed in Chapters 10 and 11.

Topological or network motifs are common structures in graph data. This area of motifs is relatively new and I have met more researchers who are skeptical about the import of these than any other. I give a slightly unusual treatment, in the sense that it is not based on graph traversals as is usually the case, of this problem in Chapter 12.

Chapter 13 is an attempt at identifying some common tools that could be utilized in most areas of pattern discovery. These include mainly structures and operations on finite sets.

The book concludes with a discussion of even more exotic pattern characterizations in the form of boolean expressions and partial orders in Chapter 14.

Part I

The Fundamentals

Chapter 2

Basic Algorithmics

The worth of an artist is measured
by the sharpness of her tools.
- based on a Chinese proverb

2.1 Introduction

To keep the book self-contained, in this chapter we review the basic mathematics and algorithmics that is required to understand and appreciate the material in the remainder of the book. To give a context to some of the abstract ideas involved, we follow a storyline. This is also an exercise in understanding an application, abstracting the essence of the task at hand, formulating the computational problem and designing and analyzing an algorithm for solving the computational problem. The last (but certainly not the least) part of the task is to implement the algorithm and analyze its performance in a real world setting.

Consider the following scenario. Professor Marwin has come across a gem of a bacterium: a *chromo-bacterium* that is identified by its color and every time a mutation occurs, it takes on a new color. Leaving her bacteria culture with a generous source of nutrition, she proceeds on a two week vacation. On her return, she is confronted with K distinct colored strains of bacteria. A closer look reveals that each bacterium has a genome of size m and with only mutational differences between the genomes. She is intrigued by this and is eager to reconstruct the evolution of the bacteria in her absence.

2.2 Graphs

A *graph* is a common abstraction that is used to model binary relationships between objects. More precisely, a graph G is a pair (V, E), consisting of a set V and a subset

$$E \subset (V \times V)$$

and we denote the graph by
$$G(V, E).$$

The elements of V are called *nodes* or *vertices* and those of E are called the *edges* of the graph. A graph is finite if its set of vertices (and hence edges) is finite. In this book all graphs considered will be finite. Further, each edge can be annotated with labels or numerical quantities. The latter is usually called the *weight* of the edge.

Professor Marwin chooses to use graphs to model her bacteria strains and their interrelationships. Quite naturally, she models the strains as the nodes in the graph and the edges between the nodes as an evolutionary distance between the two. Roughly speaking, she is seeking the simplest explanation of evolution of the bacterial strains.

However, one cannot even begin to seek a method or algorithm before defining the problem. Of course, problems in biology are usually difficult: they may not even have a precise mathematical definition. But it is essential to recognize what can and what can not be done. This goes a long way in focussing on the correct interpretations and avoiding any misrepresentation of the results.

A method or an algorithm can only be defined for a precise problem P. However P may be an approximation of the real biological problem at hand. So Professor Marwin gets down to the most important task and that is of recognizing the essence of the computational task. She identifies the following problem:

Problem 1 (Connected graph with minimum weight) *Given distinct sequences*
$$s_i, \ 1 \leq i \leq K,$$

each of size m, let d_{ij} be the number of base differences between sequences s_i and s_j. Consider a weighted graph

$$G(V, E) \ with \ |V| = K$$

and edge weights
$$wt(v_i v_j) = d_{ij}$$

where
$$v_i, v_j \in V \ and \ (v_i v_j) \in E.$$

Weight of $G(V, E)$ is defined as:

$$W_G = \sum_{(v_i v_j) \in E} wt(v_i v_j)$$

The task is to find a connected graph G^* with minimum weight (over all possible connected graphs on V).

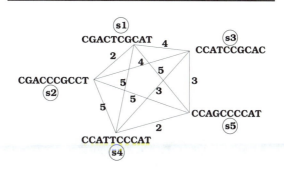

FIGURE 2.1:   The graph constructed with the data from Example (1). A vertex v_i is labeled with sequence s_i, $1 \leq i \leq 5$. The weight associated with edge $(v_i v_j)$ is the distance d_{ij}.

The edge weight d_{ij} denotes the distance between two genomes s_i, s_j, or the number of mutations that takes genome s_i to s_j or vice-versa. Clearly, all the K colored groups of bacteria that Professor Marwin observes did not evolve independently but from the one strain that she started with. Thus the connectivity of the graph G, where each node is a genome, represents how the genomes evolved over a period of time. Given different choices of graphs that explain the evolution process, the one(s) of most interest would be the ones that have the smallest weight since that would be considered as the simplest explanation of the process. Consider an instance of this problem in the following example.

Example 1 *Each genome is a sequence defined on the four nucleotides, Adenine, Cytosine, Guanine and Thymine. Consider Problem (1) with $K = 5$ and the input genomic sequences as:*

$$s_1 = CGACTCGCAT$$
$$s_2 = CGACCCGCCT$$
$$s_3 = CCATCCGCAC$$
$$s_4 = CCATTCCCAT$$
$$s_5 = CCAGCCCCAT$$

See Figure 2.1 for the graph corresponding to this data.

Before proceeding further, we pin down a few definitions to avoid ambiguity. Given a graph

$$G(V, E) \text{ and } V' \subseteq V,$$

we call

$$G(V', E'),$$

the subgraph *induced* by V' where E' is defined as

$$\{(v_1 v_2) \in E \mid v_1, v_2 \in V'\}.$$

Subgraph

$$G'(V', E')$$

is *induced* by E' when given $E' \subseteq E$, V' is defined as

$$\{v_i \in V \mid (v_i, v_j) \in E'\}.$$

A *path* in the graph $G(V, E)$ between vertices

$$v_1, v_2 \in V$$

is a sequence of vertices,

$$(w_0 = v_1), w_1, \ldots, (w_m = v_2)$$

such that for each

$$i, 0 \le i < m, \ (w_i w_{i+1}) \in E.$$

A graph is *connected* when there exists a path between any two pair of vertices. A *connected component* of a graph is an induced subgraph on a maximal set of vertices $V' \subseteq V$, such that the induced graph $G'(V', E')$ is connected.

Next, a careful study of Problem (1) reveals:

1. Each vertex $v_i \in V$ corresponds to sequence s_i.

2. $d_{ij} > 0$ for each $i \ne j$, since the sequences are *distinct*.

Determining the structure of G: Consider $G(V, E)$ with

$$E = \phi.$$

Then

$$wt(G) = 0$$

and this is the smallest possible weight of a graph. But the graph is not connected, and hence this is an incorrect solution. However this failure suggests that the solution needs to introduce as small a number of edges as possible so as to make the graph connected. Seeking the smallest (or simplest) explanation for a problem is often called the *Occam's razor principle* or the *principle of parsimony*.

Another way to look at the problem is to start with a completely connected graph

$$G(V, E),$$

i.e, each

$$(v_i v_j) \in E \text{ where } i \ne j,$$

and remove the redundant edges. Some careful thought leads to the conclusion, that this graph must have no *cycles*, i.e., no closed paths. This is because an edge can be removed from the cycle, maintaining the connectivity of the graph but reducing the number of edges. These observations lead to the following definition.

DEFINITION 2.1 **(tree, leaf node)** *A connected graph*

$$G(V, E)$$

having the property that any two vertices

$$v_1, v_2 \in V$$

has a unique path connecting them is called an acyclic *graph or a* tree. *A vertex with one incident edge is called a leaf node, all other vertices are called internal nodes.*

Does every tree have internal nodes? Consider a tree with only two vertices, all the vertices have only one incident edge. Hence this tree has no internal edge. Does every tree have a leaf node? The following lemma guarantees that it does.

LEMMA 2.1
(mandatory leaf lemma) *A tree $T(V, E)$ must have a leaf node.*

PROOF If
$$|V| \leq 2,$$
clearly all the nodes are leaves. Assume
$$|V| > 2.$$
We give a constructive argument for this claim. Let
$$V = \{v_1, v_2, v_3, \ldots, v_n\}.$$
Consider an arbitrary
$$v_{i_0} \in V.$$
If v_{i_0} is a leaf node, we are done. Assume it is not. Consider v_{i_1} where
$$(v_{i_0} v_{i_1}) \in E.$$
Again if v_{i_1} is not a leaf node, since v_{i_1} has degree at least two, there exists
$$v_{i_2} \neq v_{i_0}$$

with

$$(v_{i_1} v_{i_2}) \in E.$$

Since the graph is finite and has no cycles [1] this process must terminate with a leaf node.   ▯

We conclude that the structure of the graph we seek is a tree. Of course, this tree must denote the smallest number of changes, again by the Occam's razor principle.

2.3   Tree Problem 1: Minimum Spanning Tree

It is best to start with a naive approach to obtaining the tree with minimum weight: we simply enumerate all the trees and pick the one(s) with the least weight. It is easy to see that this algorithm of searching the entire space of spanning trees will yield the correct solution(s).

The next question to address is: How good is the algorithm? One of the important criteria is the time the algorithm takes to complete the task. There are others like the space requirements and so on, but here we focus only on the running time.

The running time is usually computed as a function of the size of the input to the problem. Here we have K sequences, each of size m. First, we focus on computing the d_{ij}'s.

Time to compute the d_{ij}'s: We assume that reading an element takes time c_r, and a comparison operation takes time c_c and any arithmetic operation takes time c_a. Each of this is a constant depending on the environment in which they are carried out. Roughly speaking, a comparison of two sequences of length m each involves the following operations:

1. initializing a count to 0,

2. reading two elements, one from each of the arrays,

3. comparing the two values and incrementing the counter by 0 or 1.

This takes time

$$c_a + m(2c_r + c_c + c_a).$$

The only factor that will change with the instance of the problem is the one that involves m, the rest being constants. The main concern is the growth

---

[1]Consider the sequence of vertices in the traversal: $v_{i_0} v_{i_1} v_{i_2} v_{i_3} \ldots v_{i_k}$. If a vertex appears multiple times in the traversal, then clearly the graph has a cycle.

of this function with m. To avoid clutter and focus on the most meaningful factor, we use an *asymptotic notation* (see Section 2.7) and the time is written as

$$c_a + m(2c_r + c_c + c_a) = \mathcal{O}(m)$$

We use a big-oh notation $\mathcal{O}(\cdot)$ which is explained in Section 2.7. In other words, all the constants are ignored. The time grows linearly with the size m and that is all that matters.

Since there are K distinct sequences,

$$\frac{K(K-1)}{2} - 1$$

comparisons are made. Again using the asymptotic notation, ignoring linear factors (K) in favor of faster growing quadratic factors (K^2), the number of comparisons is written as

$$\mathcal{O}(K^2).$$

Thus since

$$\mathcal{O}(K^2)$$

comparisons are made and each comparison takes

$$\mathcal{O}(m)$$

time, the time to compute the d_{ij}'s takes time

$$\mathcal{O}(K^2 m).$$

We first assert the following.

LEMMA 2.2
(Edge-vertex lemma) *Given a connected graph $G(V, E)$,*

$$(G \text{ is a tree}) \implies |E| = |V| - 1.$$

PROOF   This is easily shown by induction on the number of vertices n. Consider the base case, $n = 2$.

Clearly a tree with two vertices has only one edge, thus the result holds for the base case. Assume the result holds for $n > 2$, i.e., a tree with n vertices has $(n - 1)$ edges.

Consider a connected tree T with $(n + 1)$ vertices. Let T' be a tree with n vertices obtained from T by deleting a leaf node v. Such a vertex (leaf node) exists by Lemma (2.1). By the induction hypothesis, T' has $(n - 1)$ edges. v has only one edge incident on it, thus T has $(n - 1) + 1$ edges.   ⬚

COROLLARY 2.1
Given a tree $T(V, E)$, removing an edge (keeping V untouched) disconnects the tree.

Back to the algorithm. Since this algorithm involves searching the entire space of spanning trees, we next count the total number of such trees with K vertices given as N. This number N is given by Cayley's formula:

$$N = K^{(K-2)}.$$

However, this is not the end of the story, since we must devise a way to enumerate these N configurations. This is discussed in Section 2.8.2. To summarize, the time taken by the naive algorithm is

$$\mathcal{O}(K^K + mK^2),$$

a function that grows so rapidly with K that this algorithm is unacceptable for all practical purposes! Thus though the algorithm solves the problem correctly, its time complexity is too large for it to be of any use.

Recall that the component that finds the tree of the minimum weight in the naive algorithm is too inefficient to be of practical use. So we next focus on this problem identified as follows. Recall that the weight of a graph (or a tree) is the sum of the weights of all the edges on the graph (or tree).

Problem 2 *(Minimum Spanning Tree (MST))* *Given a weighted graph*

$$G(V, E)$$

with nonnegative weights on the edges, a spanning tree

$$T(V', E')$$

of $G(V, E)$ is such that

1. *$V' = V$,*

2. *$E' \subseteq E$ and*

3. *T is a tree.*

The task is to find a tree

$$T^*$$

of minimum weight amongst all possible spanning trees.

2.3.1 Prim's algorithm

The MST problem is well studied in literature [CLR90] and has a very elegant solution. We present an algorithm below based on the following observation: If a set of edges E_s disconnects the given graph, then the edge with the smallest weight in E_s must be in the MST. Formally put,

LEMMA 2.3
(Bridge lemma) *Given graph $G(V, E)$, let*

$$E_s \subseteq E$$

be such that the graph induced by

$$(E \setminus E_s)$$

on $G(V, E)$ is not connected, then an edge satisfying

$$wt(v_1 v_2) = \min_{(v_i v_j) \in E_s} wt(v_i v_j)$$

is such that

$$(v_1 v_2) \in (E_s \cap E^*)$$

where

$$T^*(V, E^*)$$

is a minimum spanning tree.

PROOF This is easily shown by contradiction. Let E_s be such that it splits the vertex set into two as

$$V = V_1 \cup V_2$$

with

$$V_1 \cap V_2 = \phi.$$

Assume the result is not true i.e.,

$$(v_1 v_2) \notin E^*.$$

Since T^* is connected, it must contain an edge

$$(v_1' v_2') \in E_s$$

with

$$wt(v_1' v_2') > wt(v_1 v_2).$$

Construct T' from T^* by deleting

$$(v_1' v_2').$$

Clearly the subgraph $T_1'(V_1, E_1)$ induced by V_1 on T^* is connected and acyclic and so is the subgraph induced by V_2,

$$T_2'(V_2, E_2), \text{ where } E_1, E_2 \subseteq E^*.$$

Next we add the edge $(v_1 v_2)$ to T' that now connects the subgraphs T_1' and T_2' without introducing cycles since $v_1 \in V_1$ and $v_2 \in V_2$, without loss of generality. As a result T' is acyclic and connected, hence a tree. But

$$wt(T') < wt(T^*),$$

which is a contradiction, hence the assumption must be wrong and

$$(v_1 v_2) \in E^*.$$

□

LEMMA 2.4
(Weakest link lemma) *The converse of Lemma (2.3) also holds true. In other words, given a minimum spanning tree*

$$T^*(V, E^*)$$

of a graph $G(V, E)$, consider an edge,

$$v_1 v_2 \in E^*,$$

and let the two connected components of T^* obtained by deleting the edge (v_1, v_2) be

$$T_k^*(V_k^*, E_k^*), \quad k = 1, 2.$$

$G^k(V^k, E^k)$ are the two subgraphs of $G(V, E)$ induced by E_k^* with

$$v_1 \in V^1 \text{ and } v_2 \in V^2.$$

Then

$$wt(v_1 v_2) = \min_{(v_i v_j) \in E, v_i \in V^1, v_j \in V^2} wt(v_i v_j)$$

From lemma to algorithm. This lemma and its converse can be used to design a straightforward algorithm (Algorithm (1)): we progressively construct an E_s in every step as we build E^* and V^*. This algorithm is also called Prim's algorithm [CLR90]. It is now straightforward to prove the correctness of the algorithm.

LEMMA 2.5
Algorithm (1) correctly computes the minimum spanning tree.

PROOF At each iteration of the algorithm

$$E_s \neq \phi$$

since

$$|V^*| \leq |V|.$$

Thus exactly one edge is added to E^*. By Lemma (2.2), T^* has

$$|V| - 1$$

edges and the algorithm is iterated $(|V| - 1)$ times. Thus T^* has $(|V| - 1)$ edges and by Lemma (2.3), each edge added to E^* is in the minimum spanning tree, hence the algorithm is correct. □

Figure 2.2 illustrates the algorithm on the graph of Example (1). Using the asymptotic notation, Step (0) takes time

$$\mathcal{O}(1).$$

Step 1 takes

$$\mathcal{O}(|E|)$$

time, Step (2) takes $\mathcal{O}(1)$ and Step 3 takes

$$\mathcal{O}(E)$$

time. Since Steps 1-3 are repeated $|V|$ times, the algorithm takes time

$$\mathcal{O}(|V||E|).$$

The running time complexity can be improved by using efficient data structures for Step (3) of the algorithm. However, for this exposition we stay content with time complexity of

$$\mathcal{O}(|V||E|).$$

Algorithm 1 *Minimum Spanning Tree Algorithm*

> *(0) $E^* \leftarrow \phi$, $V^* \leftarrow \phi$, $E_s \leftarrow E$*
>
> *FOR $i = 1 \ldots$ to $(|V| - 1)$*
>   *(1) Let $wt(v_1 v_2) = \min_{(v_i v_j) \in E_s} wt(v_i v_j)$*
>   *(2) $V^* \leftarrow V^* \cup \{v_1, v_2\}$, $E^* \leftarrow E^* \cup \{(v_1, v_2)\}$*
>   *(3) $E_s \leftarrow \{(v_i v_j) \mid (v_i \in V^*) \text{ AND } (v_j \notin V^*)\}$*

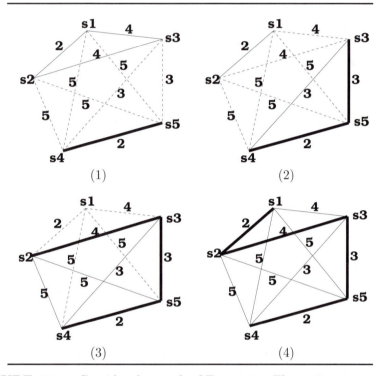

FIGURE 2.2:   Consider the graph of Figure 2.1. The minimum spanning tree (MST) algorithm on this graph is shown here. (1)-(4) denote the steps in the algorithm. At each step the edges shown in bold are the ones that constitute the spanning tree and the collection of edges E_s is shown as dashed edges. The MST is shown by the bold edges in (4).

2.4 Tree Problem 2: Steiner Tree

We now change the problem scenario of Section 2.1 slightly. After a very careful set of observations over a period of time, Professor Marwin notices that once a chromo-bacterium mutates into another color, after a while the original colored bacterium vanishes from the colonies. In other words, at any time when K distinct colors are noticed, then there had been some more K' colors that are no longer present. Problem (1) is modified as follows:

Problem 3 *(Steiner tree) Given distinct sequences*

$$s_i, \ 1 \le i \le K,$$

each of size m each, let d_{ij} be the number of base differences between sequences s_i and s_j. Consider a weighted graph

$$G(V, E)$$

with

$$|V| \ge K$$

and edge weights

$$wt(v_i v_j) = d_{ij} \ where \ v_i, v_j \in V.$$

Weight of $G(V, E)$ is defined as:

$$W_G = \sum_{(v_i v_j) \in E} wt(v_i v_j).$$

The task is to find a minimum weight tree

$$T^*(V, E)$$

with K leaf nodes corresponding to the K input sequences.

This problem requires internal nodes to correspond to new sequences that need to be constructed by the algorithm, to give a minimum weight tree. We explain this problem using Example (1) with the solution shown in Figure 2.3. The given nodes are at the leaves and the solution shows three reconstructed internal nodes.

This problem is called the Steiner tree problem and is known to be NP-hard. Section 2.9 gives a brief exposition on problem complexity. The conclusion is that the problem is hopelessly hard and it is unlikely that there exists an efficient (polynomial time) algorithm that will guarantee the optimal solution to this problem.

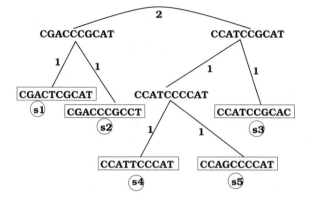

FIGURE 2.3:   The tree of minimum weight with the given vertices as leaf nodes labeled as s_1 to s_5.

Again, we first try a naive approach to solve the problem. As before, we count all possible trees that have the given K nodes as leaf nodes. This is already quite difficult so we count

$$Tb(K),$$

the number of *binary* or a bifurcating trees. In a binary tree, every internal node has exactly two children and one parent:[2] thus every internal node has degree three. If

$$Ta(K)$$

is the number of trees with K leaf nodes then

$$Tb(K) \le Ta(K).$$

We calculate the number $Tb(K)$ in Section 2.8.1.

2.5   Tree Problem 3: Minimum Mutation Labeling

We next identify the following problem. Let each node v be labeled by a character

$$\sigma \in \Sigma,$$

written as $L(v)$. In the following discussion we assume that $L(v)$ is a set of characters (instead of a single character) with the connotation that any of

---

[2]In a rooted tree, the only exception is the root which has no parents.

these characters can be assigned as a label to v to obtain an optimal assignment to the tree. Then the weight of the edge $(v_i v_j)$ is defined as:

$$d_{ij} = \begin{cases} 0 & \text{if } L(v_i) \cap L(v_j) \neq \phi, \\ 1 & \text{otherwise.} \end{cases}$$

Problem 4 *(**Minimum mutation labeling**) Given a tree $T(V, E)$ with K labeled leaf nodes, the task is to compute a labeling L so as to minimize the weight of T,*

$$W(T, L) = \sum_{v_i, v_j \in V} d_{ij}.$$

In the following discussion we assume that $L(v)$ is a set of characters (instead of a single character) and *each* of these characters can be assigned as a label to v to obtain an optimal assignment to the tree.

2.5.1 Fitch's algorithm

We use the following simple observation to design an algorithm for the problem.[3] It states that the optimal solution to the problem can be obtained from the optimal solution to the subproblems, which in this case are labeled nonoverlapping subtrees of the tree.

LEMMA 2.6
(Two-tree partition) *Let*

$$T(V, E)$$

be a tree with subtrees

$$T(V_1, E_1) \text{ and } T(V_2, E_2),$$

such that

$$V_1 \cap V_2 = \phi,$$

and

$$V = V_1 \cup V_2 \cup \{v_0\} \text{ where } v_0 \notin V_1, V_2.$$

Further

$$E = E_1 \cup E_2 \cup \{v_0 v_1, v_0 v_2\},$$

for fixed

$$v_i \in V_i, \ i = 1, 2.$$

Let the minimal weight of a tree T' be given as

$$W_{opt}(T').$$

---

[3]Strictly speaking, Fitch's algorithm was presented for a rooted bifurcating (binary) tree. We have generalized the principle here.

Then

$$W_{opt}(T_1) + W_{opt}(T_2) \leq W_{opt}(T)$$
$$\leq W_{opt}(T_1) + W_{opt}(T_2) + 1.$$

PROOF   We are given that the labeling of T_1 and T_2 are optimal. Clearly the following is not possible

$$W_{opt}(T) < W_{opt}(T_1) + W_{opt}(T_2).$$

If

$$L(v_1) \cap L(v_2) \neq \phi,$$

then by the labeling

$$L(v_0) \leftarrow L(v_1) \cap L(v_2),$$

we get

$$W_{opt}(T) = W_{opt}(T_1) + W_{opt}(T_2),$$

which is clearly the optimal, since the weight cannot be improved (reduced) any further.

If

$$L(v_1) \cap L(v_2) = \phi,$$

then by the labeling

$$L(v_0) \leftarrow L(v_1) \cup L(v_2),$$

we get

$$W_{opt}(T) = W_{opt}(T_1) + W_{opt}(T_2) + 1.$$

Again this is clearly optimal, since if it were not, then there exists a labeling of T_1, say, such that the new weight is less than the given weight, which is a contradiction.   □

The following lemma, is a more general form of Lemma (2.6) in the sense that the number of partitioning subtrees can be larger than 2.

LEMMA 2.7
(Multi-tree partition) *Let*

$$T(V, E)$$

be a tree with subtrees

$$T(V_i, E_i), \ 1 \leq i \leq p$$

such that

$$V_i \cap V_j = \phi, \ for \ 1 \leq (i \neq j) \leq p,$$

and for all i,

$$V = \bigcup_i \left(V_i \bigcup \{v_0\} \right), \text{ for } v_0 \notin V_i.$$

Also,

$$E = \bigcup_i \left(E_i \bigcup \{v_0 v_i\} \right),$$

for a fixed $v_i \in V_i$. In words, the tree T is the union of the nonoverlapping subtrees T_i and a vertex v_0 that is connected to each of the tree T_i at a vertex $v_i \in V_i$. Let

$$cnt(\sigma) = |\{v_i \mid 1 \le i \le p, \sigma \in L(v_i)\}|.$$

Then

$$\sum_{i=1}^{p} W_{opt}(T_i) \le W_{opt}(T)$$

$$\le \sum_{i=1}^{p} \left(W_{opt}(T_i) + (p - \max_{\sigma \in \Sigma}(cnt(\sigma))) \right).$$

The arguments for this lemma are along the lines of that of Lemma (2.6) and we skip the details here as they give no further insight into the problem than we already have.

From lemma to algorithm. Do the lemmas give an indication of the algorithm that can solve the problem? Actually, they do. This is a classic case of obtaining the optimal solution for a problem using optimal solutions to the subproblems. The task is to break the given tree into subtrees which can be labeled optimally and then build from there. The starting point is the collection of trees, the singleton nodes, the leaf nodes, which are optimally assigned by the given problem.

However, there is one catch. While we solve the subproblems, it is important to keep track of *all* the possible labelings of the roots, v_i of the subtrees T_i. The lemmas state that $W(T)$ is optimal but what is the guarantee that there is no other labeling of the nodes that gives the optimal solution? For example a suboptimal labeling of the internal nodes of T_1 such that

$$W(T_1) = W_{opt}(T_1) - 1$$

could give the optimal labeling for T. Let the suboptimal labeling be denoted by L'. Then

$$L'(v_1) \cap L(v_1) = \phi$$

and at least one other vertex

$$v_1' \ne v_1 \in V_1$$

is such that

$$L'(v_1') \neq L(v_1').$$

Next, we claim that a possible alternative labeling of

$$v_1' \in V_1$$

does not matter. This is because T_1 is a subtree and the only node that connects it to the remainder of the tree is v_1, and hence v_1' will never be considered in the future as well. This implies that the algorithm

1. works for a tree, but not necessarily for a graph, and,

2. gives some optimal labelings but not *all* optimal labelings of the internal nodes.

After understanding all the algorithmic implications of Lemmas (2.6), (2.7), we are now ready to present the algorithm. We first assign depth to every node as follows: Each leaf node v is assigned a

$$Depth(v) \leftarrow 0$$

and each nonleaf node v is assigned a depth $Depth(v)$ as the shortest path length to any leaf node. Let *maxdepth* be the maximum depth assigned to any vertex in the tree.

Algorithm 2 *Minimum Mutation Labeling*

> *(0-1) FOR EACH leaf node v with label σ_v,*
> $\qquad L(v) = \{\sigma_v\}$
> *(0-2) $Wt \leftarrow 0$*
>
> *FOR $d = 1, 2, \ldots, maxdepth$ DO*
> $\quad$ *FOR EACH $v \in V$ with $depth(v) = d$*
> $\quad U \leftarrow \{u \mid (uv) \in E, (depth(u) < d)\}$
> $\quad v(\sigma) = \{u \mid (u \in U) \text{ AND } (\sigma \in L(u))\}$
> $\quad L(v) \leftarrow \{\sigma' \mid v(\sigma') = \max_{\sigma \in \Sigma} |v(\sigma)|\}$
> $\quad Wt \leftarrow Wt + (|U| - \max_{\sigma \in \Sigma} |v(\sigma)|)$
> *ENDFOR*

This algorithm is simple but not necessarily the most efficient. The optimal weight is computed in the variable Wt. The depth of the nodes need not be explicitly computed and the tree can be traversed bottom-up from the leaves. For rooted bifurcating trees, this is also known as Fitch's algorithm. Figure 2.4 gives an example illustrating this simple and elegant algorithm.

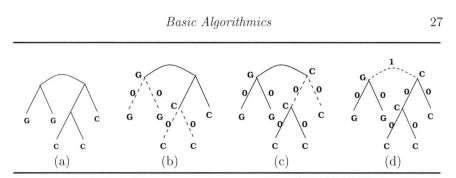

FIGURE 2.4: Labeling internal nodes so as to minimize the weight of the tree using Fitch's algorithm. The dashed lines denote the edges under consideration at that step of the algorithm. The edge is labeled with weight of 0/1 depending on the labels at the two end vertices. (a) Given tree T with leaf node labels as shown. (b) The parents of the labeled nodes are assigned labels. (c)-(d) The last step is continued until all nodes are labeled. For the given tree, $W_{opt}(T) = 1$.

2.6 Storing & Retrieving Elements

Encouraged by the success of her quest described in the previous sections, Professor Marwin decides to store the genomic data of her chromo-bacteria for future reference. For simplicity, we use only the first five bases of the genomic sequence. Her problem is formalized as follows.

Problem 5 *Let C be a collection of n elements*

$$e_{i_1} < e_{i_2} < e_{i_3} \ldots < e_{i_n}.$$

How efficiently can the existence of an arbitrary element e_i be checked (or retrieved) form C?

Supposing the elements are stored in an arbitrary order in the database. Then a search for e_i requires time

$$\mathcal{O}(n),$$

which is the time taken to linearly scan the collection.

Next, let us store the elements in the order as shown in an array A of size n. We probe the database as follows: split the n elements into two and test if e_i is in the first or the second collection by simply looking at the first element e' of the second collection, i.e.,

$$e' \leftarrow A\left[\frac{n}{2} + 1\right].$$

If

$$e_i \geq e'$$

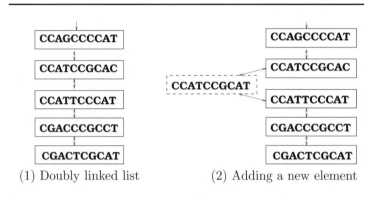

(1) Doubly linked list   (2) Adding a new element

FIGURE 2.5: Doubly liked linear list: Each element in the linked list has a pointer to the previous and the next element. The element shown in the dashed box is a new element that is added in lexicographic order as shown. This takes $\mathcal{O}(n)$ time.

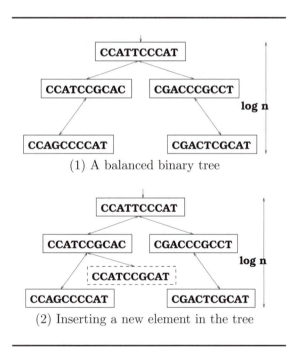

(1) A balanced binary tree

(2) Inserting a new element in the tree

FIGURE 2.6: The elements stored in a balanced binary tree. The element shown is a dashed box is the element being added to the tree, maintaining its balanced property. This takes $\mathcal{O}(\log n)$ time.

then the element is possibly in the second collection, otherwise it's possibly in the first. We repeat this process until either e_i is found or we run out of subcollections to search. The time taken for this process can be computed using the following recurrence equation

$$T(n) = \begin{cases} 1 & \text{if } n = 1 \\ T\left(\frac{n}{2}\right) + \mathcal{O}(1) & \text{if } n > 1 \end{cases}$$

Asymptotically,

$$T(n) = \mathcal{O}(\log n).$$

See Section 2.7 and Exercise 7.

Next Professor Marwin realizes that a new sequence needs to be added very often. It seem like a waste of effort to redo the whole computation every time. The problem is modified as follows.

Problem 6 (Element retrieval) *Let C be a dynamic collection of objects. How efficiently can the existence of an arbitrary element e_i be checked (or retrieved) from C?*

Since C is a dynamic collection, the elements cannot be stored in an array. The simplest alternative is to store them in a linked list as shown in Figure 2.5. But this list cannot be accessed through an index number. It is linearly traversed using the links each time. A new element is added by deleting and old link and updating the links of the previous and next elements appropriately. It is easy to see that it takes time

$$\mathcal{O}(n),$$

to insert and element. Can this time be improved?

Figure 2.6 shows a balanced binary tree. Each node stores element e_p in the tree, and has a left child that stores e_l and a right child that stores e_r where the (lexicographic) order holds:

$$e_l < e_p < e_p.$$

Balanced tree. Let the *height* of the subtree rooted at node p be the maximum number of edges it takes to reach a leaf node. For a node p let

$$H^p_{left}$$

denote the height of the tree rooted at its left child. Let the height of the right child be

$$H^p_{right}.$$

A tree is *balanced* if for every node p,

$$|H^p_{right} - H^p_{left}| \leq 1.$$

In other words a difference of at most one is allowed in the height of the left and the right subtrees of a node. In Figure 2.6, the root r (the vertex at the top) has

$$H^r_{right} = H^r_{left} = 2.$$

Every other node has a difference of one in the heights of its left and right child.

Thus when a new element is added, the tree has to remain balanced. The height of the tree is

$$\mathcal{O}(\log n).$$

In the example in Figure 2.6, the element is added in the correct lexicographic order and the tree continues to be balanced. In general, it is possible to make some local adjustments so that the tree is balanced. This balancing can be done in time

$$\mathcal{O}(\log n).$$

Thus inserting an element in a balanced binary tree takes time

$$\mathcal{O}(\log n).$$

Examples of such data structures are AVL trees (named after the authors Adelson-Velskii and Landis), 2-3 trees, B-trees, red-black trees and splay-trees [CLR90].

2.7   Asymptotic Functions

Often it is easier to represent an entity by a simpler, albeit less accurate, version that avoids clutter and helps focus on the appropriate features. For example consider a number

$$g = 205.42332122132122.$$

Let an acceptable approximation be f with only two digits after the decimal and [4]

$$|g - f| \leq 0.005.$$

The feature here is apparently the usability with acceptable monetary units. Denote this 'approximation' as Υ, then this can be written as

$$g = \Upsilon(f).$$

Note that f is an approximation to a set of g's.

We generalize this idea to a nonnegatively valued function $g(n)$. Note that this is not exactly the same notion. The similarity is only in replacing a complex function by a simpler one for further scrutiny. Here, our interest is

---

[4]If g was obtained as the interest computed for thirteen months on a sum of money and had to be paid to a client, then $f = 205.42$ is an acceptable approximation.

in studying the growth of this function with n- this is called the asymptotic behavior of the function

$$g(n).$$

In the context of algorithm analysis, various approximations are of interest, some of which are listed in Figure 2.7. We show five different forms of Υ:

(a) $\mathcal{O}(\cdot)$ read as 'big-oh',

(b) $o(\cdot)$ read as 'small-oh',

(c) $\Omega(\cdot)$ read as 'omega',

(d) $\omega(\cdot)$ read as 'small-omega' and

(e) $\Theta(\cdot)$ read as 'theta'.

Each has the following characteristics:

1. ignoring constants, since different machines would give a different number, that can be safely overlooked and

2. studying the function at infinity or very large n.

The 'big-oh' function of $f(n)$ denoted as

$$\mathcal{O}(f(n)),$$

is the most commonly used asymptotic notation and we take a closer look at its definition. $\mathcal{O}(f(n))$ is the set of functions $g(n)$ such that there exist ($\exists$) positive constants c and n_0, satisfying

$$0 \leq g(n) \leq cf(n),$$

for all ($\forall$)

$$n > n_0.$$

In other words, for sufficiently large n, there exists some constant $c > 0$ such that

$$cf(n)$$

is always larger than

$$g(n).$$

In practice, to obtain

$$\mathcal{O}(f(n)),$$

usually the highest order term, ignoring the constants suffices. For example,

$$g(n) = 507n^2 \log n + 36n + 2054$$
$$= \mathcal{O}(n^2 \log n).$$

notation		set of functions $g(n)$	$n \geq$	example $g(n)\ f(n)$
(a) $\mathcal{O}(f(n))$	$\exists\ c,\ n_0$	$0 \leq g(n) \leq cf(n)$	n_0	$10^6 n = \mathcal{O}(n)$
(b) $o(f(n))$	$\forall\ c,\ \exists\ n_c$	$0 \leq g(n) \leq cf(n)$	n_c	$n = o(n^2)$
(c) $\Omega(f(n))$	$\exists\ c,\ n_0$	$0 \leq cf(n) \leq g(n)$	n_0	$10^{-6} n = \Omega(n)$
(d) $\omega(f(n))$	$\forall\ c,\ \exists\ n_c$	$0 \leq cf(n) \leq g(n)$	n_c	$n^2 = \omega(n)$
(e) $\Theta(f(n))$	$\exists\ c_0,$ $c_1,\ n_0$	$0 \leq c_0 f(n) \leq g(n) \leq c_1 f(n)$	n_0	$10^6 n +$ $10^{10} = \Theta(n)$

FIGURE 2.7:   Different asymptotic notations of functions $g(n)$.

Sometimes, simple algebraic manipulations may help to obtain

$$\mathcal{O}(f(n))$$

in the most appropriate form. For instance,

$$\begin{aligned}
g(n) &= 507n^2 \log n + 36n \log^2 n + 2054 \\
&= \mathcal{O}(n^2 \log n + n \log^2 n) \\
&= \mathcal{O}(n \log n(n + \log n)) \\
&= \mathcal{O}(n^2 \log n).
\end{aligned}$$

For more intricate instances, refer to the definitions in Figure 2.7. A pictorial representation of some of the functions is shown in Figure 2.8 for convenience.

2.8   Recurrence Equations

A function that is defined in terms of itself is represented by a recurrence equation. For example, consider the factorial function

$$Fac(n) = n!$$

which is defined as the product of natural numbers

$$1, 2, 3, \ldots, n.$$

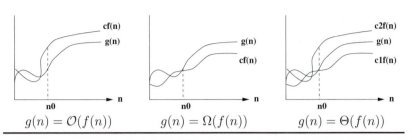

$$g(n) = \mathcal{O}(f(n)) \qquad g(n) = \Omega(f(n)) \qquad g(n) = \Theta(f(n))$$

FIGURE 3.8: The behavior of the asymptotic functions with respect to $g(n)$. Note that in each of the cases, for smaller n ($n \leq n_0$), the relative values of the functions does not matter. However, for large values of n, $g(n)$ must be consistently larger or smaller than the other functions as shown in the three cases.

The same function can also be written as:

$$Fac(n) = \begin{cases} 1 & \text{if } n = 1 \\ nFac(n-1) & \text{if } n > 1 \end{cases}$$

The recurrence equation has two parts:

1. the base case (here $n = 1$) and

2. the recurring case (here $n > 1$ case).

A useful asymptotic approximation of the factorial function is given by Stirling's formula:

$$Fac(n) = \sqrt{2\pi n} \left(\frac{n}{e}\right)^{n+o(1)} \tag{2.1}$$

In other words, for large n, $Fac(n)$ can be approximated by

$$\sqrt{2\pi n} \left(\frac{n}{e}\right)^n.$$

Consider the Fibonacci series of numbers which are generated by adding the last two numbers in the series:

$$0, 1, 1, 2, 3, 5, 8, 13, 21, \ldots.$$

The nth Fibonacci number given as

$$Fib(n),$$

can can be written as the following recurrence equation

$$Fib(n) = \begin{cases} n & \text{if } n = 0 \text{ or } n = 1, \\ Fib(n-1) + Fib(n-2) & \text{if } n > 1 \end{cases}$$

A recurrence form of this kind serves a concise way of defining the function. But a closed form is more useful for algebraic manipulations and comparison with other forms. The closed form expression for the Fibonacci number is:

$$Fib(n) = \frac{1}{\sqrt{5}} \left[\left(\frac{1 + \sqrt{5}}{2} \right)^n - \left(\frac{1 - \sqrt{5}}{2} \right)^n \right].$$

It can also be easily seen that:

$$Fib(n) = \mathcal{O}(c^n),$$

where

$$c = \left(\frac{1 + \sqrt{5}}{2} \right).$$

2.8.1   Counting binary trees

We illustrate the use of setting up recurrence equation and then obtaining a closed form solution, if possible, on a tree.

Problem 7 (Counting binary trees) *Let the leaf nodes of a binary tree be labeled from*

$$1, 2, \dots K.$$

How many such trees can be constructed?

Let

$$Tb(K)$$

denote the number of binary trees with K leaf nodes. We first compute

$$E(K),$$

the number of edges in a binary tree with K leaf nodes. Clearly,

$$E(2) = 1.$$

Given a binary tree with K leaf nodes, if a new leaf node is to be added to the tree, the leaf node with an edge has to be always attached to the middle of an existing edge. Thus the existing edge is lost and three new edges are added, effectively increasing the number of edges by 2. Thus, for $K > 2$,

$$E(K + 1) = E(K) + 2.$$

A closed form solution to this recurrent equation can be obtained by using a method called *telescoping*. We illustrate this method below.

The recurrence equation is:

$$E(K) = \begin{cases} 1 & \text{if } K = 2 \\ E(K-1) + 2 & \text{if } K > 2 \end{cases}$$

The closed form of this is computed as follows. For $K > 2$,

$$\begin{aligned} E(K) &= E(K-1) + 2 \\ &= E(K-2) + 2 + 2 \\ &= E(K-3) + 2 + 2 + 2 \\ &\;\;\vdots \\ &= E(K - (K-2)) + 2(K-2) \\ &= E(2) + 2(K-2) \\ &= 1 + 2(K-2) \\ &= 2K - 3. \end{aligned}$$

Now it should be clear why this approach is called telescoping. We now construct the next set of recurrent equations. Let

$$Tb(K)$$

denote the number of unrooted binary trees with K leaf nodes. Consider $K = 2$. Clearly, there is only one tree, thus

$$Tb(2) = 1.$$

Next, we define

$$Tb(K)$$

in terms of

$$Tb(K-1).$$

A new leaf node can be attached to any edge in a tree with $(K-1)$ leaves and this tree has $E(K-1)$ edges. Thus for $K > 2$,

$$\begin{aligned} F(K) &= E(K-1)F(K-1) \\ &= (2K-5)F(K-1). \end{aligned}$$

The recurrence equation is:

$$Tb(K) = \begin{cases} 1 & \text{if } K = 2. \\ (2K-5)Tb(K-1) & \text{if } K > 2. \end{cases}$$

The closed form is as follows:

$$Tb(K) = \frac{(2K-4)!}{2^{K-2}(K-2)!} \tag{2.2}$$

See Exercise 8 at the end of the chapter for $f(K)$, where

$$Tb(K) = \mathcal{O}(f(K)).$$

2.8.2   Enumerating unrooted trees (Prüfer sequence)

Consider the problem of enumerating trees on a collections of nodes.

Problem 8 (Enumerating unrooted trees) *Given K nodes labeled from,*

$$1, 2, \ldots, K,$$

enumerate all (unrooted) trees on the labeled nodes.

Such a tree is a spanning tree on the complete graph on K vertices. While enumerating trees, it is important to do so in a manner that avoids repetition. A moment of reflection will show that this is not as trivial a task as it seems at first glance.

The trees are unrooted, i.e., there is no particular node in the tree that is designated to be a root. Hence the process must identify identical (isomorphic) trees.

It turns out that there is a very elegant way of solving this problem using *Prüfer sequences* [Prü18]. This sequence is associated with a tree whose nodes are labeled from

$$1, 2, \ldots, K,$$

and is generated iteratively as follows:
Given a tree with labeled nodes, at step i,

1. pick the leaf node l_i with the smallest label,

2. output l_i's immediate neighbor p_i, and

3. remove l_i from tree.

This process is continued until only two vertices remain on the tree. The sequence of labels

$$p_1 p_2 \cdots p_{K-3} p_{K-2}$$

is called the Prüfer sequence. Notice that repetitions are allowed in the sequence, i.e., it is possible that for some

$$1 \leq i, j \leq K - 2,$$

the labels are such that

$$p_i = p_j.$$

Figure 2.9 shows two examples of a *Prüfer sequence* and its corresponding labeled tree.

Given a Prüfer sequence p, the construction of the tree can be done in a single scan of the sequence from left to right. The algorithm first constructs L, the set of nodes missing from p. At each step of the scan we introduce an edge between the minimum element of L and the currently scanned element

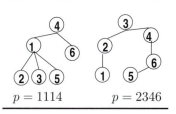

$p = 1114$ $p = 2346$

FIGURE 2.9:   Examples of trees and the corresponding Prüfer sequences.

of the sequence. L is updated by removing the minimum element and adding the scanned element, if it is not repeated any more. This process is described in Algorithm (3) and illustrated in Figure 2.10. In the algorithm description, $\Pi(s)$ denotes the set of all numbers occurring in a sequence s.

Algorithm 3 *Constructing trees from Prüfer Sequences*

> *(0-1)* $l \leftarrow |p|$
> *(0-2)* $V \leftarrow \{1, 2, \ldots, (l+2)\}$
> *(0-3)* $L \leftarrow V \setminus \Pi(p)$
>
> *FOR* $i = 1, 2, \ldots, l$ *DO*
> *(1)* $m \leftarrow \min L$
> *(2)*   *Introduce edge* $(m, p[i])$ *in* E
> *(3)* $L \leftarrow L \setminus \{m\}$
> *(4)*   *If* $p[i] \notin \Pi(p[i+1 \ldots l])$ $L \leftarrow L \cup \{p[i]\}$
> *ENDFOR*
> *(5) Introduce edge* (m_1, m_2) *in* E *where* $m_1, m_2 \in L$

Proof of correctness of algorithm (3).   We first show that the graph constructed by Algorithm (3) is connected. At the start of the algorithm, there are K connected components corresponding to each vertex. Assign a distinct color to each component. Every time a vertex from component i is connected to a vertex from component $j(> i)$, all the vertices in component j are assigned the color of component i. The very first time line (2) of the iteration is executed, assume the first connected component of the graph is under construction with the color of $p[1]$ as color 1.

Next it is easy to see that at the end of the i-th iteration of the algorithm, the vertex $p[i]$ (which belongs to the connected component containing the last edge added), must occur again as incident on a subsequently added edge (either as $p[j]$ with $j > i$ or as $\min(L)$). Every vertex gets added to L exactly once, either at line (0-3) or at line (4) and at each iteration a vertex gets removed from L (line (3)). Hence at the end of $l = K - 2$ iterations at line (5), $|L| = 2$ and there can be at most two connected components remaining corresponding to the two vertices in L and these get connected in the last step. Thus the constructed graph is connected.

Further, an edge is constructed in each iteration and the number of iterations is $K-1$, hence the graph has $K-1$ edges. A connected graph with $K-1$ edges on K nodes is a tree (see Exercise 2). This proves that the algorithm is correct.   □

The theorem below follows directly from Algorithm (3) and its proof of correctness.

THEOREM 2.1
(Prüfer's theorem) *There is a bijection [5] from the set of Prüfer sequences of length*

$$K - 2$$

on

$$V = \{1, 2, \ldots, K\}$$

to the set of trees T with vertex set V. In other words, a Prüfer sequence corresponds to a unique tree and vice-versa.

COROLLARY 2.2
The number of trees on K nodes is (Cayley's number):

$$K^{K-2}.$$

PROOF   Since there is a one-to-one correspondence between sequences of length

$$K - 2$$

on

$$\{1, 2, \ldots, K\}$$

and trees with K nodes, the number of such trees is the same as number of distinct sequences of length

$$K - 2,$$

which is

$$K^{K-2}.$$

                                                                                    ⧠

Back to enumerating all trees. Now, it is easy to see that to enumerate all trees on K nodes, we need to enumerate all Prüfer sequences of length

---

[5] Given sets X, Y, a function $f : X \to Y$ is *injective* or *one-to-one* if $(f(x) = f(y)) \implies (x = y)$ and is *surjective* or *onto* if for every $y \in Y$, there is an $x \in X$ with $f(x) = y$. A function f that is both injective and surjective is called a *bijection* or a *bijective* function.

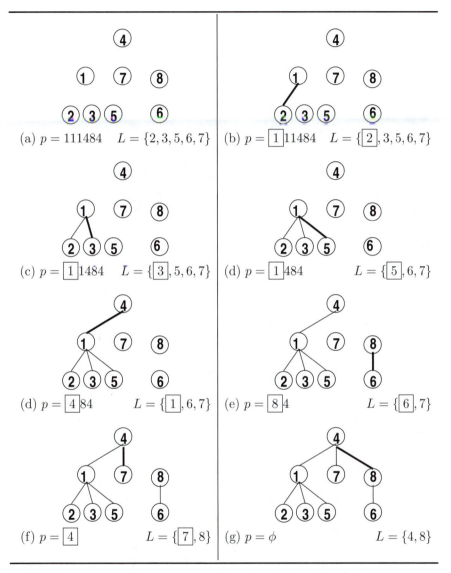

FIGURE 2.10: Illustration of Algorithm (3) on Prüfer sequence $p = 111484$. (a) The tree has $|p| + 2 = 8$ nodes. (b)-(f) Each iteration of the algorithm: The boxed element in p is being scanned and the minimum label of L is boxed. Consequently the bold edge in the graph is the one that is constructed at this step. (g) The very last edge is on the two vertices of L.

$K - 2$ and a unique labeled tree can be constructed from each sequence in linear time,

$$\mathcal{O}(K),$$

using Algorithm (3).

2.9   NP-Complete Class of Problems

We saw earlier in the chapter that the minimum spanning tree problem (Problem (2)) has a polynomial time solution. We were unable to devise such an algorithm for a very similar problem, the Steiner tree problem (Problem (3)). Usually, a problem that has a polynomial time algorithm,

$$\mathcal{O}(n^c),$$

is called *tractable*, where n is the size of the input and c is some constant. So, is the Steiner tree problem really not tractable or were we not smart enough to find one?

Theoretical computer scientists study an interesting class of problems, called the NP-complete [6] problems, whose tractability status is unknown. No polynomial time algorithm has been discovered for any problem in this class, to date. However, the general belief is that the problems in this class are intractable. Needless to mention, this is the most perplexing open problem in the area.

Notwithstanding the fact that the central problem in theoretical computer science remains unresolved, techniques have been devised to ascertain the tractability of a problem using relationships between problems in this class. Suppose we encounter Problem X. First we need to check if this problem has been studied before. A growing compendium of problems in the class of NP-complete problems exist and it is very likely that a new problem one encounters is identical to one of this collection. For example, our Problem (3) was identified to be the Steiner tree problem.

If Problem X cannot be identified with a known problem, then the next step is to *reduce* a problem, Problem A, from the NP-complete class in polynomial time to Problem X. The reduction is a precise process that demonstrates that a solution to Problem A can be obtained from a solution to Problem X in polynomial time. This proves that Problem X also belongs to the class of NP-complete problems.

---

[6]NP stands for 'Non-deterministic Polynomial' and any further discussion requires a fair understanding of formal language theory and is beyond the scope of this exposition.

Once it is established that Problem X is in the class of NP-complete problems, it is futile to seek a polynomial time algorithm.[7] The next course of action may be to either redefine the problem depending on the application, or design an approximation algorithm. Loosely speaking, an approximation algorithm guarantees the computed solution to be within an ϵ factor of the optimal.

Summary

The chapter introduces the reader to a very basic abstraction, trees. The link between pattern discovery and trees will become obvious in the later chapters. To understand and appreciate the issues involved with trees, we elaborate on three problems: (1) Minimum spanning tree, (2) Steiner tree and (3) Minimum mutation labeling. The first and the third have a polynomial time solution. The reader is also given a quick introduction to using the same abstraction as a data structure (balanced binary trees). Recurrence equation is a simple yet powerful tool that helps in counting and Prüfer sequences are invaluable in enumerating trees.

This was a brief introduction to the exciting world of algorithmics. In my mind, this is the foundation for a systematic subject such as bioinformatics. The beauty of this field is that some very powerful statements (that will be used repeatedly elsewhere in the book), such as *the number of internal nodes in a tree is bounded by the number of leaf nodes*, are consequences of very simple ideas (see Exercise 4). The intent of the chapter has been to introduce the reader to basic concepts used elsewhere in the book as well as influence his or her thought processes while dealing with computational approaches to biological problems.

2.10   Exercises

Exercise 1 *(mtDNA) DNA in the mitochondria (mtDNA) of a cell traces the lineage of a mother to child. A health center that gathers mtDNA data for families to help trace and understand hereditary diseases accidentally mixes up the mtDNA data losing all lineage information for a family with seven*

---

[7]Though in theory it is possible since the central tractability question is still unresolved, in practice, it is extremely unlikely.

*generations. The entire sequence information of the mtDNA for each member
is accessible from a database.*

1. *Assuming the only change in the mtDNA of a daughter is at most one
   mutation at a locus of the mother's mtDNA, can the lineage information
   be recovered?*

2. *What are the issues to be kept in mind in designing a recovery system
   for this scenario?*

Exercise 2 (Tree edge)   *Show the converse of Lemma (2.2).   In other
words, prove that given a connected graph $G(V, E)$,*

$$(G \text{ is a tree}) \Leftarrow (|E| = |V| - 1)$$

Hint: Use induction on $|V|$, the number of vertices.

Exercise 3 (Tree leaf nodes)   *Show the a tree $T(V, E)$ with $|V| > 1$ must
have at least $l = 2$ leaf nodes.*
Does the statement hold for $l = 3$ (assume $|V| > 2$)?

Hint: Fix a vertex $v_f \in V$. Define a notion of distance from v_f to any

$$(v \neq v_f) \in V.$$

Note that v with the largest distance from v_f must be a leaf node.

Exercise 4 (Linear size of trees)   *Given a tree $T(V, E)$, let l be the number
of leaf nodes. Show that*

$$|V| \leq 2l.$$

Exercise 5   *In the optimally (and completely) labeled Steiner tree, a node v
is always optimally assigned with respect to its immediate neighbors.*
Prove or disprove this statement.

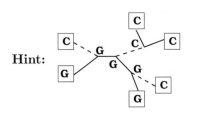

Hint:

The leaf node assignments, that cannot be relabeled, are shown enclosed in boxes. Notice that the above statement holds true for this labeling of the tree. The cost of this tree is 3 corresponding to the three dashed edges. Can the cost be improved with a different labeling of the internal nodes?

Exercise 6 *Show that Equation (2.2) is the solution to the recurrence equation:*

$$F(K) = \begin{cases} 1 & \text{if } K = 2 \\ (2K - 5)F(K - 1) & \text{if } K > 2 \end{cases}$$

Hint: Use the telescoping technique to obtain

$$F(K) = 1 \times 3 \times 5 \times \ldots \times (2K - 5),$$

for $K > 2$. Next, use induction on K.

Exercise 7 *Show that asymptotically*

$$T(n) = \mathcal{O}(\log n),$$

where

$$T(n) = \begin{cases} 1 & \text{if } n = 1, \\ T(n/2) + \mathcal{O}(1) & \text{if } n > 1. \end{cases}$$

Exercise 8 *Show that asymptotically*

$$Tb(k) = (\mathcal{O}(k))^k,$$

where

$$Tb(k) = \frac{(2k - 4)!}{2^{k-2}(k - 2)!}$$

Hint: Use Stirling's approximation (Equation (2.1) for the factorials in $Tb(k)$.

Exercise 9 ** *(1) Design an algorithm to add an element to a balanced tree. (2) Design an algorithm to delete an element from a balanced tree.*

Hint: An efficient algorithm can be designed by maintaining extra information at each node. See a standard text on algorithms and data structures such as [CLR90].

Exercise 10 (MaxMin path problem) *Let $G(V, E)$ be a weighted connected graph with*

$$wt(v_1 v_2) \geq 0 \text{ for each } (v_1 v_2) \in E.$$

A path, $P(v_1, v_2)$, from v_1 to v_2 is given as:

$$P(v_1, v_2) = (v_1 = v_{i_1}) \, v_{i_2} \ldots v_{i_{k-1}} (v_{i_k} = v_2),$$

for some k. The path length is defined as follows:

$$Len(P(v_1, v_2)) = Len((v_1 = v_{i_1}) \, v_{i_2} \ldots v_{i_{k-1}} (v_{i_k} = v_2))$$
$$= \min_{j=2\ldots k} wt(v_{i_{j-1}} v_{i_j}).$$

Given two vertices $v_1, v_2 \in V$, the distance between the two is given as the maximum path length over all possible paths between v_1 and v_2, i.e.,

$$Dis(v_1, v_2) = \max_{P(v_1, v_2)} Len(P(v_1, v_2)).$$

1. *Design an algorithm to compute*

$$Dis(u, v) \text{ for each } v \in V,$$

   where $u \in V$ is some fixed vertex.

2. *What is the time complexity of the algorithm?*

Hint: 1. Is the following algorithm correct? Why?

1. For each $v \in V$, $dis[v] \leftarrow 0$
   $dis[u] \leftarrow \infty$

2. Let $S \leftarrow V$

3. WHILE ($S \neq \phi$)

   (a) Let $x \in S$ be such that

   $$dis[x] = \max_{v \in S} (dis[v]).$$

   (b)

   $$S \leftarrow S \setminus \{x\}$$

(c) For each $(v \in S \text{ AND } (xv) \in E)$

     i. $dis[v] \leftarrow \max\{dis[v], \min(dis[x], wt(xv))\}$

     ii. Keep track of the path here, by storing backpointer to x, if the second term contributed to the max value

4. For each v, $dis[v]$ is the required distance.

2. This works in time

$$\mathcal{O}(|V|^2|E|).$$

Is it possible to improve the time complexity to

$$\mathcal{O}(|V|\log|V||E|)?$$

Exercise 11 (Expected depth on an execution tree) *Now we address the problem of computing the average path depth (number of vertices in the path from the root, or, depth from the root) in the MaxMin Problem (Exercise 10), when each edge weight $wt(v_1, v_2)$ is distributed uniformly over $(0,1)$. Let T be the execution tree generated by the algorithm: this has k nodes.*

1. *What is the total number of such trees?*

2. *What is the average distance of a node from the root?*

Hint: Let

$$M(k, d), \quad d \le k,$$

be the total number of vertices at depth d in ALL rooted trees with k nodes.

$$
\begin{aligned}
M(0, 1) &= 1, \\
M(1, 1) &= 1, \\
M(d, k) &= \begin{cases} M(d-1, k-1) + kM(d, k-1) & \text{for } 0 \le d \le k, \\ 0 & \text{otherwise.} \end{cases}
\end{aligned}
$$

1. $M(0, k)$, $k \ge 1$, is the total number of all possible trees (not necessarily distinct):

$$M(0, k) = k!$$

2. Average distance, $\mu(k)$, of a node from the root on the tree with k nodes:

$$
\begin{aligned}
\mu(k) &= \frac{\sum_{d=1}^{k} dM(d, k)}{kM(0, k)} \\
&= \frac{\sum_{d=1}^{k} dM(d, k)}{k \times k!}.
\end{aligned}
$$

Constructing the functions $M(d,k)$, $\mu(k)$:

$k\backslash d$	-1	0	1	2	3	4	5	6	-	-	$\mu(k)$
1	0	1	1	0							1
2	0	2	3	1	0						1.25
3	0	6	11	6	1	0					1.44
4	0	24	50	35	10	1	0				1.60
5	0	96	274	225	85	15	1	0			2.17
6	0	-	-	-	735	175	21	1	0		-
7	0	-	-	-	-	-	322	13	1	0	-

Chapter 3

Basic Statistics

Statistics is like your cholesterol health,
it is not just what you do, but more importantly,
what you do not do, that matters.
- anonymous

3.1 Introduction

To keep the book self-contained, in this chapter we review some basic statistics. To navigate this (apparent) complex maze of ideas and terminologies, we follow a simple story-line.

Professor Marwin, who kept us busy in the last chapter with her algorithmic challenges, also tends a Koi [1] pond with four types of this exquisite fish: *Asagi* (A), *Chagoi* (C), *Ginrin Showa* (G) and *Tancho Sanshoku* (T). The fish have multiplied tremendously for the professor to keep track of their exact number but she claims that the four types have an equal representation in her large pond.

She is introduced to a scanner (a kind of fish net or trap) that catches no more than one at each trial i.e., zero or one koi. The manufacturer sells the scanners in a turbo mode, as a k-scanner where $k(\geq 1)$ scanners operate simultaneously yielding an ordered sequence of k results. The professor further asserts that each fish in the pond is equally healthy, agile and alert to avoid a scanner, thus having equal chance of being trapped (or not) in the scanner.

We study the relevant important concepts centered around this koi pond scenario. We give a quick summary below of the questions that motivate the different concepts.

---

[1] The scientific name for this beautiful fish is *Cyprinus carpio*.

The koi pond setting with A, C, G, T types.	Foundations, Terminology $(\Omega, \mathcal{F}, P)$
k-scanner: What is the probability of having a nonhomogenous scan?	Bernoulli trials Binomial distribution Multiple events
What is the probability of an unbalanced scan?	Random variables Expectations
∞-scanner with type counts	Poisson distribution
∞-scanner with type mass	Normal distribution
Uniform model: Is the scanner fair (random)?	Statistical significance p-value Central Limit Theorem

3.2 Basic Probability

Probability is a branch of applied mathematics that studies the effects of *chance*. More precisely, probability theory is the study of probability spaces and random variables.

3.2.1 Probability space foundations

We adopt the Kolmogorov's axiomatic approach to probability and define a *probability space* as a triplet

$$(\Omega, \mathcal{F}, P) \tag{3.1}$$

satisfying certain axiom stated in the following paragraphs. Once a given problem is set in this framework, interesting questions (which are usually about probabilities of specific events) can be answered.

3.2.1.1 Defining Ω and $\mathcal{F}$

We go back to the koi pond to define the different terms. In this scenario, a *random trial* [2] is the process of fishing in the pond with a k-scanner. For example when a 1-scanner is used, the outcome of the random trial is the type

---

[2]Historically this has been called an *experiment*.

it yields: A, C, G, T or zero (denoted by -). The *sample space* of a random trial is the set of all possible outcomes. This is a nonempty set and is usually denoted as Ω. In this example,

$$\Omega = \{\text{A, C, G, T, -}\}. \tag{3.2}$$

An element of Ω is usually denoted by ω.

An *event* is a set of outcomes of a random trial, thus a subset of the sample space Ω. When the event is a single outcome, it is often called an *elementary event* or an *atomic event*. Thus events are subsets of Ω and the set of events is usually denoted by $\mathcal{F}$. An element of $\mathcal{F}$ is usually denoted by E. Sometimes

$$\omega \in \Omega,$$

is not distinguished from the singleton set

$$\{\omega\} \in \mathcal{F}$$

and is referred to as an elementary event.

3.2.1.2 Measurable spaces**

In mathematics, a σ-algebra over a set Ω is a collection, $\mathcal{F}$, of subsets of Ω that is closed under countable set operations. In other words,

1. if $E \in \mathcal{F}$, then its complement $\bar{E} \in \mathcal{F}$ and

2. if $E_1, E_2, E_3, \ldots \in \mathcal{F}$ then $(\cup_i E_i), (\cap_i E_i) \in \mathcal{F}$.

When Ω is discrete (finite or countable), then the σ-algebra is the whole power set 2^Ω (set of all subsets of Ω). However, when Ω is not discrete, care needs to be taken for probability P to be meaningfully defined. In case

$$\Omega = \mathbb{R}^n$$

it is convenient to take the class of *Lebesgue measurable* subsets of $\mathbb{R}^n$ (which form a σ-algebra) for $\mathcal{F}$. Since the construction of a non-Lebesgue measurable subset usually involves the *Axiom of Choice*, one does not encounter such a set in practice. So, we do not belabor this point. But it is still important to note that not every set of outcomes is an event.

3.2.1.3 Defining probability P

Assuming that the pair Ω and $\mathcal{F}$ is a measurable space, we now define a *probability measure* (or just *probability*) denoted by P. It is a function from $\mathcal{F}$ to the nonnegative real numbers, written as,

$$P : \mathcal{F} \to \mathbb{R}$$

satisfying the *Kolmogorov Axioms*:

1. For each $E \in \mathcal{F}$,

$$P(E) \geq 0.$$

2. It is certain that some atomic element of Ω will occur. Mathematically speaking,

$$P(\Omega) = 1.$$

3. For *pairwise disjoint sets* $E_1, E_2, \ldots \in \mathcal{F}$,

$$P(E_1 \cup E_2 \cup \ldots) = \sum_i P(E_i).$$

We leave it as an exercise for the reader (Exercise 12) to show that under these conditions, for each $E \in \mathcal{F}$,

$$0 \leq P(E) \leq 1. \tag{3.3}$$

Usually the probability measure is also called a *probability distribution function* (or simply the *probability distribution*).

3.2.2 Multiple events (Bayes' theorem)

In practice, we almost always deal with multiple events, so the next natural topic is to understand the delicate interplay between these multiply (in conjunction) occurring events.

Bayes' rule. The Bayesian approach is one of the most commonly used methods in a wide variety of applications ranging from bioinformatics to computer vision. Roughly speaking, this framework exploits multiply occurring events in observed data sets by using the occurrence of one or mote events to (statistically) guess the occurrence of the other events. Note that there can be no claim on an event being either the *cause* or the *effect*.

The simplicity of the Bayesian rule is very appealing and we discuss this below.

Joint probability is the probability of two events in conjunction. That is, it is the probability of both events together. The joint probability of events E_1 and E_2 is written as

$$P(E_1 \cap E_2)$$

or just

$$P(E_1 E_2).$$

Going back to the foundations, Kolmogorov axioms lead to the natural concept of *conditional probability*. For E_1 with

$$P(E_1) > 0,$$

the *probability of E_2 given E_1* denoted by

$$P(E_2|E_1),$$

is defined as follows:

$$P(E_2|E_1) = \frac{P(E_1 \cap E_2)}{P(E_1)}.$$

In other words, conditional probability is the probability of some event E_2, given the occurrence of some other event E_1.

In this context, the probability of an event say E_1 is also called the *marginal probability*. It is the probability of E_1, regardless of event E_2. The marginal probability of E_1 is written $P(E_1)$.

Bayes' theorem relates the conditional and marginal probability distributions of random variables as follows:

THEOREM 3.1
(Bayes' theorem) *Given events E_1 and E_2 in the same probability space, with*

$$P(E_2) > 0,$$

the following holds:

$$P(E_1|E_2) = \frac{P(E_2|E_1)}{P(E_2)} P(E_1). \tag{3.4}$$

The proof falls out of the definitions and the result is often interpreted as:

$$\text{Posterior} = \frac{\text{Likelihood}}{\text{normalization factor}} \text{Prior}.$$

We will pick up this thread of thought in a later chapter on maximum likelihood approach to problems.

3.2.3 Inclusion-exclusion principle

Mutually exclusive vs independent events. Recall that an event, in a sense, is synonymous with a set. Two nonempty sets E_1 and E_2 are *mutually exclusive* if and only if they have an empty intersection. Using the probability measure P, two events E_1 and E_2 are mutually exclusive if and only if

1. $P(E_1) \neq 0$, $P(E_2) \neq 0$, and,

2. $P(E_1 \cap E_2) = 0$.

Mutually exclusive events are also called *disjoint*. It follows that if E_1 and E_2 are mutually exclusive, then the conditional probabilities are zero. i.e.,

$$P(E_1|E_2) = P(E_2|E_1) = 0.$$

What can we say when $E_1 \cap E_2$ is not empty, i.e,

$$E_1 \cap E_2 \neq \emptyset \ ?$$

There is a very important concept called the *independence of events*, that comes into play here. It is a subtle concept and has the same connotation as the natural meaning of the word 'independence'. Loosely speaking, when two events are independent, it means that knowing about the occurrence of one of them does not yield any information about the other. In case the events are dependent, usually great care needs to be taken to account for the interplay arising from their dependence. Note that if

$$E_1 \cap E_2 = \emptyset$$

and

$$P(E_1)P(E_2) \neq 0,$$

then the events are (very) dependent.

Mathematically speaking, two events E_1 and E_2 are *independent* if and only if the following hold:

1. $P(E_1) \neq 0$, $P(E_2) \neq 0$, and,

2. $P(E_1 \cap E_2) = P(E_1)P(E_2)$.

An alternative view of the same is as follows. Let E_1 and E_2 be two events with

$$P(E_1), P(E_2) > 0.$$

If the marginal probability of E_1 is the same as its conditional (with E_2) probability, then E_1 and E_2 are independent. In other words, if

$$P(E_1) = P(E_1|E_2),$$

then E_1 and E_2 are said to be independent events.

Union of events.   Using the Kolmogorov axioms one can deduce the following:

$$P(E_1 \cup E_2) = P(E_1) + P(E_2) - P(E_1 \cap E_2).$$

For a natural generalization to

$$P(E_1 \cup E_2 \cup \ldots \cup E_n),$$

define the following quantities: S_l is the sum of the probabilities of the intersection of all possible l out of the n events. Thus

$$S_1 = \sum_i P(E_i),$$

$$S_2 = \sum_{i<j} P(E_i \cap E_j),$$

$$S_3 = \sum_{i<j<k} P(E_i \cap E_j \cap E_k),$$

and so on.

THEOREM 3.2
(Inclusion-exclusion principle)

$$P(E_1 \cup E_2 \cup \ldots \cup E_n) = S_1 - S_2 + S_3 - S_4 + S_5 \ldots + (-1)^{n+1} S_n$$
$$= \sum_{i=1}^n (-1)^{i+1} S_i.$$

It can be seen that

$$S_1 \geq S_2 \geq \ldots \geq S_n,$$

hence,

$$(-1)^k S_k + (-1)^{k+1} S_{k+1} \begin{cases} \geq 0 & \text{for } k \text{ odd,} \\ \leq 0 & \text{for } k \text{ even.} \end{cases} \qquad (3.5)$$

This implies that (the proof is left as an exercise for the reader):

$$P(E_1 \cup E_2 \cup \ldots \cup E_n) \begin{cases} \leq \sum_{i=1}^k (-1)^{i+1} S_i & \text{for } k \text{ odd,} \\ \geq \sum_{i=1}^k (-1)^{i+1} S_i & \text{for } k \text{ even.} \end{cases} \qquad (3.6)$$

Inequalities (3.6) are often referred to as *Bonferroni's Inequalities*. They are useful in practice in order to obtain quick upper and lower estimates on the probabilities of unions of events. For instance, when $k = 1$,

$$P(E_1 \cup E_2 \cup \ldots \cup E_n) \leq S_1.$$

The above is expanded as follows, which is also known as *Boole's Inequality*.

$$P(E_1 \cup E_2 \cup \ldots \cup E_n) \leq P(E_1) + P(E_2) + \ldots + P(E_n). \qquad (3.7)$$

3.2.4   Discrete probability space

When Ω is finite or countable, it is called a *discrete* space. For such cases, P induces the following function,

$$M_P : \Omega \to \mathbb{R}_{\geq 0}$$

given by

$$M_P(\omega) = P(\{\omega\}). \tag{3.8}$$

In other words, M_P assigns a probability to each element $\omega \in \Omega$. The function M_P is often called a *probability mass function*. It can be verified that if P satisfies the Kolmogorov's axioms (Section (3.2.1.3)) then M_P must satisfy the following (called the probability mass function conditions or *probability axioms*):

1. $0 \leq M_P(\omega) \leq 1$, for each $\omega \in \Omega$, and,

2. $\sum_{\omega \in \Omega} M_P(\omega) = 1$.

Note also that M_P in turn determines P. Thus when Ω is finite, it is simpler to specify M_P instead of P. Also, in this case

$$\mathcal{F} = 2^{\Omega}.$$

Thus for a discrete setting the probability space is specified by the triplet:

$$(\Omega, 2^{\Omega}, M_P) \tag{3.9}$$

A concrete example.   We go back to the professor's koi pond. She decides to use a 2-scanner and then asks the following question:

> *What is the probability of having a nonhomogenous scan, i.e., different types in the same scan?*

The specification of Ω and $\mathcal{F}$ is usually determined from the nature of the random event, which in this case is the result of a 2-scanner. Let this be denoted by uv where u is the outcome of the first and v the outcome of the second of the 2-scanner. Since the two scanners operate simultaneously, we define Ω as the following set of 25 elements:

$$\Omega = \begin{Bmatrix} \text{-- } & \text{A-} & \text{C-} & \text{G-} & \text{T-} \\ \text{-A} & \text{AA} & \text{CA} & \text{GA} & \text{TA} \\ \text{-C} & \text{AC} & \text{CC} & \text{GC} & \text{TC} \\ \text{-G} & \text{AG} & \text{CG} & \text{GG} & \text{TG} \\ \text{-T} & \text{AT} & \text{CT} & \text{GT} & \text{TT} \end{Bmatrix} \tag{3.10}$$

Next, we need some more information before we can specify P or M_P. We gather the following intelligence:

1. The manufacturer asserts the following:

    (a) each scanner in the k-scanner operates simultaneously, thus the scan of each scanner is independent of the other, and,

    (b) the odds of having an empty scan, in each scanner, is just one in nine and the scanner is not partial to any particular koi type.

2. The professor asserts that each type in her pond is equally likely to be caught in a scanner (we call this the *uniform model*).

Bernoulli trials. We model each scanner separately as a (*multi-outcome*) *Bernoulli trial*. Usually a Bernoulli trial has two outcomes: 'success' and 'failure'. However, we use a generalization where each trial has N possible outcomes and N is some integer larger than one. The probability of each outcome x_i is given as $pr(x_i)$ which are so defined so that the following holds:

$$\sum_{i=1}^{N} pr(x_i) = 1$$

Using (1b) and (2), we have $N = 5$ and we model $pr(x)$ as follows:

$$pr(x) = \begin{cases} 2/9 & \text{if } x = A \\ 2/9 & \text{if } x = C \\ 2/9 & \text{if } x = G \\ 2/9 & \text{if } x = T \\ 1/9 & \text{otherwise} \end{cases} \tag{3.11}$$

From (1a) above, for each $uv \in \Omega$,

$$M_P(uv) = pr(u)pr(v) \tag{3.12}$$

Next, there is a need to show that the probability function P (as derived from M_P) satisfies the probability measure conditions. This is left as an exercise for the reader (Exercise 15).

Back to the query. Let E denote the event that the outcome of the 2-scanner is homogenous. We claim the following:

$$E = \{--, A-, -A, -C, C-, -G, G-, -T, T-, AA, CC, GG, TT\}$$

We specify nonhomogeneity to be the presence of at least two distinct types in the scan. We treat the absence of nonhomogeneity to be homogenous and with this view (subjective interpretation), the first nine elements in the set E are considered to be homogenous. The last four elements are clearly homogenous.

Treating E as the union of singleton sets and since all singleton intersections are empty, we get

$$P(E) = \sum_{\{uv\} \in E} P(\{uv\}) \qquad \text{(probability measure cond (2))}$$

$$= \sum_{\{uv\} \in E} M_P(uv) \qquad \text{(probability mass function defn (3.8))}$$

$$= \sum_{\{uv\} \in E} pr(u)pr(v) \qquad \text{(using Eqn (3.12))}$$

$$= 33/81 \qquad \text{(using Eqn (3.11))}$$

Let $\bar{E}$ denote the event that the scan is *not* homogenous. Then $P(\bar{E})$ is given by:

$$\begin{aligned}
P(\bar{E}) &= P(\Omega \setminus E) &&\text{(since } E \cup \bar{E} = \Omega) \\
&= P(\Omega) - P(E) &&\text{(probability measure cond (2), and,} \\
& &&\text{since } E \cap \bar{E} = \emptyset) \\
&= 1 - P(E) &&\text{(since } P(\Omega) = 1) \\
&= 48/81
\end{aligned}$$

Alternative (CG-rich) model.   Note that Ω (and thus $\mathcal{F}$) were specified in the treatment by studying the nature of the query, which in turn defined the random trial. However, P was specified by taking the input from the scanner manufacturer and the professor.

Professor Marwin realizes that her pond has more of the C and G type than A and T. We call this the *CG-rich model*. She realizes that some quantitative information is required and does some testing of her own. She concludes that the C and G type is three times more ponderous than the A and T. At this point, we do not question how she gets these numbers (that is a different topic of discussion), but we must build this conclusion into our model.

Note that the specification of Ω (and thus $\mathcal{F}$) does not change for the CG-rich model. But $pr(x)$ is redefined as follows:

$$pr(x) = \begin{cases} 1/9 & \text{if } x = \text{A} \\ 3/9 & \text{if } x = \text{C} \\ 3/9 & \text{if } x = \text{G} \\ 1/9 & \text{if } x = \text{T} \\ 1/9 & \text{otherwise} \end{cases} \qquad (3.13)$$

We leave the arguments that $pr(x)$ leads to a probability measure as an exercise (Exercise 15) for the reader. The computation of the probability of the nonhomogenous event under this new model is also left as an exercise (Exercise 16) for the reader.

3.2.5 Algebra of random variables

Motivation. Using the CG-rich model defined in the last scenario, consider the questions:

> *What is the average number of C's in a scan?*
> *What is the variance?*
> *What is the average number of A's and C's?*

To answer a question of this kind conveniently, we need to expand our vocabulary. A *random variable* is a variable that is associated with the outcome of a random trial or an event and takes real values.

In the algebraic axiomatization of probability theory, the primary concept is that of a random variable.[3] The measurable space Ω, the event space $\mathcal{F}$ and the probability measure P arise from random variables and expectations by means of *representation theorems of analysis*. However, we do not explore this line of thought in detail here but just appeal to the intuition of the reader.

A random variable is not a variable in the usual sense that a single value may not be assigned to it. However, a probability distribution P [4] may be assigned to a random variable X, written as

$$X \sim P.$$

More precisely, given a probability space $(\Omega, \mathcal{F}, P)$, a random variable, X is a measurable function defined as:

$$X : \Omega \to \mathbb{R},$$

i.e., X maps Ω to real numbers. See Section (3.2.1.2) for the notion of measurability of functions. Often

$$P(\{\omega \in \Omega \mid X(\omega) \leq x_0\})$$

is abbreviated as

$$P(X \leq x_0).$$

Further, since random variables are simply *functions*, they can be manipulated as easily as functions. Thus, we have the following.

1. A real constant c is a random variable.

2. Let X_1 and X_2 be two random variables on the same probability space.

---

[3] Recall that in the Kolmogorov's axiomatic approach the event and its probability is the primary concept.

[4] Usually P is specified by the distribution parameters mean μ and variance σ^2 and written as $P(\mu, \sigma)$.

(a) The sum of X_1 and X_2 is a random variable denoted

$$X_1 + X_2.$$

(b) The product of X_1 and X_2 is a random variable denoted

$$X_1 X_2.$$

Note that addition and multiplication of random variables are commutative.

3.2.6 Expectations

The *mathematical expectation* (or simply *expectation*) of X, denoted by $E[X]$, is defined as follows:

$$E[X] = \int_\Omega X \, dP \tag{3.14}$$

Recall that X is a measurable function.[5] $E[X]$ can be interpreted to be the average value of X. When Ω is finite or countable, X is a *discrete random variable*. Recall that the probability space is

$$(\Omega, 2^\Omega, M_P)$$

and

$$E[X] = \sum_{\omega \in \Omega} X(\omega) M_P(\omega). \tag{3.15}$$

The expected values of the powers of X are called the *moments* of X. The l-th moment of X is defined by

$$E[X^l] = \int_\Omega X^l \, dP. \tag{3.16}$$

Similarly, the *central moments* are expected values of powers of $(X - E[X])$.

3.2.6.1 Properties of expectations

Using the definition of expectation, we can show the following properties. Given two random variables X_1 and X_2 defined on the same probability space, the following properties hold.

---

[5]In this chapter we have been denoting an event with E. Expectation is also denoted with an E, and it should be clear from the context what we mean.

1. *Inequality property:*

$$\text{If } X_1 \leq X_2 \text{ then } E[X_1] \leq E[X_2].$$

2. *Linearity of expectations:* For any $a, b \in \mathbb{R}$,

$$E[aX_1 + bX_2] = aE[X_1] + bE[X_2].$$

3. *Nonmultiplicative:* In general,

$$E(X_1 X_2) \neq E[X_1]E[X_2].$$

However, when X_1 and X_2 are *independent*,

$$E[X_1 X_2] = E[X_1]E[X_2].$$

These results are straightforward to prove and follow from the properties of integration (or summations in the discrete scenario).

3.2.6.2 Back to the concrete example

Consider our running example of the koi pond: *What is the average number of C's in a scan? What is the variance? What is the average numbers of A's and what is the average number of A's and C's?*

To answer this query, two random variables

$$X_C, X_A : \Omega \rightarrow \mathbb{R}$$

may be defined as follows (where Z is C, A):

$$X_Z(\omega) = l,$$

where l is the number of Z's in ω. Then the average number of C's in ω is given by the expected value of X_C. Using Ω defined in Equation (3.10) and the probabilities in Equation (3.13) we obtain:

$$
\begin{aligned}
E[X_C] &= \sum_{\omega \in \Omega} X_C(\omega) M_P(\omega) \\
&= \sum_{uv \in \Omega} X_C(uv) M_P(uv) \\
&= \sum_{(u \neq v) \in \Omega} M_P(uv) + \sum_{(u = v) \in \Omega} M_P(uv) \\
&= 1(3/81 + 3/81 + 3/81 + 3/81 + 3/81 + 3/81 + 9/81 + 3/81) + 2(9/81) \\
&= 48/81
\end{aligned}
$$

Similarly the average number of A's in ω is given by

$$E[X_A] = 18/81.$$

By linearity of expectations, the average number of A's and C's is given by

$$E[X_C + X_A] = E[X_C] + E[X_A]$$
$$= 66/81.$$

The variance of X_C, $V[X_C]$, is the second moment about the mean, and is given by

$$V[X_C] = E\left[(X_C - E[X_C]^2\right]$$
$$= E\left[X_C^2 + E[X_C]^2 - 2X_C E[X_C]\right]$$
$$= E[X_C^2] + E[X_C]^2 - 2E[X_C]E[X_C] \quad \text{(using linearity of } E\text{)}$$
$$= E[X_C^2] - E[X_C]^2.$$

As before using Ω defined in (3.10) and Equation (3.13) we obtain:

$$E[X_C^2] = \sum_{\omega \in \Omega} X_C^2(\omega) M_P(\omega)$$
$$= 1(M_P(\text{-C}) + M_P(\text{C-}) + M_P(\text{AC}) + M_P(\text{GC})$$
$$+ M_P(\text{TC}) + M_P(\text{CA}) + M_P(\text{CG}) + M_P(\text{CT}))$$
$$+ 4(M_P(\text{CC}))$$
$$= 1(3/81 + 3/81 + 3/81 + 3/81 + 3/81 + 3/81 + 9/81 + 3/81)$$
$$+ 4(9/81)$$
$$= 66/81.$$

Thus

$$V[X_C] = (66/81) - (48/81)^2$$
$$= 3042/6561.$$

3.2.7   Discrete probability distribution (binomial, Poisson)

Let X be a random variable, then the *probability density function* of X is the function f satisfying

$$P(x \leq X \leq x + dx) = f(x)dx.$$

This distribution can also be uniquely described by its *cumulative distribution function* $F(x)$, which is defined for any $x \in \mathbb{R}$ as

$$F(x) = P(X \leq x).$$

Random variable $X \sim P(\mu, \sigma)$ (domain)	Probability mass function f (discrete)	mean μ $E[X] =$ $\bar{X}$	variance σ^2 $V[X] =$ $E[(X\text{-}\bar{X})^2]$
(a) Binomial $k = 0, 1, \ldots, n$	$\binom{n}{k} p^k (1-p)^{n-k}$	np	$np(1-p)$
(b) Poisson $k \in \mathbb{N}$	$\frac{e^{-\lambda}\lambda^k}{k!}$	λ	λ
	Probability density function f (continuous)		
(c) Normal $x \in \mathbb{R}$	$\frac{1}{\sigma\sqrt{2\pi}} \exp\left(-\frac{(x-\mu)^2}{2\sigma^2}\right)$	μ	σ^2
(c') Standard Normal $x \in \mathbb{R}$	$\frac{1}{\sqrt{2\pi}} \exp\left(-\frac{x^2}{2}\right)$	0	1

FIGURE 3.0: Probability distributions and their parameters.

Note that the density function f and the cumulative distribution function F are related by

$$F(x) = \int_{-\infty}^{x} f(t)dt.$$

The binomial and Poisson distribution are among the most well-known discrete probability distributions. The normal or the Gaussian is the most commonly used continuous distribution.

Binomial distribution is the discrete probability distribution that expresses the probability of the number of 1's in a sequence of n independent 0/1 experiments, each of which yields 1 with probability p. Thus the mass function

$$pr_{binomial}(k; n, p) = \binom{n}{k} p^k (1-p)^{n-k}$$

gives the probability of seeing exactly k number of 1's for a fixed n and p. It can be verified that

$$\sum_{k=0}^{n} pr_{binomial}(k; n, p) = 1.$$

The mean and variance are as follows:

$$\bar{X}_{binomial} = E[X]$$
$$= \sum_{k=0}^{n} k \; pr_{binomial}(k; n, p)$$
$$= np$$
$$V[X] = E[(X - \bar{X}_{binomial})^2]$$
$$= np(1 - p)$$

Poisson distribution is a discrete probability distribution that expresses the probability of a number of events occurring in a fixed period of time if these events occur with a known average rate λ, and are independent of the time since the last event. The mass function is given by

$$pr_{poisson}(k; \lambda) = \frac{e^{-\lambda}\lambda^k}{k!}$$

and denotes the probability of seeing exactly k events for a fixed λ. It can be verified that

$$\sum_{k=0}^{\infty} pr_{poisson}(k; \lambda) = 1.$$

Note the convention that

$$0! = 1.$$

The mean and variance are as follows:

$$\bar{X}_{poisson} = E[X]$$
$$= \sum_{k=0}^{n} k \; pr_{poisson}(k; \lambda)$$
$$= \lambda$$
$$V[X] = E[(X - \bar{X}_{poisson})^2]$$
$$= \lambda$$

The following lemma gives a convenient way of modeling additive influences of phenomena. It says that sum of two binomial (respectively Poisson) is also a binomial (respectively Poisson) random variable. The proofs of the claims require some intricate but straightforward algebraic manipulations as shown below. We say that

$$X \sim f$$

if X is a discrete (respectively continuous) random variable with probability mass (respectively density) function f.

LEMMA 3.1
Let X_1 and X_2 be two independent random variables.

1. *If $X_1 \sim Binomial(n_1, p)$ and $X_2 \sim Binomial(n_2, p)$, then*

$$X = X_1 + X_2 \sim Binomial(n_1 + n_2, p)$$

2. *If $X_1 \sim Poisson(\lambda_1)$ and $X_2 \sim Poisson(\lambda_2)$, then*

$$X = X_1 + X_2 \sim Poisson(\lambda_1 + \lambda_2)$$

PROOF   Let $X = X_1 + X_2$. Since the domain for both X_1 and X_2 are nonnegative integers, we have

$$pr(X = k) = pr\left((X_1 + X_2) = k \mid X_1 = i\right)$$
$$= \sum_{i=0}^{k} f_1(i) f_2(k - i).$$

f_i is the probability mass function for X_i, $i = 1, 2$. We now deal with the two distributions separately.

Binomial distribution:

$$pr(k) = \sum_{i=0}^{k} \binom{n_1}{i} p^i (1-p)^{n_1-i} \binom{n_2}{k-i} p^{k-i} (1-p)^{n_2-k+i}$$

Separating the factors independent of i from the summation

$$pr(k) = p^k (1-p)^{n_1+n_2-k} \sum_{i=0}^{k} \binom{n_1}{i} \binom{n_2}{k-i}$$

The following combinatorial identity, also called the *Vandermonde's identity*,

$$\sum_{i=0}^{k} \binom{n_1}{i} \binom{n_2}{k-i} = \binom{n_1 + n_2}{k},$$

can be verified by equating the coefficients of x^k of both sides in

$$(1+x)^{n_1}(1+x)^{n_2} = (1+x)^{n_1+n_2}.$$

Thus,

$$pr(k) = \binom{n_1 + n_2}{k} p^k (1-p)^{n_1+n_2-k}.$$

Hence

$$X \sim Binomial(n_1 + n_2, p).$$

Poisson distribution:

$$pr(k) = \sum_{i=0}^{k} \left(\frac{e^{-\lambda_1}\lambda_1^i}{i!}\right)\left(\frac{e^{-\lambda_2}\lambda_2^{k-i}}{(k-i)!}\right)$$

$$= e^{-(\lambda_1+\lambda_2)} \sum_{i=0}^{k} \frac{\lambda_1^i \lambda_2^{k-i}}{i!(k-i)!}$$

Multiplying by $k!/k!$, we get

$$pr(k) = \frac{e^{-(\lambda_1+\lambda_2)}}{k!} \sum_{i=0}^{k} \frac{k!}{i!(k-i)!}\lambda_1^i \lambda_2^{k-i}$$

It can be verified that

$$\sum_{i=0}^{k} \frac{k!}{i!(k-i)!}\lambda_1^i \lambda_2^{k-i} = \sum_{i=0}^{k} \binom{k}{i}\lambda_1^i \lambda_2^{k-i}$$

$$= (\lambda_1 + \lambda_2)^k.$$

Thus

$$pr(k) = \frac{e^{-(\lambda_1+\lambda_2)}(\lambda_1 + \lambda_2)^k}{k!}$$

Hence

$$X \sim Poisson(\lambda_1 + \lambda_2).$$

$\Box$

3.2.8 Continuous probability distribution (normal)

The normal or the Gaussian distribution arises in statistics in the following way. The sampling distribution of the mean is approximately normal, even if the distribution of the population, the sample is taken from, is not necessarily normal (we show this in the *Central Limit Theorem* is a later section). In fact, a variety of natural phenomena like heights of individuals, or exam scores of students in a large undergraduate calculus class, or photon counts can be approximated with a normal distribution. While the underlying mechanisms may be unknown or little understood, if it can be justified that the errors or effects of independent (but small) causes are *additive*, then the likelihood of the outcomes may be approximated with a normal distribution.

The probability density function of a normal distribution, with mean μ and variance σ^2 is

$$pr_{normal}(X\mu, \sigma^2) = \frac{1}{\sigma\sqrt{2\pi}} \exp\left(-\frac{(X-\mu)^2}{2\sigma^2}\right).$$

It can be verified that

$$\int_{-\infty}^{\infty} pr_{normal}(X; \mu, \sigma^2) dX = 1.$$

The mean and variance are

$$\mu = \int_{-\infty}^{\infty} X \; pr_{normal}(X; \mu, \sigma^2) dX$$

$$\sigma^2 = \int_{-\infty}^{\infty} (X - \mu)^2 \; pr_{normal}(X; \mu, \sigma^2) dX$$

A random variable X is standardized using the theoretical mean and standard deviation:

$$Z = \frac{X - \mu}{\sigma},$$

where $\mu = E[X]$ is the mean and $\sigma^2 = V[X]$ is the variance of the probability distribution of X.

The lemma below follows from using the *characteristic functions* discussed in Section (3.3.3) which we leave as an exercise for the reader (Exercise 20).

LEMMA 3.2
If $X_1 \sim Normal(\mu_1, \sigma_1^2)$ and $X_2 \sim Normal(\mu_2, \sigma_2^2)$, are two independent random variables, then

$$X = X_1 + X_2 \sim Normal(\mu_1 + \mu_2, \sigma_1^2 + \sigma_2^2)$$

Binomial, Poisson & normal distributions. Figure 3.1 summarizes the different probability distributions. Before we conclude this section, we relate the three probability distributions that we studied here. Binomial is in a sense the simplest discrete probability distribution. It gets its name from its resemblance to the coefficients of a binomial

$$(x + y)^n.$$

The function values are indeed the coefficients of this polynomial. Putting

$$x = p$$

and

$$y = 1 - p$$

in fact shows the values add up to 1.

As n approaches ∞ and p approaches 0 while np remains fixed at $\lambda > 0$ or np approaches $\lambda > 0$, the

$$Binomial(n, p)$$

distribution approaches the Poisson distribution with expected value λ,

$$Poisson(\lambda).$$

As n approaches ∞ while p remains fixed, the distribution of

$$\frac{X - np}{\sqrt{np(1-p)}}$$

approaches the normal distribution with expected value 0 and variance 1,

$$Normal(0, 1).$$

3.2.9 Continuous probability space (Ω is $\mathbb{R}$)

We go back to the running example using the professor's koi pond. She is lured by the manufacturer to try their turbo ∞-scanner, which they claim is an extremely long scanner. Then she asks the following question:

> *What is the probability of an unbalanced scan, i.e., having at least three times as many as C's and G's than A's and T's in a scan?*

The professor asserts that the pond is large enough for these huge scanners. Now, we must work closely with the in-house scientists of the manufacturer to model this problem appropriately. After a series of carefully controlled experiments, they make the following observations for the ∞-scanners.

1. The chance of getting an empty outcome in an individual scanner in the ∞-scanner, has gone up significantly with large k's, giving rise to sparsely filled ones at each scan.

2. The average number, λ, of a type scanned is independent of the actual length of the scanner. In the professor's koi pond the average numbers for the four types C, G, A and T are observed to be

$$\lambda_C = \lambda_G = .054,$$
$$\lambda_T = \lambda_A = .018,$$

   respectively.

3. For a fixed i, the chance of having a nonzero outcome in the ith of the ∞-scanner in multiple scans (trials) is zero.

Due to independence of the occurrence of each event, the joint probability of the quadruple is given as:

$$P(X_A, X_C, X_G, X_T) = P(X_A)P(X_C)P(X_G)P(X_T).$$

To construct the probability space, let the outcome of a trial be denoted by a quadruple

$$(i_A, i_C, i_G, i_T)$$

where i_z,

$$z = \{A, C, G, T\},$$

is the number of type z in the scan. Then we define the probability space

$$(\Omega, 2^\Omega, M_P)$$

as follows:

$$\Omega = \{(i_A, i_C, i_G, i_T) \mid 0 \le i_A, i_C, i_G, i_T\}.$$

The occurrence of each type is independent of the other (see observation 3), and we model each as a Poisson distribution. Thus:

$$M_P(i_A, i_C, i_G, i_T) = \prod_{z=\{A,\ C,\ G,\ T\}} Poisson(i_z; \lambda_z). \tag{3.17}$$

and

$$\lambda_C = \lambda_G = .054,$$
$$\lambda_T = \lambda_A = .018.$$

Next, does this probability distribution satisfy the Kolmogorov's axioms? This is left as an exercise for the reader (Exercise 17).

However, the model can be simplified using the two following observations:

1. $\lambda_A = \lambda_T$, $\lambda_C = \lambda_G$, and

2. the professor lumps the C's and G's together, and, A's and T's together in her query.

Under this 'condensed' model, let the outcome of a trial be denoted by a pair

$$(i_{AT}, i_{CG}).$$

i_{AT} is the number of type A or T and i_{CG} is the number of type C or G in the scan. We define the probability space

$$(\Omega, 2^\Omega, M_P)$$

as follows:

$$\Omega = \{(i_{AT}, i_{CG}) \mid 0 \le i_{AT}, i_{CG}\}.$$

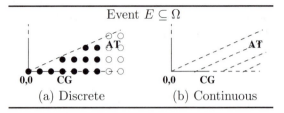

FIGURE 3.1:   The Ω space in Poisson and Normal distributions.   The shaded region corresponds to $x_{CG} \geq 3y_{AT}$ as shown.

Also we use the fact the sum of two Poisson random variables (parameters λ_1, λ_2) is another Poisson random variable (parameter $\lambda_1 + \lambda_2$), by Lemma (3.1). Thus

$$M_P : \Omega \to \mathbb{R}$$

is defined as follows:

$$M_P(i_{AT}, i_{CG}) = Poisson(i_{AT}; 2\lambda_A)Poisson(i_{CG}; 2\lambda_C). \tag{3.18}$$

However, the professor's question is quite a curveball. But we have set up the probability space quite conveniently to handle queries of such complex flavors. Let E denote the event that there are at least thrice as many as C+G's than A+T's in a scan. Our interest is in the Ω space shown in Figure 3.1(a). The shaded region corresponds to

$$i_{CG} \geq 3i_{AT}$$

as shown. Then

$$P(E) = \sum_{i_{CG}=0}^{\infty} \sum_{i_{AT}=0}^{3i_{CG}} Poisson(i_{AT}; 2\lambda_A)Poisson(i_{CG}; 2\lambda_C)$$

$$= \sum_{i_{CG}=0}^{\infty} \sum_{i_{AT}=0}^{3i_{CG}} Poisson(i_{AT}; 2 \times .018)Poisson(i_{CG}; 2 \times .054)$$

$$= 0.968.$$

Continuous probability distribution.   Let's go back to the koi pond and change the scenario as follows. The scanner manufacturer noticed that each type of koi has a characteristic size (i.e., mass or weight) and since the professor's interest is in separating the types, they provide her a scanner that works along the lines of a *mass spectrometer.*

In physical chemistry, a mass spectrometer is a device that measures the mass-to-charge ratio of ions in a given sample. This is done by ionizing the sample and passing them through an electric and/or magnetic field, thus separating the ions of differing masses. The relative abundance is deduced by measuring intensities of the ion flux.

The scanner provides the relative masses of the four types in a catch. The average mass of a type scanned has been provided by the manufacturer as

$$\mu_A = \mu_T = 15$$

and

$$\mu_C = \mu_G = 36$$

and a standard deviation of

$$\sigma = 5\sqrt{2}$$

for each type.

Going back to our running example and using the 'condensed' model the fraction of the number of A+T's is denoted by the random variable X which follows a normal distribution, using Lemma (3.2),

$$X \sim Normal(x : 2\mu_C, 2\sigma^2).$$

Similarly, the fraction of the number of C+G's is denoted by the random variable Y, and

$$Y \sim Normal(y : 2\mu_A, 2\sigma^2).$$

So we define the probability space as:

$$\Omega = [0, \infty) \times [0, \infty).$$

Let E denote the event that there are thrice as many as C+G's than A+T's in a scan. Our interest is in the shaded portion of the Ω space shown in Figure 3.1(b). The shaded region corresponds to

$$i_{CG} \geq 3i_{AT}$$

as shown. Then

$$P(E) = \int_0^\infty \int_0^{3y} XY\,dY\,dX$$
$$= 0.7154.$$

3.3 The Bare Truth about Inferential Statistics

Simply put, inferential statistics is all about the study of *deviation from chance events*. In other words, is a given observation a result of

1. chance phenomenon (called the *null hypothesis* or in the genetics parlance *neutral model*), or,

2. is there an underlying nonrandom phenomenon?

The latter is usually more interesting and may actually be an evidence of some biological basis and a possible candidate for a new discovery (and yet another research paper in the sea of such findings).

Interestingly, the deviation of the observations from a random model can be measured (called the *p-value*). In the context of hypothesis testing this is possible due to an incredibly simple, yet powerful result called the *Central Limit Theorem*. In this section, we will develop the ideas that lead to this theorem.

3.3.1 Probability distribution invariants

What is common across all probability distributions? Recall that we define probability distribution as any function that satisfies Kolmogorov's axioms.

We begin by a very interesting theorem called the *Chebyshev's Inequality Theorem* which says that in any probability distribution, almost all values of the distribution lie close to the mean. The distance from the mean is in terms of standard deviation of the distribution and the number of values are a fraction of all the possible values. For example, no more than $1/4$ of the values are more than 2 standard deviations away from the mean, no more than $1/9$ are more than 3 standard deviations away, no more than $1/16$ are more than 4 standard deviations away, and so on.

The theorem is formally stated below. The proof is surprisingly straightforward and is based on *Markov's Inequality* which is discussed first. This proof simply uses the properties of integrals.

THEOREM 3.3
(Markov's inequality) *If X is a random variable and $a > 0$, then*

$$P(X \geq a) \leq \frac{E[|X|]}{a}.$$

PROOF   Let $(\Omega, \mathcal{F}, P)$ be the probability space and let

$$X : \Omega \to \mathbb{R}$$

be a random variable. Then Ω can be partitioned into disjoint sets Ω_1 and Ω_2, where

$$\Omega_1 = \{\omega \in \Omega \mid |X(\omega)| < a\} \text{ and}$$
$$\Omega_2 = \{\omega \in \Omega \mid |X(\omega)| \geq a\}.$$

By the definition of expectation,

$$E[|X|] = \int_\Omega |X(\omega)| dP$$

$$= \int_{\Omega_1} |X(\omega)| dP + \int_{\Omega_2} |X(\omega)| dP.$$

Thus,

$$E[|X|] \geq \int_{\Omega_2} |X(\omega)| dP,$$

since

$$\int_{\Omega_1} |X(\omega)| dP \geq 0.$$

But,

$$\int_{\Omega_2} |X(\omega)| dP \geq a \int_{\Omega_2} dP,$$

since $|X(\omega)| \geq a$ for all $\omega \in \Omega_2$ and

$$\int_{\Omega_2} dP = P\left(|X| \geq a\right).$$

Hence,

$$E[|X|] \geq aP\left(|X| \geq a\right).$$

$\square$

Now we are ready to prove Chebyshev's inequality which is a direct application of Markov's inequality.

THEOREM 3.4
(Chebyshev's inequality theorem) *If X is a random variable with mean μ and (finite) variance σ^2, then for any real $k > 0$,*

$$P\left(|X - \mu| \geq k\sigma\right) \leq \frac{1}{k^2}.$$

PROOF   Define a random variable Y as $(X - \mu)^2$. Setting a as

$$a = (k\sigma)^2,$$

and using Markov's inequality we obtain

$$P(|X - \mu| \geq k\sigma) \leq \frac{E\left[|X - \mu|^2\right]}{k^2\sigma^2}$$

$$= \frac{\sigma^2}{k^2\sigma^2} = \frac{1}{k^2}.$$

$\square$

3.3.2   Samples & summary statistics

Consider a very large population. Assume we repeatedly take samples of size n from the population and compute some *summary statistics*. We briefly digress here to discuss our options here.

Summary statistics.   Statistics is the branch of applied mathematics that deals with the analysis of a large collection of data. The two major branches of this discipline are *descriptive statistics* and *inferential statistics*. Descriptive statistics is the branch that concerns itself mainly with summarizing a set of data.

In this section, our focus is on *summary statistics* which are numerical values that summarize the data. One is to the study the *central tendency*, usually via the mean, median or the mode and the other is to study the *variability*, usually via range or variance.

Let's go back to the professor's pond. Let x_i be the number of T types in a scan labeled i. Let

$$x_1, x_2, \ldots, x_n$$

be the collection of such real-valued numbers i.e, the counts of type T in n scans. Let us call this collection S.

The *mean* μ of S is:

$$\mu = \frac{1}{n} \sum_{i=1}^{n} x_i.$$

The *mode* of S is the element x_i that occurs the most number of times in the collection. Sorting the elements of S as

$$x_{i_1} \le x_{i_2} \le \ldots \le x_{i_n},$$

the median is the element x_{i_l} in this sorted list, where

$$l = \begin{cases} \lfloor \frac{n}{2} \rfloor & \text{for } n \text{ odd,} \\ \frac{n}{2} \text{ and } \frac{n}{2} + 1 & \text{for } n \text{ even.} \end{cases}$$

The range of S is $(x_{\max} - x_{\min})$, where

$$x_{\min} = \min_i x_i,$$
$$x_{\max} = \max_i x_i.$$

Next, σ is the *standard deviation* and the *variance* σ^2 of S is given as:

$$\sigma^2 = \frac{1}{n} \sum_{i=1}^{n} (x_i - \mu)^2.$$

Back to samples. Recall that we repeatedly take samples of size n from a large population. We compute the mean μ and variance σ^2, as discussed in the last section, for each sample of size n. Clearly, the different samples will lead to different sample means. The distribution of these means is the *sampling distribution of the sample mean for n.*

What can we say about this distribution? What can we say about the population by studying only the samples of size n?

It is not immediately apparent from what we have seen so far. It turns out that we can say the following:

1. Let the mean of the sample be

$$\mu_{n-\text{sample}}.$$

   Let the standard deviation be

$$\sigma_{n-\text{sample}}$$

   (this is sometimes also called the *standard error*). If

$$\mu_{\text{population}}$$

   is the mean and

$$\sigma_{\text{population}}$$

   the standard deviation of the (theoretical) population that the sample is derived from, then

   - $\mu_{\text{population}}$ is usually estimated to be $\mu_{n-\text{sample}}$, and,
   - $\sigma_{\text{population}}$ is estimated to be $\sqrt{n}\,\sigma_{n-\text{sample}}$.

   Thus if we need to halve the standard error, the sample size must be quadrupled.

2. The sampling distribution of the sample mean is in fact a normal distribution

$$Normal(\mu_{n-\text{sample}}, \sigma_{n-\text{sample}})$$

   (assuming the original population is 'reasonably behaved' and the sample size n is sufficiently large). A weaker version of this is stated in the following theorem.

THEOREM 3.5

(Standardization theorem) *If*

$$X_i \sim Normal(\mu, \sigma^2),$$

$1 \leq i \leq n$, *and*

$$\bar{X} = \frac{1}{n} \sum_{i=1}^{n} X_i,$$

then

$$Z = \frac{\bar{X} - \mu}{\sigma/\sqrt{n}} \sim Normal(0, 1).$$

This follows from the *Central Limit Theorem*.

3. Consider the sample median from the sample data. It has a different sampling distribution which is usually not normal.

This is just to indicate that not all summary statistics have a normal behavior.

Now, we look at the theorems that lie behind the observations above. The first result on the sample mean and variance is easily obtained by simply using the definition of expectations.

THEOREM 3.6
(Sample mean & variance theorem) *If*

$$X_1, X_2, X_3, \ldots,$$

is an infinite sequence of random variables, which are pairwise independent and each of whose distribution has the same mean μ and variance σ^2 and

$$\bar{X}_n = \frac{X_1 + \ldots + X_n}{n},$$

then

$$E[\bar{X}_n] = \mu$$

and

$$V[\bar{X}_n] = \frac{\sigma^2}{n},$$

i.e., the mean of the distribution of random variable $\bar{X}_n$ is μ and its standard deviation $\sigma/n^{\frac{1}{2}}$.

PROOF By linearity of expectation we have,

$$E(\bar{X}_n) = E\left[\frac{X_1 + X_2 + \ldots + X_n}{n}\right]$$

$$= \frac{E[X_1] + E[X_2] + \ldots + E[X_n]}{n}$$

$$= \mu.$$

Next,

$$E[\bar{X}_n^2] = E\left[\left(\frac{\sum_i X_i}{n}\right)^2\right]$$

$$= \frac{1}{n^2}E\left[\sum_i X_i^2 + \sum_{i \neq j} 2X_i X_j\right]$$

$$= \frac{1}{n^2}\left(\sum_i E[X_i^2] + 2\sum_{i \neq j} E[X_i X_j]\right).$$

The last step above is due to linearity of expectations, The random variables are independent thus for each i and j,

$$E[X_i X_j] = E[X_i]E[X_j].$$

Further, for each i,

$$E[X_i^2] = \sigma^2 + \mu^2.$$

Hence we have,

$$E[\bar{X}_n^2] = \frac{1}{n^2}\left(\sum_i (\sigma^2 + \mu^2) + 2\sum_{i \neq j} \mu^2\right)$$

$$= \frac{1}{n^2}\left(n(\sigma^2 + \mu^2) + 2\frac{n^2 - n}{2}\mu^2\right).$$

Recall that

$$V[\bar{X}_n] = E[\bar{X}_n^2] - E[\bar{X}_n]^2.$$

Thus,

$$V[\bar{X}_n] = \left(\frac{\sigma^2}{n} + \mu^2\right) - \mu^2$$

$$= \frac{\sigma^2}{n}.$$

$\Box$

In the last theorem, we saw that the sample mean is μ. *Law of Large Numbers* gives a stronger result regarding the distribution of the random variable $\bar{X}_n$. It says that as n increases the distribution concentrates around μ. The formal statement and proof is given below.

THEOREM 3.7
(Law of large numbers) *If*

$$X_1, X_2, X_3, \ldots,$$

is an infinite sequence of random variables, which are pairwise independent and each of whose distribution has the same mean μ and variance σ^2, then for every $\epsilon > 0$,

$$\lim_{n\to\infty} P(|\bar{X}_n - \mu| < \epsilon) = 1.$$

where

$$\bar{X}_n = \frac{X_1 + X_2 + \ldots + X_n}{n}$$

is the sample average.

PROOF   By Theorem (3.6),

$$E[\bar{X}_n] = \mu$$

and

$$V[\bar{X}_n] = \frac{\sigma}{\sqrt{n}}.$$

For any $k > 0$, Chebyshev's inequality on the random variable $\bar{X}_n$ gives

$$P\left(|\bar{X}_n - \mu| \geq k\frac{\sigma}{\sqrt{n}}\right) \leq \frac{1}{k^2}.$$

Thus letting

$$\epsilon = k\frac{\sigma}{n},$$

we get

$$P\left(|\bar{X}_n - \mu| \geq \epsilon\right) \leq \frac{\sigma^2}{\epsilon^2 n}.$$

In other words, since P is a probability distribution,

$$P\left(|\bar{X}_n - \mu| < \epsilon\right) \leq 1 - \frac{\sigma^2}{\epsilon^2 n}.$$

Note that since σ is finite and ϵ is fixed,

$$\lim_{n\to\infty} \frac{\sigma^2}{\epsilon^2 n} = 0.$$

Thus

$$\lim_{n\to\infty} P(|\bar{X}_n - \mu| < \epsilon) = 1.$$

□

3.3.3 The central limit theorem

A much stronger version of the Large Number Theorem is the Central Limit Theorem which also gives the distribution of the sample mean as the sample size grows. Assuming the random variables have finite variances, regardless of the underlying distribution of the random variables, the standardized difference between the sum of the random variables and the mean of this sum, converges in distribution to a standard normal random variable.

In this section we give only an indication of the proof of the Central Limit Theorem, using methods from *Fourier analysis*. For a more rigorous proof the reader must consult a standard textbook on probability theory [Fel68].

We first define a characteristic function and state four important properties that we use in the proof of the theorem.

Characteristic function & its properties. Let F_X be the probability density function of a random variable X. The characteristic function Φ_X of X is defined by:

$$\Phi_X(t) = \int_{-\infty}^{\infty} e^{itx} F(x) dx.$$

Note that here we mean

$$i = \sqrt{-1}.$$

For those of you familiar with Fourier analysis, this is just the Fourier transform of the probability density function of X. The four crucial properties of the characteristic function we will be using are as follows.

1. For any constant $a \neq 0$ and random variable X,

   $$\Phi_{aX}(t) = \Phi_X(at).$$

   This is easy to verify from the definition.

2. The characteristic functions of random variables also have the nice property that the characteristic function of the sum of two independent random variables is the product of the characteristic functions of the individual random variables. More precisely, if $X = X_1 + X_2$ then

   $$\Phi_X = \Phi_{X_1} \Phi_{X_2}.$$

   We leave the proof of this as an exercise for the reader.

3. Using differentiation under the integral sign, we have

   $$\Phi_X(t) = \int_{-\infty}^{\infty} e^{itx} F(x) dx$$

   $$\Phi'_X(t) = \int_{-\infty}^{\infty} ix e^{itx} F(x) dx$$

   $$\Phi''_X(t) = \int_{-\infty}^{\infty} (ix)^2 e^{itx} F(x) dx$$

Using the above and Taylor's formula, we have that for all small enough $t > 0$,

$$\Phi_X(t) = \Phi_X(0) + \frac{\Phi'_X(0)}{1!}t + \frac{\Phi''_X(0)}{2!}t^2 + \ldots$$

$$= 1 + i\frac{E[X]}{1}t + i^2\frac{E[X^2]}{2!}t^2 + o(t^2).$$

4. If

$$X \sim Normal(0,1)$$

implying that

$$F_X(x) = \frac{1}{\sqrt{2\pi}}e^{-x^2/2},$$

then one can show using complex integration that

$$\Phi_X(t) = e^{-t^2/2}.$$

THEOREM 3.8
(Central limit theorem) *If*

$$X_1, X_2, X_3, \ldots,$$

is an infinite sequence of random variables, which are pairwise independent and each of whose distribution has the same mean μ and variance σ^2, then

$$\lim_{n\to\infty} S_n \sim Normal(n\mu, \sigma^2 n),$$

where random variable S_n is defined by,

$$S_n = X_1 + \ldots + X_n.$$

Normalizing S_n as Z_n, we have the following. If

$$Z_n = \frac{S_n - n\mu}{\sigma\sqrt{n}},$$

then

$$\lim_{n\to\infty} Z_n \sim Normal(0,1).$$

PROOF   Let

$$Y_i = \frac{X_i - \mu}{\sigma}.$$

Then Y_i is a random variable with mean 0 and standard deviation 1 and

$$Z_n = \frac{1}{\sqrt{n}}\sum_{i=1}^{n} Y_i.$$

Applying properties (1) and (2)

$$\Phi_{Z_n}(t) = \prod_{i=1}^{n} \Phi_{Y_i}\left(\frac{t}{\sqrt{n}}\right). \tag{3.19}$$

Using Property 3, for all $i > 0$,

$$\Phi_{Y_i}(t) = 1 - \frac{t^2}{2} + o(t^2). \tag{3.20}$$

It follows from the Equations (3.19) and (3.20),

$$\Phi_{Z_n}(t) = \left(1 - \frac{t^2}{2n} + \frac{\gamma(t,n)}{n}\right)^n,$$

where

$$\lim_{n\to\infty} \gamma(t,n) = 0.$$

Hence

$$\lim_{n\to\infty} \Phi_{Z_n}(t) = \lim_{n\to\infty} \left(1 - \frac{t^2}{2n} + \gamma(t,n)\right)^n$$
$$= e^{-t^2/2}.$$

Thus the characteristic function of Z_n approaches $e^{-t^2/2}$ as n approaches ∞. By Property 4, we know that

$$e^{-t^2/2}$$

is the characteristic function of a normally distributed random variable with mean 0 and standard deviation 1. This implies (with some more work, which we omit) that the probability distribution of Z_n converges to

$$Normal(0,1)$$

as n approaches ∞. ⬜

This is a somewhat astonishing and a strong result, particularly when you realize that each X_i can have *any* distribution, as long as the mean is μ and the variance is σ^2. Under these conditions, the sample sum S_n approaches the normal distribution

$$Normal(n\mu, \sigma^2 n)$$

as n approaches ∞. Very often, it is simpler to use the standard normal distribution

$$Normal(0,1)$$

and Z_n converges to this distribution as n approaches ∞.

3.3.4   Statistical significance (p-value)

An observed event is significant *if it is unlikely to have occurred by chance.*
The significance of the event is also called its *p-value*, a real number in the
interval $[0, 1]$. The smaller the p-value, the more significant the occurrence of
the event.

Theoretical framework & practical connotations.   Note that in our
framework of the probability space

$$(\Omega, \mathcal{F}, P),$$

the p-value of an event $E \in \mathcal{F}$ is simply the probability of the event

$$P(E).$$

The probability distribution P models the 'chance' phenomenon (note that
this model does not allow supporting an alternative phenomenon or explanation). It is important to understand this for a correct interpretation of the
p-value.

However, in practice often a binary question needs to be answered:

Is the observed event statistically significant?

To deal with this question, certain thresholds are widely used and more or
less universally accepted.

When p-value (probability) of an event E is

$$P(E) = p,$$

then the level of significance is defined to be $p \times 100\%$. Fixing a threshold at
α (generally $0 \leq \alpha \ll 0.5$), an event E is significant at

$$\alpha \times 100\%$$

level if

$$P(E) \leq \alpha.$$

Or, the event is simply *statistically significant* or *interesting*.

The following thresholds (also known as *significance level*), are generally
used in practice:

1. ($\alpha = 0.10$): denoted by α and the level of significance is 10%.

2. ($\alpha = 0.05$): denoted by α^* and the level of significance is 5%.

3. ($\alpha = 0.01$): denoted by α^{**} and the level of significance is 1%.

Clearly the result is most impressive when threshold α^{**} is used and the least
impressive when α is used.

An alternative view of *significance level* (usually in the context of traditional
hypothesis testing) is as follows. If the null hypothesis is true, the significance
level is the (largest) probability that the observation will be rejected in error.

Concrete example. We go back to the professor's koi pond under the uniform model where we treat the C and G types as one *kind* and the A and T types as the other. We wish to test this hypothesis:

> *The chance of scanning each kind is equal and the 1-scanner is not favorable towards any particular kind.* This is our null hypothesis.

We study this in four steps.

1. Experiment: We will carry out 20 scans using the 1-scanner and count the number of C+G (to be interpreted as either C or G) kind in each scan.

2. Probability Space: The discrete probability space

$$(\Omega, 2^{\Omega}, M_P)$$

   is defined as follows:

$$\Omega = \{0, 1, 2, \ldots, 20\}.$$

   By our null hypothesis, M_P is defined by the distribution

$$Binomial(20, 1/2).$$

   Thus,

$$M_P(\omega \in \Omega) = \binom{20}{\omega} p^{\omega} (1-p)^{20-\omega}.$$

3. Outcome of the experiment: Let X denote the number of C+G types in a scan. Suppose in our experiment of 20 scans we see 16 C+G types. Then

$$P(X \geq 16) = \sum_{k=16}^{20} M_P(k)$$

$$= \sum_{k=16}^{20} \binom{20}{k} \left(\frac{1}{2}\right)^{20}$$

$$= 0.0573$$

   In other words,

$$\text{p-value} = 0.0573$$

4. Significance evaluation: We use α^* and α below to evaluate the statistical significance.

   - Using a threshold of α^*, since p-value $> \alpha^*$, we say: '*Observing 16 C+G kind is not statistically significant at the 5% level*'. In other words, at the 5% significance level, the null hypothesis is true.

- Using a threshold of α, since p-value $< \alpha$, we say: '*Observing 16 C+G kind is statistically significant at the 10% level.*' In other words, at the 10% significance level, the null hypothesis is false.

Our conclusion is as follows: At the 5% significance level, the observation is not 'surprising'. But at the 10% significance level, the observation is 'surprising'.

3.4   Summary

The chapter was a whirlwind tour of basic statistics and probability. It is worth realizing that this field has taken at least a century to mature, and what it has to offer is undoubtedly useful as well as beautiful. A correct modeling of real biological data will require understanding not only of the biology but also of the probabilistic and statistical methods.

The chapter has been quite substantial in keeping with the amount of work in bioinformatics literature that use statistical and probabilistic ideas. *Information theory* is an equally interesting field that deserves some discussion here. Since we have already introduced the reader to random variables and expectations, a few basic ideas in information theory are explored through Exercise 21.

3.5   Exercises

Exercise 12 *Consider the Kolmogorov's axioms of Section (3.2.1.3). Show that under these conditions, for each $E \in \mathcal{F}$,*

$$0 \leq P(E) \leq 1.$$

Exercise 13 (alternative axioms) *We call the following the* SB *Axioms on*

$$P : \mathcal{F} \to \mathbb{R}_{\geq 0}$$

where Ω is finite:

1. $P(\emptyset) = 0.$

2. $P(\Omega) = 1.$

3. *For each $E_1, E_2 \in \mathcal{F}$,*

$$P(E_1 \cup E_2) = P(E_1) + P(E_2) - P(E_1 \cap E_2).$$

Prove the following statements.

1. *Kolomogorov Axiom 1 follows from the SB Axioms.*

2. $P(\emptyset) = 0$ *follows from Kolomogorov Axioms.*

3. *The Kolmogorov and SB Axioms are equivalent.*

Exercise 14 (Exclusive-inclusive principle)

1. *Show that*
$$S_1 \geq S_2 \geq \ldots \geq S_n,$$
where the S_i's are defined as in Theorem (3.2).

2. *Show that*
$$P(E_1 \cup E_2 \cup \ldots \cup E_n) \begin{cases} \leq \sum_{i=1}^{k}(-1)^{i+1}S_i & \text{for } k \text{ odd}, \\ \geq \sum_{i=1}^{k}(-1)^{i+1}S_i & \text{for } k \text{ even}. \end{cases}$$

Hint: Use Equation (3.5).

Exercise 15 1. *Let $pr(x)$ be defined as follows (Eqn (3.11)),*
$$pr(x) = \begin{cases} 2/9 & \text{if } x = A, \\ 2/9 & \text{if } x = C, \\ 2/9 & \text{if } x = G, \\ 2/9 & \text{if } x = T, \\ 1/9 & \text{otherwise.} \end{cases}$$

For each $uv \in \Omega$, let
$$M_P(uv) = pr(u)pr(v).$$

Show that M_P is a probability mass function.

2. *Let $pr(x)$ be defined as follows (Eqn (3.13)),*
$$pr(x) = \begin{cases} 1/9 & \text{if } x = A \\ 3/9 & \text{if } x = C \\ 3/9 & \text{if } x = G \\ 1/9 & \text{if } x = T \\ 1/9 & \text{otherwise} \end{cases}$$

For each $uv \in \Omega$, let
$$M_P(uv) = pr(u)pr(v).$$

Show that M_P is a probability mass function.

Hint: What is $\sum_{uv \in \Omega} p(u)p(v)$?

Exercise 16 *Consider a k-scanner, CG-rich model of the koi pond scenario used in this chapter.*

1. *What is the probability of having a nonhomogenous scan?*

2. *What is the average number of A's in a scan?*

Exercise 17 *Show that M_P as defined in Equation (3.17) as*

$$M_P(i_A, i_C, i_G, i_T) = \prod_{z=\{A,\ C,\ G,\ T\}} Poisson(i_z; \lambda_z)$$

satisfies the Kolmogorov's axioms.

Exercise 18 (Inequality property of expectations) *Based on the definition of $E[X]$, prove the following: If*

$$X_1 \leq X_2$$

then

$$E[X_1] \leq E[X_2].$$

Exercise 19 *Let a function*

$$f_\lambda(k)$$

defined on the integers $\mathbb{Z}$ i.e., on

$$\ldots, -1, 0, 1, \ldots,$$

for a given $\lambda > 0$, be as follows:

$$
\begin{aligned}
f_\lambda(k) &= e^{-\lambda}, & \text{if } k = 0, \\
&= \frac{e^{-\lambda}\lambda^{|k|}}{2|k|!}, & \text{otherwise.}
\end{aligned}
$$

1. *Show that f_λ is a probability mass function.*

2. *If the probability mass function of X is f_λ, we say a random variable*

$$X \sim pseudoPoisson(\lambda).$$

If two independent random variables X_1 and X_2 are such that

$$X_1 \sim pseudoPoisson(\lambda) \text{ and}$$
$$X_2 \sim pseudoPoisson(\lambda),$$

and

$$X = X_1 + X_2,$$

then determine $pr(X = k)$.

Hint:

1. Show that

$$\sum_{k=-\infty}^{\infty} f(k; \lambda) = 1.$$

2. The *convolution* of functions f_1 and f_2 is written $f_1 * f_2$ and is given by

$$f_1 * f_2(x) = \int_{-\infty}^{\infty} f_1(t) f_2(x - t) dt.$$

   We assume that the functions f_1 and f_2 are defined on the whole real line, extending them by 0 outside their domains of definition otherwise. For the intrepid reader: can you prove the following?

   (a)

   $$f_1 * f_2 = f_2 * f_1.$$

   (b) The probability distribution of the sum of two independent random variables is the convolution of each of their distributions.

   Back to our problem: What is the convolution of f_λ with itself?

Exercise 20 (Normal random variables) *If*

$$X_1 \sim Normal(\mu_1, \sigma_1^2) \text{ and}$$
$$X_2 \sim Normal(\mu_2, \sigma_2^2),$$

are two independent random variables, then show that

$$X = X_1 + X_2 \sim Normal(\mu_1 + \mu_2, \sigma_1^2 + \sigma_2^2)$$

Hint: Use convolution (see Exercise 19) of the two functions:

$$f_X(x) = \int_{-\infty}^{\infty} f_{X_1}(x - y) f_{X_2}(y) dy.$$

Exercise 21 (Information theory) *In information theory, the* entropy *of a random variable X is a measure of the amount of 'uncertainty' of X. In the following, we consider discrete random variables.*

1. Let $pr(k)$ denote the probability

$$pr(X = k).$$

Then the entropy of X is defined to be

$$\begin{aligned} H(X) &= \sum_k pr(k) \log \frac{1}{pr(k)} \\ &= -\sum_k pr(k) \log pr(k). \end{aligned}$$

(a) Using the definition of entropy, show that

$$H(X) = E[-\log(pr(x))].$$

(b) Show that the entropy $H(X)$ is maximized under the uniform distribution i.e, each outcome is equally likely.

2. Let X_1 and X_2 be two discrete random variables and let

$$pr(k_i)$$

denote $pr(X_i = k_i)$, $i = 1, 2$. Let

$$pr(k_1, k_2)$$

denote the joint probability of $X_1 = k_1$ and $X_2 = k_2$.

(a) The joint entropy is defined to be

$$\begin{aligned} H(X_1, X_2) &= \sum_{k_1, k_2} pr(k_1, k_2) \log \frac{1}{pr(k_1, k_2)} \\ &= -\sum_{k_1, k_2} pr(k_1, k_2) \log pr(k_1, k_2). \end{aligned}$$

(b) The conditional entropy of X_1 given X_2 is defined to be

$$H(X_1 | X_2) = \sum_{k_1, k_2} pr(k_1, k_2) \log \frac{pr(k_2)}{pr(k_1, k_2)}.$$

Show that

$$H(X|Y) = H(X, Y) - H(Y).$$

3. Mutual information is a measure of information that can be obtained about one random variable by observing another. This is given as:

$$I(X_1; X_2) = \sum_{k_1, k_2} pr(k_1, k_2) \log \frac{pr(k_1, k_2)}{pr(k_1) pr(k_2)}.$$

Show that

$$(1) \quad I(X_1; X_2) = H(X_1) - H(X_1|X_2)$$
$$(2) \quad I(X_1; X_2) = I(X_2; X_1).$$

Hint: To show (1)(b) above use induction on k (base case $k = 1$).
A few interesting nuggets about entropy and random variables:

1. Note that when X is a continuous random variable with density function $f(x)$, $H(X)$ is defined as

$$H(X) = \int_{-\infty}^{\infty} f(x) \log \frac{1}{f(x)} dx.$$

2. The normal distribution $Normal(\mu, \sigma^2)$ has the maximum entropy over all distributions on reals with mean μ and variance σ^2. Thus if we know only the mean and variance of a distribution, it is very reasonable to assume it to be a normal distribution.

3. The *uniform distribution* on the interval $[a, b]$ is defined by the probability density function $f(x)$ as

$$f(x) = \begin{cases} \frac{1}{b-a} & \text{when } a \le x \le b, \\ 0 & \text{otherwise.} \end{cases}$$

The uniform distribution is a maximum entropy distribution among all continuous distributions defined on $[a, b]$ and 0 elsewhere.

Exercise 22 (Expected minimum, maximum values)

1. *Show that the expected value of the minimum of n independent random variables, uniformly distributed over $(0, 1)$, is*

$$\frac{1}{n+1}.$$

2. *Show that the expected value of the maximum of n independent random variables, uniformly distributed over $(0, 1)$, is*

$$\frac{n}{n+1}.$$

3. *Consider the MaxMin Problem described in Exercise 10.*
   If each edge weight $wt(v_1, v_2)$ is distributed uniformly over $(0,1)$, then what is the expected path length?
   (Recall that the path length is the maximum over all paths and in each path the length is the minimum of all the edge weights.)

Hint: 1. & 2. Let $n = 2$. The value of the min and max functions respectively, is constant along the dashed lines shown below. We need to integrate these two functions over the unit area. Note that for the min function, in the lower triangle $min(x, y) = y$ and in the upper triangle $min(x, y) = x$. The reverse is true for the max function.

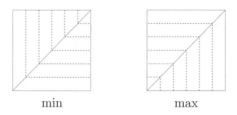

| min | max |

3. Use 1 and 2. and the expected path depth (see Exercise 11).

Comments

My experience with students (and even some practicing scientists) has been that there is a general haziness when it comes to interpreting or understanding statistical computations. I prefer Kolmogorov's axiomatic approach to probability since it is very clean and elegant. In this framework, it is easy to formulate the questions, identify the assumptions and most importantly convince yourself and others about the import and the correctness of your models.

A small section of this chapter belabors on the proofs of some fundamental theorems in probability and statistics. This is just to show that most of the results we need to use in bioinformatics are fairly simple and the basis of these foundational results can be understood by most who have a nodding acquaintance with elementary calculus.

In my mind the most intriguing theorem in this area is the *Central Limit Theorem* (and also the *Large Number Theorem*). While the latter, quite surprisingly, has an utterly simple proof, the former requires some familiarity with function convergence to appreciate the underlying ideas. In a sense it is a relief to realize that the proof of the most used theorem in bioinformatics is not too difficult.

Chapter 4

What Are Patterns?

> *Patterns lie in the eyes of the beholder*
> - anonymous bioinformaticist

4.1   Introduction

Patterns haunt the mind of a seeker. And, when the seeker is equipped with a computer with boundless capabilities, limitless memory, and easily accessible data repositories to dig through, there is no underestimating the care that should be undertaken in defining such a task and interpreting the results. For instance, the human genome alone is 3,000,000,000 nucleotides long! A sequence of 3 billion characters defined on an alphabet size of four (A, C, G, T) offers endless possibilities for useful as well as possibly useless nuggets of information.[1]

The genomes are so large in terms of nucleotide lengths that a database [2] reports the sizes of the genomes by its weight in picograms (1 picogram = one trillionth of a gram). It is also important to note that genome size does not correlate with organismal complexity. For instance, many plant species and some single-celled protists, have genomes much larger than that of humans. Also, genome sizes (called *C-values*) vary enormously among species and that relationship of this variation to the number of genes in the species is unclear. The presence of noncoding regions settles the question to a certain extent in eukaryotes. The human genome, for example, comprises only about 1.5% protein-coding genes, with the remaining being various types of noncoding DNA, especially transposable elements.[3]

Hence our tools of analysis or discovery must be very carefully designed. Here we discuss models that are mathematically and statistically 'sensible' and hopefully, this can be transposed to biologically 'meaningful'.

---

[1] Also, Ramsey theory states that even random sequences when sufficiently large will contain some sort of pattern (or order).

[2] Gregory, T.R. (2007). Animal Genome Size Database. http://www.genomesize.com

[3] As reported by the International Human Genome Sequencing Consortium 2001.

4.2   Common Thread

What is a pattern? In our context, a pattern is a nonrandom entity. We need a characterization that is amenable to objective scrutiny or an automatic discovery. We define a pattern to be a nonunique phenomenon.[4] A pattern could be rare or rampant. But, a unique event is not a pattern. The term *pattern* and *motif* is used interchangeably in the book. The choice of one over the other in different contexts is mainly to be consistent with published literature.

We start with the premise that we do not know the questions, but believe the answers to be embedded in the observations or a given body of data. What must the question be?

The hope is that if a phenomenon appears multiple times it is likely to tell a tale of interest. In nature, if a pattern is seen in a structure, or in a DNA coding segment, then perhaps it is a sign that nature is reusing the 'objects' or 'modules' and could be of potential interest.

Is it possible to elicit invariant features independent of the precise definition of the 'phenomenon' ?

4.3   Pattern Duality

Let the input data be I. A *pattern* is a nonunique phenomenon observed in I. Let this pattern, say p, *occur* k distinct times in the data. The k *occurrences* are represented as an occurrence list or a *location list*, $\mathcal{L}_p$, written as

$$\mathcal{L}_p = \{i \mid p \text{ occurs at } location \ i \text{ in } I\}. \tag{4.1}$$

If I were m-dimensional, then a *location* in the input could be specified by an m-tuple

$$i = (i_1, i_2, \ldots, i_m).$$

We call the description of p as a combinatorial feature and its location list $\mathcal{L}_p$ as a statistical feature. This is so called since the former offers a description of the pattern and the latter denotes its frequency in the data.

Statistical definition (using $\mathcal{L}_p$).   We next address the following curiosity:

Given I, does $\mathcal{L}$, a collection of positions on I define a pattern p?

---

[4]See Exercises 32, 33 and 34 for possible alternative models.

There are at least two issues with this:

1. Firstly, there may be no pattern p that occurs at the positions in $\mathcal{L}$.

2. Secondly, if there is indeed a p that occurs at each position $i \in \mathcal{L}$, it is still possible that p may occur at

$$j \notin \mathcal{L}.$$

Thus this definition of a pattern is 'well-defined' or 'nontrivial' since any arbitrary collection of positions does not qualify as a pattern specification.

Back to duality. It is interesting to note that given an input I, a pattern is *completely* specified by

1. either its combinatorial description, p, written as (I, p),

2. or its statistical description, $\mathcal{L}_p$, written as $(I, \mathcal{L}_p)$.

In other words, (I, p) is determined by $(I, \mathcal{L}_p)$ and conversely.

Fuzziness of an occurrence $i \in \mathcal{L}$. Consider the following example where the pattern is defined as an uninterrupted sequence such as

$$p = C \, G \, T \, T \, T.$$

The occurrence of this pattern in an input is shown as a box below:

$$I = \text{A A} \boxed{\text{C G T T T}} \text{C G}$$
$$\phantom{I = \text{A A }}\uparrow \uparrow \uparrow \uparrow \uparrow$$
$$i = 1 \; 2 \; 3 \; 4 \; 5 \; 6 \; 7 \; 8 \; 9$$

What is the occurrence position, i, of p ? Should i be 3 or 4 or 5 or 6 or 7? It might be easier to follow some convention, such as the left most position ($i = 3$). But this has an unreasonable bias to the left. Perhaps a more unbiased choice of i is 5. Thus it is merely a convention and it is important to bear that in mind.

To keep the generality, without being bogged down with technicalities, we assume that two positions

$$i \in \mathcal{L}_{p_1} \text{ and } j \in \mathcal{L}_{p_2}$$

are equal, written as

$$i =_\delta j,$$

when the following holds

$$|i - j| \leq \delta(i, j),$$

for some appropriately chosen $\delta(i, j) \geq 0$. To avoid clutter, we omit the subscript δ in $=_\delta$. Thus, for $i \in \mathcal{L}_{p_1}$ and $j \in \mathcal{L}_{p_2}$,

$$(i = j) \Leftrightarrow (|i - j| \leq \delta(i, j)).$$

This is also discussed in the two concrete examples of Section 4.4.4.

Back to pattern specification. To summarize, a pattern is uniquely defined by

1. specifying the form of p (such as a string or as a permutation or a partial order or a boolean expression and so on) and

2. its occurrence (i.e., what condition must be satisfied to say that p occurs at position i in the input).

Thus a pattern can be specified in many and varied ways, but its location list, mercifully, is always a set. This gives a handle to extract the common properties of patterns, independent of its specific form.

4.3.1   Operators on p

What operators should (or can) be defined on patterns? Can we identify them even before assigning a specific definition to a pattern?

Here we resort to the duality of the patterns. So, what are the operations defined on location lists? Since sets have union and intersection operations defined on them, then it is meaningful to define two operators as follows. The plus operator ($\oplus$):

$$p = p_1 \oplus p_2 \Leftrightarrow \mathcal{L}_p = \mathcal{L}_{p_1} \cap \mathcal{L}_{p_2}.$$

The times operator ($\otimes$):

$$p = p_1 \otimes p_2 \Leftrightarrow \mathcal{L}_p = \mathcal{L}_{p_1} \cup \mathcal{L}_{p_2}.$$

Further, the following properties hold which again are a direct consequence of the corresponding set operation.

$$p_1 \oplus p_2 = p_2 \oplus p_1$$
$$p_1 \otimes p_2 = p_2 \otimes p_1$$
$$(p_1 \oplus p_2) \oplus p_3 = p_1 \oplus (p_2 \oplus p_3)$$
$$(p_1 \otimes p_2) \otimes p_3 = p_1 \otimes (p_2 \otimes p_3)$$
$$(p_1 \otimes p_2) \oplus p_3 = (p_1 \oplus p_3) \otimes (p_2 \oplus p_3)$$
$$(p_1 \oplus p_2) \otimes p_3 = (p_1 \otimes p_3) \oplus (p_2 \otimes p_3)$$

4.4   Irredundant Patterns

Let $P(I)$ be the set of *all* patterns in an input I.

Patterns by their very nature have a large component of repetitiveness. A very desirable requirement is to identify (and eliminate) these duplications.[5] So a very natural question arises:

Is there some redundancy in $P(I)$?

In other words, is it possible to trim $P(I)$ of its belly-fat so to say, without losing its essence?

This is the motivation for the next definition. For $p \in P(I)$ and a set of patterns

$$\{p_1, p_2, \ldots, p_l\} \subset P(I),$$

with each $p_i \neq p$, let

$$\mathcal{L}_p = \mathcal{L}_{p_1} \cup \mathcal{L}_{p_2} \cup \ldots \cup \mathcal{L}_{p_l}.$$

Then p is *redundant* with respect to (w.r.t.) these l patterns which we write as

$$\{p_1, p_2, \ldots, p_l\} \hookrightarrow p.$$

Also the pattern p is defined as

$$p_1 \otimes p_2 \otimes \ldots \otimes p_l = p.$$

If p is such that for each $l \geq 1$ there exists no $p_1, p_2, \ldots, p_l \in P(I)$, with

$$\{p_1, p_2, \ldots, p_l\} \hookrightarrow p,$$

then p is called *irredundant*.

Note that redundancy and irredundancy of patterns is always with respect to I. There is no notion of irredundancy or redundancy for an isolated form p.

Also, since redundancy is defined in terms of sets (location lists), this holds across the board for patterns of all specifications.

4.4.1   Special case: maximality

The reader may already have an intuitive sense for maximality, at least for string patterns. For a concrete example, see the three boxed occurrences of

$$p = \text{A T T G C},$$

on an input string I:

$$I = \text{G A} \boxed{\text{A T T G C}} \text{G G} \boxed{\text{A T T G C}} \text{C A C} \boxed{\text{A T T G C}} \text{C T}$$

---

[5]The reader may be familiar with the the use of maximality which is widely used to remove redundancies.

and

$$\mathcal{L}_p = \{3, 10, 18\}.$$

Intuitively, what are the nonmaximal patterns? Assume, for convenience, that a string pattern must have at least two elements. Then all the nonmaximal patterns $p_1, p_2, \ldots, p_9$ are shown below:

$$p = \text{A T T G C} \quad p_1 = \text{A T T G} \quad p_2 = \text{A T T} \quad p_3 = \text{A T}$$
$$p_4 = \text{T T G C} \quad p_5 = \text{T T G} \quad p_6 = \text{T T}$$
$$p_7 = \text{T G C} \quad p_8 = \text{T G}$$
$$p_9 = \text{G C}$$

Note that if the input were I', then the occurrences of p_8 is shown below:

$$I' = \text{G A} \;\boxed{\text{A T}\; \boxed{\text{T G}}\; \text{C}}\; \text{G G} \;\boxed{\text{A T}\; \boxed{\text{T G}}\; \text{C}}\; \text{C A C} \;\boxed{\text{A T}\; \boxed{\text{T G}}\; \text{C}}\; \text{C} \;\boxed{\text{T G}}$$

Here p_8, a substring of p, is considered maximal [6] since it has an independent fourth occurrence.

However, here we define maximality as a special case of irredundancy. If l is restricted to 1, this is a special case and p is called *nonmaximal* w.r.t. p_1. If p is such that there exists no $p_1 \in P(I)$ with

$$p_1 \hookrightarrow p,$$

then p is called *maximal*.

In our concrete example, we observe that (using the 'fuzzy' definition of position i), for $1 \le j \le 9$,

$$\mathcal{L}_p = \mathcal{L}_{p_i}.$$

This says that at every occurrence of p, there is also an occurrence of p_j for each $1 \le j \le 9$. Thus one of the patterns must be more informative than all others and this pattern is the maximal pattern p.

Thus, if for two distinct patterns $p_1 \ne p_2$,

$$\mathcal{L}_{p_1} = \mathcal{L}_{p_2},$$

then one of them must be nonmaximal. This agrees with the intuitive notion that if a multiple motifs occur in the same positions, then the largest (or most informative) is the maximal motif.

Note that maximality is always in terms of an input I. There is no notion of maximality on an isolated string p.

---

[6]It is possible to have an alternative view and this is discussed as Exercise 84.

4.4.2 Transitivity of redundancy

Let $P(I)$ be the set of all patterns on an input I. Let $p \in P(I)$ be redundant w.r.t. $p_1, p_2, \ldots, p_l \in P(I)$, i.e.,

$$\{p_1, p_2, \ldots, p_l\} \hookrightarrow p.$$

Further, let each of p_i, $1 \le i \le l$, be redundant as follows

$$P_1 = \{p_{11}, p_{12}, \ldots, p_{1l_1}\} \hookrightarrow p_1,$$
$$P_2 = \{p_{21}, p_{22}, \ldots, p_{2l_2}\} \hookrightarrow p_2,$$
$$\vdots \quad \vdots$$
$$P_l = \{p_{l1}, p_{l2} \ldots, p_{ll_l}\} \hookrightarrow p_l.$$

with

$$\left.\begin{array}{c} p_{11}, \ p_{12}, \ \ldots, \ p_{1l_1}, \\ p_{21}, \ p_{22}, \ \ldots, \ p_{2l_2}, \\ \vdots \quad \vdots \quad \quad \vdots \\ p_{l1}, \ p_{l2}, \ \ldots, \ p_{ll_l}. \end{array}\right\} \in P(I).$$

Then p is redundant as follows

$$\left\{\begin{array}{c} p_{11}, \ p_{12}, \ \ldots, \ p_{1l_1}, \\ p_{21}, \ p_{22}, \ \ldots, \ p_{2l_2}, \\ \vdots \quad \vdots \quad \quad \vdots \\ p_{l1}, \ p_{l2}, \ \ldots, \ p_{ll_l} \end{array}\right\} \hookrightarrow p$$

In other words, if if p is redundant with respect to

$$p_1, p_2, \ldots, p_l,$$

and each p_i in turn is redundant w.r.t. set P_i, then p is redundant with respect to the union of sets, i.e.,

$$\bigcup_i P_i \hookrightarrow p.$$

4.4.3 Uniqueness property

A direct consequence of transitivity of redundancy is the following important theorem. See Exercise 23 for an illustration.

THEOREM 4.1
Let $P(I)$ be the set of all patterns that occur in a given input I. Then the collection of irredundant patterns,

$$P_{irredundant}(I) \subset P(I),$$

is unique.

COROLLARY 4.1
Let $P(I)$ be the set of all patterns that occur in a given input I. Then the collection of maximal patterns,

$$P_{maximal}(I) \subset P(I),$$

is unique.

Since the construction of the irredundant patterns result in a unique collection, i.e., they are independent of the order of the construction. This set is also called a *basis* since any pattern

$$p \notin P(I),$$

can be written as

$$p_1 \otimes p_2 \otimes \ldots \otimes p_l = p,$$

for some

$$p_1, p_2, \ldots, p_l \in P(I).$$

Thus the basis is defined as follows:

$$\begin{aligned} P_{basis}(I) &= P_{irredudant}(I) \\ &= \{p \in P(I) \mid p \text{ is irredundant}\}. \end{aligned}$$

Note that

$$P_{basis}(I) \subseteq P_{maximal}(I) \subseteq P(I).$$

4.4.4 Case studies

Here we sketch two concrete examples. The focus is on studying the relation between p and $\mathcal{L}_p$.

Concrete example 1 (string pattern). Let the input be defined on Σ. An element in Σ is called a solid character. The dot character, '.', is a wild card and stands for any character in that position. A string pattern is a sequence on

$$\Sigma \cup \{\text{'.'}\}.$$

A pattern of length one or one that does not start or end with a solid character is called a *trivial* pattern. If for $1 \leq i \leq l$, the following holds

$$p \hookrightarrow p_i,$$

then we write

$$p \hookrightarrow p_1, p_2, \ldots, p_l$$

For ease of exposition, we consider the input to be four sequences as shown below and $P(I)$ is the set of all nontrivial patterns.

$$
\begin{aligned}
s_1 &= \text{A B C D E F,} \\
s_2 &= \text{A G C D E G,} \\
s_3 &= \text{A B G D E H,} \\
s_4 &= \text{A B C G E A.}
\end{aligned}
\qquad
P(I) = \left\{
\begin{aligned}
&\text{AB, BC, CD, DE,} \\
&\text{A.C, B.D, C.E,} \\
&\text{ABC, CDE,} \\
&\text{AB.D, A.CD, A..D,} \\
&\text{BC.E, B.DE, B..E,} \\
&\text{AB..E, A..DE, A.C.E,} \\
&\text{ABC.E, A.CDE, AB.DE,} \\
&\text{A.CDE, AB.DE, ABC.E}
\end{aligned}
\right\}.
$$

Patterns & location lists. In this example a position is a two-tuple (i, j) which denotes position j in the ith sequence. Following the convention the that position represents the leftmost position of the pattern in the input, we have the following:

$$
\begin{aligned}
\mathcal{L}_{\text{ABC.E}} &= \{(1,1), (2,1), (3,1), (4,1)\}, \\
\mathcal{L}_{\text{BC.E}} &= \{(1,2), (2,2), (3,2), (4,2)\}, \\
\mathcal{L}_{\text{ABC}} &= \{(1,1), (2,1), (3,1), (4,1)\}, \\
\mathcal{L}_{\text{BC}} &= \{(1,2), (2,2), (3,2), (4,2)\}.
\end{aligned}
$$

But each location in $\mathcal{L}_{\text{BC.E}}$ is one position away from a position in $\mathcal{L}_{\text{ABC.E}}$.

$$
\mathcal{L}_{\text{BC.E}} = \{(i, j + \delta) \mid (i, j) \in \mathcal{L}_{\text{ABC.E}}\},
$$

where $\delta = 1$. The same holds for $\mathcal{L}_{\text{BC}}$. Intuitively, the lists should be the same since they are capturing the same common segments in the input with a phase shift. Thus, the location lists are 'fuzzily' equal, i.e.,

$$
\mathcal{L}_{\text{ABC.E}} = \mathcal{L}_{\text{BC.E}} = \mathcal{L}_{\text{ABC}} = \mathcal{L}_{\text{BC}}.
$$

Similar arguments hold for the following equalities:

$$
\begin{aligned}
\mathcal{L}_{\text{A.CDE}} &= \mathcal{L}_{\text{CD}} = \mathcal{L}_{\text{CDE}} = \mathcal{L}_{\text{A.CD}}, \\
\mathcal{L}_{\text{AB.DE}} &= \mathcal{L}_{\text{B.D}} = \mathcal{L}_{\text{AB.D}} = \mathcal{L}_{\text{B.DE}}, \\
\mathcal{L}_{\text{AB..E}} &= \mathcal{L}_{\text{AB}} = \mathcal{L}_{\text{B..E}}, \\
\mathcal{L}_{\text{A..DE}} &= \mathcal{L}_{\text{DE}} = \mathcal{L}_{\text{A..D}}, \\
\mathcal{L}_{\text{A.C.E}} &= \mathcal{L}_{\text{A.C}} = \mathcal{L}_{\text{C.E}}.
\end{aligned}
$$

Also,

$$
\mathcal{L}_{\text{A..DE}} = \mathcal{L}_{\text{AB.DE}} \bigcup \mathcal{L}_{\text{A.CDE}},
$$

$$
\mathcal{L}_{\text{A.C.E}} = \mathcal{L}_{\text{A.CDE}} \bigcup \mathcal{L}_{\text{ABC.E}},
$$

$$\mathcal{L}_{\text{AB..E}} = \mathcal{L}_{\text{AB.DE}} \bigcup \mathcal{L}_{\text{ABC.E}},$$

$$\mathcal{L}_{\text{A...E}} = \mathcal{L}_{\text{A.CDE}} \bigcup \mathcal{L}_{\text{AB.DE}} \bigcup \mathcal{L}_{\text{ABC.E}}.$$

What are the redundancies ? The redundancies are as shown below (redundancy with restriction $l = 1$ gives the maximal elements):

$$
\left.
\begin{array}{l}
\text{AB..E} \hookrightarrow \text{AB, B..E} \\
\text{A..DE} \hookrightarrow \text{DE, A..D} \\
\text{A.C.E} \hookrightarrow \text{A.C, C.E} \\
\text{ABC.E} \hookrightarrow \text{BC, ABC, BC.E} \\
\text{A.CDE} \hookrightarrow \text{CD, CDE, A.CD} \\
\text{AB.DE} \hookrightarrow \text{B.D, AB.D, B.DE}
\end{array}
\right\} \text{maximal}
\qquad
\begin{array}{l}
\{\text{AB.DE, A.CDE}\} \hookrightarrow \text{A..DE} \\
\{\text{A.CDE, ABC.E}\} \hookrightarrow \text{A.C.E} \\
\{\text{AB.DE, ABC.E}\} \hookrightarrow \text{AB..E} \\
\{\text{A.CDE, AB.DE, ABC.E}\} \hookrightarrow \text{A...E}
\end{array}
$$

Thus

$$P_{maximal}(I) = \{\text{AB..E, A..DE, A.C.E, ABC.E, A.CDE, AB.DE}\},$$
$$P_{basis}(I) = \{\text{A.CDE, AB.DE, ABC.E}\}.$$

Concrete example 2 (permutation pattern).

A permutation pattern is a set that appears in any order in the input. A pattern of length one is called a *trivial* pattern.

For ease of exposition, we consider the input to be two sequences and $P(I)$ is the set of all nontrivial patterns.

$$
\begin{array}{ll}
s_1 = \text{d e a b c,} & \\
s_2 = \text{c a b e d.} & P(I) = \{\{a,b\}, \{a,b,c\}, \{e,d\}, \{a,b,c,d,e\}\}.
\end{array}
$$

Patterns & location lists. In this example a position is a two-tuple (i, j) which denotes position j in the ith sequence. Following the convention the that position represents the leftmost position of the pattern in the input, we have the following:

$$\mathcal{L}_{\{a,b\}} = \{(1,3), (2,2)\},$$
$$\mathcal{L}_{\{a,b,c\}} = \{(1,3), (2,1)\},$$
$$\mathcal{L}_{\{e,d\}} = \{(1,1), (2,4)\},$$
$$\mathcal{L}_{\{a,b,c,d,e\}} = \{(1,1), (2,1)\}.$$

Each pattern is in the same two collection of segments in the input. Thus by using the sizes of the patterns, i.e.,

$$|\{a,b,c,d,e\}| = 5 \text{ and } |\{a,b\}| = 2,$$

it is possible to guess if one pattern is contained in the other (in this example the occurrence is without gaps). Hence

$$\mathcal{L}_{\{a,b\}} = \left\{ (i, j + \delta(i,j)) \mid (i,j) \in \mathcal{L}_{\{a,b,c,d,e\}} \right\},$$

where

$$\delta(1,1) = 2 \text{ and } \delta(2,1) = 1.$$

Thus the value of the δ's is clear from the context. Using similar arguments, we get the following:

$$\mathcal{L}_{\{a,b\}} = \mathcal{L}_{\{a,b,c\}} = \mathcal{L}_{\{e,d\}} = \mathcal{L}_{\{a,b,c,d,e\}}.$$

The redundancies are shown below.

$$\left. \begin{array}{l} \{a,b,c\} \hookrightarrow \{a,b\}, \\ \{a,b,c,d,e\} \hookrightarrow \{a,b,c\}, \\ \{a,b,c,d,e\} \hookrightarrow \{d,e\}. \end{array} \right\} \text{ maximal}$$

Thus

$$P_{maximal}(I) = P_{basis}(I) = \{\{a, b, c, d, e\}\}.$$

4.5 Constrained Patterns

The reality of the situations may sometimes demand a constrained version of the patterns. A *combinatorial constraint* is one that is applied on the form of the pattern. For instance, a pattern of interest may have to follow some *density constraint* or *size constraint* (see Chapter 6). A pattern specified as a boolean expression may have to have a specific form such as a *monotone form* (see Chapter 14).

A *statistical constraint* is one that is applied on the location list of the pattern. It is possible to impose a quorum constraint, k and for a pattern p to meet the quorum constraint, the following must hold

$$|\mathcal{L}_p| \geq k.$$

Thus the pattern must occur at least k times in the input to meet this quorum constraint.

4.6 When is a Pattern Specification Nontrivial?

Often in real life, confronted with a daunting task of understanding or mining on a very large set, there is a need to specify a pattern, whose definition can then be used to extract patterns objectively from the data. Then it is an extremely valid question to ask:

Is the specification nontrivial?

Here, we suggest one such condition.

The specification should be such that there must exist some collections of positions in the input, $\mathcal{L}$, that do not correspond to a pattern. See Section 4.3 for a detailed discussion. For an input I, let $\mathbf{L}$ denote all possible such sets, i.e.,

$$\mathbf{L} = \{\mathcal{L} \mid \mathcal{L} \text{ is a collection of positions } i \text{ of } I\} .$$

Consider the collection of sets that correspond to patterns by the specification:

$$\mathbf{L'} = \{\mathcal{L} \in \mathbf{L} \mid \mathcal{L} = \mathcal{L}_p \text{ for some pattern } p\} .$$

The following property is desirable (ϵ is small constant):

$$\frac{|\mathbf{L'}|}{|\mathbf{L}|} < \epsilon.$$

The specification of a pattern should not be so lax that any arbitrary collection of segments in the input qualify to be a pattern. In other words, the probability of a list of positions to correspond to a pattern p should be fairly low i.e.,

$$pr(\mathcal{L} \in \mathbf{L'}) < \epsilon.$$

4.7   Classes of Patterns

We conclude the chapter by giving the reader a taste of different possible classes of patterns in the following table. Note that the *pattern form* and the *pattern occurrence* must be specified for an unambiguous definition. For brevity, we omit the details and appeal to the reader's intuition in the following examples.

	pattern p	input I (strings)	occurrence
	string pattern		
(1a)	C T T	$s_1 = $ A $\boxed{\text{C T T}}$ C G $s_2 = $ C C G T C $s_3 = \boxed{\text{C T T}}$ C C G	appears exactly
(1b)	C . T C	$s_1 = $ A $\boxed{\text{C T T C}}$ G $s_2 = $ C $\boxed{\text{C G T C}}$ $s_3 = \boxed{\text{C T T C}}$ C G	appears with wild card
(1c)	C T T-2,3-G	$s_1 = $ A $\boxed{\text{C T T C G}}$ $s_2 = $ C C G T C $s_3 = \boxed{\text{C T T C C G}}$	appears with gap of 2 or 3 wild cards
(1d)	G T C	$s_1 = $ A $\boxed{\text{C|G T C}}$ C G $s_2 = $ C C $\boxed{\text{G T C}}$ $s_3 = \boxed{\text{G|T T C}}$ C C G	appears exactly (but input has multiple elements in places)
	permutation pattern		
(2a)	$\{g_1, g_2, g_3, g_4\}$	$s_1 = \boxed{g_2\ g_4\ g_1\ g_3}\ g_6$ $s_2 = g_7\ g_9\ \boxed{g_1\ g_2\ g_3\ g_4}$ $s_3 = g_8\ \boxed{g_3\ g_1\ g_4\ g_2}$	appears together in any order
(2b)	$\{g_1, g_2, g_3\}$	$s_1 = \boxed{g_2\ g_5\ g_1\ g_6\ g_3}$ $s_2 = g_4\ g_9\ \boxed{g_1\ g_2\ g_3}\ g_7$ $s_3 = g_6\ \boxed{g_3\ g_1\ g_8\ g_2}$	appears together in any order, with at most 1 gap
(2c)	$\{g_1, g_2, g_4\}$	$s_1 = \boxed{g_2\ g_5\ g_1\ g_4}\ g_6$ $s_2 = g_4\ g_9\ \boxed{g_1\ g_2\ g_3\ g_4}$ $s_3 = g_6\ \boxed{g_3\ g_1\ g_4\ g_2}$	appears together in any order, in fixed window size 4
(2d)	$\{m_1(2), m_2\}$	$s_1 = \boxed{m_2\ m_1\ m_2}\ m_5\ m_6$ $s_2 = m_4\ m_9\ \boxed{m_1\ m_2\ m_2}\ m_4$	appears together in any order, but m_2 appears 2 times

	pattern p	input I	occurrence
	partial order motif	strings	
(3a)		$q_1 = 7\ \boxed{1\ 2\ 3\ 4}$ $q_2 = \boxed{3\ 1\ 4\ 2}\ 8$ $q_3 = 9\ \boxed{4\ 1\ 3\ 2}$	in cluster, 1 precedes 2 (without gap)
(3b)		$q_1 = 7\ \boxed{1\ 2\ 6\ 3\ 4}$ $q_2 = \boxed{3\ 7\ 1\ 4\ 2}\ 8$ $q_3 = 9\ \boxed{4\ 1\ 3\ 8\ 2}$	in cluster, 1 precedes 2 (with gap)
(3c)	sequence pattern $2 \rightarrow 3$	$q_1 = 1\ 7\ \boxed{2\ 6\ 3}\ 4$ $q_2 = \boxed{2\ 3}\ 1\ 4\ 2\ 7$ $q_3 = 9\ 4\ 1\ \boxed{2\ 8\ 3}$	2 precedes 3 at each occurrence
(4)	topological motif	graph	
			isomorphic to a subgraph
	bicluster motif	2D array	
(5a)	cols 1,3 rows 2,3,5	$\boxed{\begin{matrix} .20 & .50 & .20 & .50 \\ \mathbf{.80} & .10 & \mathbf{.60} & .70 \\ \mathbf{.75} & .60 & \mathbf{.55} & .80 \\ .10 & .40 & .30 & .60 \\ \mathbf{.78} & .30 & \mathbf{.57} & .90 \end{matrix}}$	values along col within $\delta = 0.1$
(5b)	cols 1,3,4 rows 1,2,3,5	$\boxed{\begin{matrix} \mathbf{.80} & .50 & \mathbf{.59} & \mathbf{.50} \\ \mathbf{.60} & .10 & \mathbf{.60} & \mathbf{.55} \\ \mathbf{.75} & .60 & .35 & \mathbf{.57} \\ .10 & .40 & .30 & .60 \\ \mathbf{.78} & .30 & \mathbf{.57} & \mathbf{.58} \end{matrix}}$	most values along col within $\delta = 0.1$
(6)	boolean expression	incidence matrix	
	$m_1 \wedge (m_2 \vee m_3)$	$\begin{matrix} m_1 & m_2 & m_3 & m_4 \\ \hline \mathbf{1} & \mathbf{0} & \mathbf{1} & 0 \\ \mathbf{1} & \mathbf{1} & \mathbf{1} & 1 \\ \mathbf{1} & \mathbf{1} & \mathbf{1} & 0 \\ 1 & 0 & 0 & 1 \end{matrix}$	expression evaluates to TRUE (in rows 1, 2, 3)

4.8 Exercises

Exercise 23 (Nonuniqueness) *The chapter discussed a scheme, using re-dundancy, to trim P, the set of all patterns in the data to produce a unique set, say P'.*

Construct a specification of a pattern and an elimination scheme where the resulting P' is not unique.

Hint: Let the pattern p be an interval (l, u) on reals. If two intervals overlap and at least half of one interval is in the overlap region, then the smaller interval is eliminated from the set of intervals P. Concrete example:

$$P = \{p_1 = (1, 3), p_2 = (2, 6), p_3 = (4, 12)\}$$

1. Consider p_1 and p_2; p_1 is eliminated. Then between p_2 and p_3, p_2 is removed. Thus

$$P' = \{p_3\}.$$

2. Consider p_3 and p_2; p_2 is removed. Then between p_1 and p_3, none is removed. Thus

$$P' = \{p_1, p_3\}.$$

Exercise 24 (Trend patterns) *Let s be a sequence of real values. Discuss how the problem can be framed to extract common pattern of (additive) trends in the data?*

Hint: Use forward differences.

Exercise 25 (Nonmaximal partial order motif) *Consider a partial order motif B on n elements, which captures all the common order information of the elements across all its occurrences and there are no gaps in the occurrences (See (3a) of the table in Section 4.7).*

Give a specification of a nonmaximal partial order motif.

Hint: A nonmaximal partial order motif is on a subset of the n elements where the subsets occur without gaps in the input. How can these elements be identified in the partial order? Chapter 14 discusses partial order motifs.

Exercise 26 (Sequence motif) *Let s be an input sequence defined on an alphabet Σ. Let*

$$m = \sigma_1\,\sigma_2\,\ldots\sigma_l,$$

where $\sigma_i \in \Sigma$, then

$$\Pi(m) = \{\sigma_1, \sigma_2, \ldots, \sigma_l\}.$$

m is a sequence motif and at each occurrence this sequence may be interrupted with at most g elements from

$$\Sigma \setminus \Pi(m).$$

For example, occurrence of

$$m = \sigma_1\,\sigma_2\,\sigma_3$$

with gap, $g = 2$, is shown below:

$$\sigma_5\,\sigma_4\ \boxed{\sigma_1\,\sigma_6\,\sigma_2\,\sigma_4\,\sigma_5\,\sigma_3}\ \sigma_2.$$

Define redundancy on sequence motifs

$$m_1, m_2, \ldots, m_r,$$

where for each $i \neq j$, $1 \leq i < j \leq r$, the following holds

$$\Pi(m_i) \cap \Pi(m_j) = \emptyset.$$

Hint: Define the binary $\oplus$ operator. See also Chapter 14.

Exercise 27 (Nontrivial, nonmaximal boolean expressions) *See (6) of the table in Section 4.7 for an example of boolean expression motifs.*

1. *Note that given a an array of n boolean values corresponding to n boolean variables, there always exists an expression on the n variables that evaluates to TRUE for exactly these assignments.*

   Discuss possible specifications for nontrivial boolean expression motifs.

2. *Discuss the possible interpretation of a nonmaximal boolean expression.*

Hint: 1. See monotone boolean expressions in Chapter 14 2. See *redescriptions* in Chapter 14. What is the relationship between the two, if any?

Exercise 28 *Assuming the following*

$$p = p_1 \oplus p_2 \Leftrightarrow \mathcal{L}_p = \mathcal{L}_{p_1} \cap \mathcal{L}_{p_2},$$
$$p = p_1 \otimes p_2 \Leftrightarrow \mathcal{L}_p = \mathcal{L}_{p_1} \cup \mathcal{L}_{p_2},$$

show that the following statements hold:

1. $p_1 \oplus p_2 = p_2 \oplus p_1$.
2. $p_1 \otimes p_2 = p_2 \otimes p_1$.
3. $(p_1 \oplus p_2) \oplus p_3 = p_1 \oplus (p_2 \oplus p_3)$.
4. $(p_1 \otimes p_2) \otimes p_3 = p_1 \otimes (p_2 \otimes p_3)$.
5. $(p_1 \otimes p_2) \oplus p_3 = (p_1 \oplus p_3) \otimes (p_2 \oplus p_3)$.
6. $(p_1 \oplus p_2) \otimes p_3 = (p_1 \otimes p_3) \oplus (p_2 \otimes p_3)$.

Hint: Use the properties of sets

Exercise 29 (Operators on string patterns) *Consider (1a)-(1c) in the table of classes of patterns of Section 4.7 where each is a string pattern.*

Define operators $\oplus$ and $\otimes$ in terms of the pattern descriptions for the three cases.

Exercise 30 (Homologous sets) *Consider (1d) in the table of patterns of Section 4.7 where the pattern is defined on a string of sets (not just singleton elements), called multi-sets.*

1. *What are the issues of prime concern in this scenario?*

2. *Can the pattern also be defined on sets as characters? What should the constraints be?*

Hint: 1. If we convert these multiset input sequences to multiple sequences as follows:

$$A|C\,G\,G\,T\,C|T\,G \Leftrightarrow \begin{cases} A\,G\,G\,T\,C\,G \\ A\,G\,G\,T\,T\,G \\ C\,G\,G\,T\,C\,G \\ C\,G\,G\,T\,T\,G \end{cases}$$

How many such multiple sequences result in general? Is there any escape from this explosion?

2. It is better to allow for multiple sets in patterns, as long as each multi-set S in the pattern satisfies

$$S \subseteq S'_i,$$

where S'_i is the multi-set in the corresponding position of its ith occurrence in the input.

Exercise 31 (Nontrivial vs flexibility) *Consider the different classes of patterns shown in Section 4.7. Notice that different definitions of occurrences* *on the same (or similar) pattern forms give rise to different pattern specifi-cations. Usually the more flexible a pattern, the more likely it is to pick up false-positives. However, biological reality may demand such a flexibility.*

Discuss which of the patterns are more nontrivial than the others in each of the groups below. Why?

1. *string patterns (1a)-(1d),*

2. *permutation patterns (2a)-(2d),*

3. *partial order motifs (3a)-(3c),*

4. *bicluster motifs (5a)-(5b).*

Exercise 32 (Recombination patterns or haplotype blocks) *Patterns sometimes may not fit into the 'repeating phenomenon' framework. Instead they may be so defined so as to optimize a global function. One such exam-ple is a* haplotype block *or sometimes also called the* recombination pattern. *Consider a simplified definition of a haplotype block here.*

The input, I, is an $[n \times m]$ array, where each column j, denotes a SNP (Single Nucleotide Polymorphism) and each row i denotes a sample or an individual. The order of the columns represents the order in which the SNPs appear in a chromosome, although they may not be adjacent.

An interval, $[j_1, j_2]$, $1 \leq j_1 \leq j_2 \leq m$, is a block, b, represents the submatrix of I consisting of rows from 1 to n and columns from j_1 to j_2, and is written as

$$I_b = I_{[j_1 - j_2]}.$$

Let the length of the block be

$$l = j_2 - j_1 + 1.$$

The block pattern, b_p, *is defined to be a vector of length l where,*

$$b_p[i] = x, \text{ where the majority entries in column } i \text{ of } I_b \text{ is } x,$$

for $1 \leq i \leq l$. Given some constant,

$$0 < \alpha \leq 1,$$

a row i in block b is compatible *if the Hamming distance (number of differ-ences) between row i of block b and b_p is no more than*

$$\alpha (j_2 - j_1 + 1).$$

Given I and some α, the task is to find a minimum *number of block patterns that partition I into blocks. In other words, every column, j, must belong to exactly one block*

1. *Does this problem always have a solution? If it does, is it unique?*

2. *Note that α is defined for a row i. Is it meaningful to define such a constant for a column j?*

3. *Is there a necessity to define maximal blocks (patterns)?*

4. *Is the problem so defined 'nontrivial' ? Why?*

Hint: 1. When $\alpha = 1$, it is easy to construct an I that has no solution. Possible block partitions for an input I and $\alpha = 0.5$ are shown below.

1 2	3 4 5 6 7	8 9
1 0	1 0 1 1 0	0 1
1 0	0 0 1 0 0	0 0
1 0	1 0 1 0 1	1 0
1 1	0 1 1 0 0	0 1
1 0	0 0 1 0 1	1 0
b_1	b_2	b_3
1 0	0 0 1 0 0	0 0
p_{b_1}	p_{b_2}	p_{b_3}

(1) Block Partition 1.

1 2 3	4 5 6 7 8 9
1 0 1	0 1 1 0 0 1
1 0 0	0 1 0 0 0 0
1 0 1	0 1 0 1 1 0
1 1 0	1 1 0 0 0 1
1 0 0	0 1 0 1 1 0
b_1	b_2
1 0 0	0 1 0 0 0 0
p_{b_1}	p_{b_2}

(2) Block Partition 2.

1 2 3 4	5 6 7 8 9
1 0 1 0	1 1 0 0 1
1 0 0 0	1 0 0 0 0
1 0 1 0	1 0 1 1 0
1 1 0 1	1 0 0 0 1
1 0 0 0	1 0 1 1 0
b_1	b_2
1 0 0 0	1 0 0 0 0
p_{b_1}	p_{b_2}

(3) Block Partition 3.

1 2 3 4 5 6 7 8 9
1 0 1 0 1 1 0 0 1
1 0 0 0 1 0 0 0 0
1 0 1 0 1 0 1 1 0
1 1 0 1 1 0 0 0 1
1 0 0 0 1 0 1 1 0
b_1
1 0 0 0 1 0 0 0 0
p_{b_1}

(4) Block Partition 4.

Exercise 33 (Haplogroup patterns) *Here is another example where patterns do not fit into the 'repeating phenomenon' framework. Consider the following scenario.*

The input, I, is an $[n \times m]$ array, where each column j, denotes a SNP and each row i denotes a sample or an individual.

The task is to find patterns that divide the rows into at most K (possibly overlapping) groups.

1. *Discuss possible formalizations of this problem.*

2. *Is there a necessity to define maximal patterns?*

3. *For a 'nontrivial' formalization, devise a method to solve the optimization task.*

4. *How do the solutions with $K' < K$ groups compare?*

Hint: 1. See a possible scheme with the example below.

$$s_1 = 0\ 1\ 0\ 1\ 0\ 1\ 1\ 1\ 1\ 0$$
$$s_2 = 1\ 1\ 0\ 1\ 1\ 1\ 0\ 0\ 1\ 1$$
$$s_3 = 0\ 1\ 1\ 1\ 1\ 1\ 0\ 0\ 1\ 0$$
$$s_4 = 1\ 0\ 0\ 0\ 1\ 0\ 0\ 1\ 1\ 1$$
$$s_5 = 1\ 0\ 1\ 0\ 1\ 0\ 0\ 1\ 1\ 1$$

pattern 1 = * 1 * 1 * 1 * * * * for group $\{s_1, s_2, s_3\}$
pattern 2 = * * * * 1 * * * 1 * for group $\{s_2, s_3, s_4, s_5\}$

4. This is sometimes known as the issue of *model selection*.

Exercise 34 (Anti-patterns, forbidden words) *The subject of this chapter (and the book) is repeating phenomenon in data. However, a unique or an absent phenomenon is also of interest sometimes and is called an anti-pattern. An absent string pattern is also called a* forbidden word *[BMRS02] in literature.*

The following table shows some characteristic properties of patterns. Note the use of the idea of 'minimality' (instead of maximality).

Discuss the essential difference between patterns and anti-patterns.

p	# of occurrences, k	length	property
		pattern	
pattern	≥ 2	*maximal*	If p' is a substring p, then p' is a also a pattern
unique	$= 1$	*minimal*	If p is a substring p', and p' occurs in the input, then p' is also unique
		anti-pattern	
forbidden	$= 0$	*minimal*	If p is a substring p', then p' is also forbidden

Hint: What and why is the dramatic shift in paradigm as k changes from 0 to 1 to ≥ 2.

Comments

While we get get down and dirty in the rest of the chapters, where the focus is on a specific class of patterns, in this chapter we indulge in the poetry of abstraction. One of the nontrivial ideas in the chapter, is that of *redundancy* of patterns, which I introduced in [PRF$^+$00], by taking a frequentist view of patterns in strings, rather than the usual combinatorial view. Also, a nonobvious consequence of this abstraction is the identification of maximality in permutations as recursive substructures (PQ trees; see Chapter 10).

As an aside, out of curiosity, we take a step back and pick a few definitions of the word 'pattern' from the dictionary:

1. natural or chance marking, configuration, or design: *patterns of frost markings.*

2. a combination of qualities, acts, tendencies, etc., forming a consistent or characteristic arrangement: *the behavior patterns of ambitious scientists.*

3. anything fashioned or designed to serve as a model or guide for something to be made: *a paper pattern for a dress.*

In fact none of these, that I picked at random, actually suggest repetitiveness. The etymology of the word is equally interesting, confirming a Latin origin, and I quote from Douglas Harper's online etymology dictionary:

> *1324, 'the original proposed to imitation; the archetype; that which is to be copied; an exemplar' [Johnson], from O.Fr. patron, from M.L. patronus (see patron). Extended sense of 'decorative design' first recorded 1582, from earlier sense of a 'patron' as a model to be imitated. The difference in form and sense between patron and pattern wasn't firm till 1700s. Meaning 'model or design in dressmaking' (especially one of paper) is first recorded 1792, in Jane Austen. Verb phrase pattern after 'take as a model' is from 1878.*

Part II

Patterns on Linear Strings

Chapter 5

Modeling the Stream of Life

Hypotheses are what we lack the least.
- attributed to J. H. Poincare

5.1 Introduction

In 1950, when it appeared that the capabilities of computing machines were boundless, Alan Turing proposed a litmus test of sorts to evaluate a machine's intelligence by measuring its capability to perform human-like conversation. The test with a binary PASS/FAIL result, called the *Turing Test*, is described as follows: A human judge engages in a conversation via a text-only channel, with two parties one of which is a machine and the other a human. If the judge is unable to distinguish the machine from the human, then the machine passes the Turing Test.

In the same spirit, can we produce a stream or string of nucleotides 'indistinguishable' from a human DNA fragment (with the *judge* of this test possibly being a bacteria or a virus)?

In this chapter we discuss the problem of modeling strings: DNA, RNA or protein sequences. We use basic statistical ideas without worrying about structures or functions that these biological sequences may imply. Such modeling is not simply for amusing a bacteria or a virus but for producing in-silico sample fragments for various studies.[1]

5.2 Modeling a Biopolymer

Biopolymer is a special class of polymers produced by living organisms. Proteins, DNA and RNA are all examples of biopolymers. The monomer unit in proteins is an amino acid and in DNA and RNA it is a nucleic acid.

---

[1] A quick update on the capabilities of computing machines: More than half a century later, it is quite a challenge to engage a machine in an 'intelligent' conversation.

5.2.1   Repeats in DNA

This is just to remind the reader that DNA displays high repetitiveness. Here we briefly discuss some classes of repeats seen in the human DNA.

Tandem repeats and *variable number tandem repeats* (VNTR) in DNA occur when two or more nucleotides is repeated and the repetitions are directly adjacent to each other. For example,

$$T\ T\ A\ C\ G\ T\ T\ A\ C\ G\ T\ T\ A\ C\ G\ T\ T\ A\ C\ G$$

is a tandem repeat where

$$T\ T\ A\ C\ G$$

is repeated four times in the sequence. A VNTR is a short nucleotide sequence ranging from 14-100 base pairs that is organized into clusters of tandem repeats, usually repeated in the range of between 4-40 times per occurrence.

Microsatellites consist of repeating units of 1-4 base pairs in length. They are polymorphic, in the sense that they can be characterized by the number of repeats. However, they are typically neutral, and can be used as molecular markers in the field of genetics, including population studies. One common example is the

$$(C\ A)\ n$$

repeat, where n varies across samples providing different alleles. This repeat in fact is very frequent in human genome.

A *short tandem repeat* (STR) in DNA is yet another class of polymorphisms repeating units of 2-10 base pairs in length and the repeated sequences are directly adjacent to each other. For example

$$(C\ A\ T\ G)\ n$$

is an STR. These are usually seen in the noncoding intron region, hence 'junk' DNA. The A C repeat is seen on the Y chromosome.

Short interspersed nuclear elements (SINEs) and Long Interspersed Elements (LINEs) are present in great numbers in many eukaryote genomes. They are repeated and dispersed throughout the genome sequences. They constitute more than 20% of the genome of humans and other mammals. The most famous SINE family are the Alu repeats typical of the human genome. SINEs are very useful as markers for phylogenetic analysis whereas STRs and VNTRs are useful in forensic analysis.

Moral of the story: Although this section discussed the human DNA, repeats are known to exist in nonhuman DNA including that of bacteria and archaea. Thus DNA in a sense is highly nonuniform and it is perhaps best to model segments of the DNA separately, than a 'single-size-fits-all' model.

5.2.2 Directionality of biopolymers

The DNA, RNA and even protein sequences, can be viewed as linear strings. Nature also endows an orientation to these polymers i.e., either a left-to-right or a right-to-left order.

The backbone of the DNA strand is made from alternating phosphate and sugar residues. The sugars are joined together by phosphate groups that form phosphodiester bonds between the third and fifth carbon atoms of adjacent sugar rings. It is these asymmetric bonds that give a strand of DNA its direction. The asymmetric ends of a strand of DNA are referred to as the *5' end* and the *3' end*.

The protein polymer is built from twenty different amino acids. Due to the chemical structure of the individual amino acids, the protein chain has directionality. The end of the protein with a free carboxyl group is known as the C-terminus (*carboxy terminus*), and the end with a free amino group is known as the N-terminus (*amino terminus*).

Moral of the story: The directionality in the biopolymers, in a sense, makes the modeling a simpler task, i.e., they can be viewed as left-to-right strings. Contrast this with the scenario of aligning multiple double stranded DNA segments, marked only by the restriction enzyme cut sites, in the absence of the direction information [AMS97].[2]

Back to modeling. Having noted the nature of some of the biopolymers, we get down to the task of modeling a generic biopolymer under most simplistic assumptions.

Consider the task of producing a *n*-nucleotide DNA fragment. Here we make our first assumption about the underlying model that produces this DNA fragment.

Independent & Identical: One of the simplest models is the i.i.d model: a sequence is <u>i</u>ndependent and <u>i</u>dentically <u>d</u>istributed (i.i.d.)[3] if each element of the sequence has the same probability distribution as the other and all are mutually independent. Let

$$pr_A, pr_C, pr_G, pr_T$$

be the probabilities of occurrence of the nucleotide base A, C, G and T respectively. Since these are the only bases possible in the DNA fragment,

$$pr_A + pr_C + pr_G + pr_T = 1.$$

[2]Optical mapping is a single molecule method and the reader is directed to [SCH+97, Par98, CJI+98, Par99, PG99] for an introduction to the methods and the associated computational problems.

[3]A statistician's love for alliteration is borne out by the use of terms such as iid and MCMC (Markov Chain Monte Carlo).

5.2.2.1   Random number generator

A random number generator is designed to generate a sequence of numbers that lack any pattern. In other words the sequence is 'random'. Most computer programming languages include functions or library routines that perform the task of a random number generator. These routines usually provide an integer or a floating point (real) number uniformly distributed over an interval.

For ease of exposition, in the rest of the chapter, assume the availability of a function RANDOM($\cdot$) that returns a random integer r, uniformly distributed over the interval

$$[0 \dots 1000).$$

In other words,

$$0 \leq r = \text{RANDOM}(\cdot) < 1000,$$

and each integer

$$0 \leq r < 1000,$$

has the same probability of being picked by the function RANDOM($\cdot$).

Back to the model.   Under a further simplifying assumption, that the chance of occurrence of each nucleotide is the same, we have,

$$pr_A = pr_C = pr_G = pr_T = \frac{1}{4}.$$

Then each $s[i]$, is generated as follows:

$$s[i] = \begin{cases} A & \text{if } 0 \leq r < 250, \\ C & \text{if } 250 \leq r < 500, \\ G & \text{if } 500 \leq r < 750, \\ T & \text{otherwise.} \end{cases}$$

Nonequiprobable case.   It is easy to modify this to a more general scenario where the probability of occurrence of each character (nucleotide) is not necessarily the same but

$$pr_A + pr_C + pr_G + pr_T = 1.$$

Let

$$r_A = 1000\, pr_A,$$
$$r_C = r_A + 1000\, pr_C,$$
$$r_G = r_C + 1000\, pr_G.$$

Then

$$s[i] = \begin{cases} A & \text{if } 0 \leq r < r_A, \\ C & \text{if } r_A \leq r < r_C, \\ G & \text{if } r_C \leq r < r_G, \\ T & \text{otherwise.} \end{cases}$$

A general scenario. This can be further generalized to an alphabet of size N where each character

$$a_j, \ 1 \le j \le N,$$

occurs with probability pr_{a_j} and

$$\sum_{j=1}^{N} pr_{a_j} = 1.$$

Assuming (see Section 5.2.2.1)

$$0 \le r = \text{RANDOM}(\cdot) < 1000,$$

let

$$r_0 = 0,$$
$$r_j = r_{j-1} + 1000 \, pr_{a_j}, \ \text{for each } 0 < j \le N.$$

Then, $s[i]$ is constructed as follows:

$$s[i] = a_j, \text{ if } r_{j-1} \le r < r_j, \tag{5.1}$$

where r is the random integer picked by function $\text{RANDOM}(\cdot)$. It can be verified that for

$$0 \le r < 1000,$$

there always exists a j such that the following holds:

$$r_{j-1} \le r < r_j.$$

We leave the proof of this statement as an exercise for the reader (Exercise 35).

5.2.3 Modeling a random permutation

A permutation is an *arrangement* of the members (elements) of a finite alphabet Σ where each element appears exactly once. The alphabet Σ could be a cluster of genes that is common to two different organisms (see Chapter 11 for details).

Given a finite alphabet Σ, what is meant by a *random permutation* of size $|\Sigma|$?

We give a precise meaning to this term in two different ways. The fact that the two models are equivalent is left as an exercise for the reader (Exercise 38).

Model 1 (Sample-space model). Recall the definitions of the terms used below from Section 3.2.1.1.

1. The *sample space*, Ω, is the collection *all* possible permutations of Σ. In other words,

$$\Omega = \{s \mid s \text{ is a permutation of the elements of } \Sigma\}.$$

2. An *event* is the outcome of a *random trial*, which in this case is the process of drawing exactly one permutation from the sample space.

3. It is easy to see that

$$|\Omega| = |\Sigma|!.$$

Each permutation $s \in \Omega$ has an equal chance of being drawn. Thus the *probability mass function*, $M_P(s)$, is defined as follows. For each $s \in \Omega$,

$$M_P(s) = \frac{1}{|\Sigma|!}.$$

Notice that

$$\sum_{s \in \Omega} M_P(s) = 1.$$

Having set the stage, a *random permutation* is the outcome of a random trial of the probability space defined above.

Model 2 (Constructive model).   Yet another way of viewing a random permutation is as follows. A permutation s is produced under the following scheme. Assume an urn with Σ number of balls, each labeled by a distinct element of

$$\sigma_i \in \Sigma.$$

Each ball has an equal chance of being drawn from the urn. At each iteration, a ball is drawn from the urn and its label is noted. Note that this ball is not replaced in the urn and the size of the urn reduces by 1 at each iteration.

How do we generate a random permutation using the RANDOM($\cdot$) (see Section 5.2.2.1) function? Recall

$$0 \leq r = \text{RANDOM}(\cdot) < 1000.$$

We construct the permutation s in $|\Sigma|$ iterations as described below. At each iteration k, $1 \leq k \leq |\Sigma|$, $s[k]$ is constructed as described below.

1. Initialize $\sigma' = \phi$, the empty symbol.

2. At iteration $k, 1 \leq k \leq |\Sigma|$,

   (a) Let

$$\Sigma_k = \Sigma \setminus \{\sigma'\}$$
$$= \{\sigma_1, \sigma_2, \ldots, \sigma_j, \ldots, \sigma_{N_k}\},$$

   where

$$N_k = |\Sigma_k|.$$

(b) For $1 \leq j \leq N_k$, define the following:

$$r_0 = 0,$$
$$pr(\sigma_j \in \Sigma_k) = \frac{1}{|\Sigma_k|},$$
$$r_j = r_{j-1} + 1000 \, pr_\sigma.$$

(c) Then, $s[k]$ is constructed as follows:

    i. Let r be the integer returned by RANDOM($\cdot$).

    ii. If $r_{j-1} \leq r < r_j$, then $s[k]$ is set to σ_j.

(d) Then set σ' to $s[k]$ and go to Step 2 for the next iteration.

Wrapping up. The second model provides a way of constructing (or simulating) a random permutation, which has the precise property of the first model.

5.2.4 Modeling a random string

Given

1. a finite alphabet Σ and

2. an integer n,

what is meant by a *random string* of size n?

This is an overloaded term and we explore its meaning here. As before, we give a precise meaning to this term in two different ways. The fact that the two models are equivalent is left as an exercise for the reader (Exercise 39).

Model 1 (Sample-space model). Recall the definitions of the terms used below from Section 3.2.1.1.

1. The *sample space*, Ω, is the collection *all* possible strings on Σ of length n each. In other words,

$$\Omega = \{s \mid s \text{ is a string on } \Sigma \text{ of length } n\}.$$

2. An *event* is the outcome of a *random trial*, which in this case is the process of drawing exactly one string from the sample space.

3. It is easy to see that
$$|\Omega| = |\Sigma|^n.$$

Each string $s \in \Omega$ has an equal chance of being drawn. Thus the *probability mass function*, $M_P(s)$, is defined as follows. For each $s \in \Omega$,

$$M_P(s) = \frac{1}{|\Sigma|^n}.$$

Notice that
$$\sum_{s\in\Omega} M_P(s) = 1.$$

Having set the stage, a *random string* is the outcome of a random trial of the probability space defined above.

Model 2 (Constructive model).   Yet another way of viewing a random string is as follows. A string s produced under the following scheme,

1. i.i.d. model and

2. equi-probable alphabet, i.e.,

$$pr_{a_i} = pr_{a_j}, \text{ for all } i \text{ and } j,$$

and
$$\sum_i pr_{a_i} = 1.$$

is a *random string*. This construction has already been discussed in the earlier part of this section.

Wrapping up.   The second model provides a way of constructing (or simulating) a random string, which has the precise property of the first model.

In a sense the random string is the simplest model and other more sophisticated schemes, that are possibly better models of biological sequences, are discussed in the following sections.

5.3   Bernoulli Scheme

A *stochastic process* is simply a random function. Usually, the domain over which the function is defined is a time interval (referred to as a *time series* in applications). In our case the domain is a line, each integral point being referred to as a *position*. A *stationary process* is a stochastic process whose probability distribution at a fixed time or position is the same for all times or positions.

The *Bernoulli scheme* is a sequence of independent random variables

$$X_1, X_2, X_3, \ldots$$

where each random variable is a *generalized Bernoulli trial* (see Section 3.2.4), i.e., it takes one possible value from a finite set of states S, with outcome

$x_i \in S$ occurring with probability pr_i such that

$$\sum_{x_i \in S} pr_i = 1.$$

In fact the model discussed in Section 5.2 is a Bernoulli scheme with n random variables where n is the length of the fragment under construction.

5.4   Markov Chain

A *Markov chain* is a sequence of random variables

$$X_1, X_2, X_3, \ldots$$

that satisfy the *Markov property*:

$$P(X_{t+1}=x|X_0=x_0, X_1=x_1, \ldots, X_t=x_t) = P(X_{t+1}=x|X_t=x_t) \qquad (5.2)$$

The possible values of X_i form a countable set S is called the *state space* of the chain. We will concern ourselves only with finite S (often written as $|S| < \infty$). Also by Bayes' rule (Equation (3.4)), we get:

$$P(X_{t+1}=x|X_0=x_0, X_1=x_1, \ldots, X_t=x_t)$$

$$= P(X_0=x_0)\; P(X_1=x_1|X_0=x_0)\; \ldots\; P(X_{t+1}=x|X_t=x_t)$$

In other words, the Markov property states that:

> *In a Markov chain, given the present state, the future as well as the past states are independent.*

Hence, Markov chains can be described by a directed graph with $|S|$ vertices or nodes where

(1) each vertex is labeled by a *state* described by a one-to-one mapping $M : S \rightarrow V$, and

(2) the edges are labeled by the probabilities of transitions from one state to the other states, called the *transition probabilities*.

(3) Further, since the weights on the edge labels are interpreted as probabilities, the sum of the weights of the outgoing edges, for each vertex, must add up to 1.

Such a graph is also referred to as a *finite state machine* (FSM). See Figure 5.1(a) for a simple example.

Yet another way of stating the Markov property is as follows:

Markov process $(S, \mathbf{P})$ with $S = \{C,G\}$

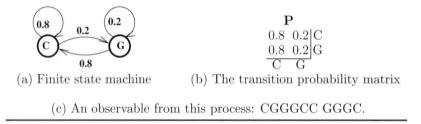

(a) Finite state machine   (b) The transition probability matrix

(c) An observable from this process: CGGGCC GGGC.

FIGURE 5.1:   (a) Specification of a Markov process: A finite state machine (or a directed graph) with two states given as $S = \{C, G\}$. Each transition (edge) is labeled with the probability of going from one state to another. (b) The corresponding transition probability matrix $\mathbf{P}$.

> *If the FSM is in state x at time t, then the probability that it moves to state y at time $t+1$ depends only on the current state and does not depend on the time t.*

The FSM can also be represented by a matrix (see Figure 5.1(b)), called the *transition (probability) matrix* usually denoted as $\mathbf{P}$. We follow the convention that

$$\mathbf{P}[i, j]$$

represents the probability of going from state x_i to state x_j, also written as:

$$\mathbf{P}[i, j] = P(X_{t+1}=x_j | X_t=x_i).$$

Thus matrix $\mathbf{P}$ has the following properties:

1. $\mathbf{P}[i, j] \geq 0$ for each i and j since it represents a probability. In other words $\mathbf{P}$ is a *nonnegative matrix*.

2. The entries in each row must add up to 1. In other words, *the row vector is a probability vector*. We call such a matrix a *stochastic matrix*.

   A vector with nonnegative values whose entries add up to 1 is called a *stochastic vector* or simply *stochastic*.

Stochastic matrix $\mathbf{P}$ is a transition matrix for a single step and the k-step transition probability can be computed as the kth power of the transition matrix, $\mathbf{P}^k$.

Any stochastic matrix can be considered as the transition matrix of some finite Markov chain. When all rows as well as all columns are probability vectors, we call such a matrix *doubly stochastic*. To summarize, we specify a Markov process as the tuple

$$(S, \mathbf{P}), \tag{5.3}$$

where

1. S is a finite set of states and

2. $\mathbf{P}$ is a stochastic matrix.

This is also conveniently represented by the FSM (graph) G.

5.4.1 Stationary distribution

We next discuss an important concept associated with Markov chains, which is the idea of the probability vector of the state space.

For example consider the Markov process shown in Figure 5.1. A possible string of length 10 generated by this process is

$$C\ G\ G\ G\ C\ C\ G\ G\ G\ C.$$

In fact the process can generate all possible strings of C and G. Then, how is this different from another Markov process with the same state space S, capable of generating all possible strings of C and G, but with a different transition matrix $\mathbf{P}$?

So a natural question arises from this:

Given the specification of a Markov process (say as an FSM), what can be said about the probability of the occurrences of each state (C, G in our example) in sufficiently long strings generated by this process?

Mathematically speaking, we seek a probability row vector π that satisfies the following:

$$\pi = \pi\mathbf{P}. \tag{5.4}$$

π is called the *stationary distribution*.[4] The two following questions are of paramount interest.

1. (Question 1): Does π always exist?

2. (Question 2): If it does, is it unique?

We address the two questions in the discussion below.

Question 1 (Existence).   Consider a state space

$$S = \{A,T,C,G\},$$

---

[4]Note that π is a left *eigenvector* of $\mathbf{P}$ associated with the *eigenvalue* 1.

and the following stochastic matrix, which is the transition probability matrix of a Markov process with

$$
\mathbf{P} = \begin{bmatrix} 0.3 & 0.3 & 0.4 & 0 \\ 0 & 0.4 & 0 & 0.6 \\ 0.25 & 0.25 & 0.25 & 0.25 \\ 0 & 0.5 & 0 & 0.5 \end{bmatrix} \begin{matrix} A \\ T \\ C \\ G \end{matrix} \tag{5.5}
$$
$$
\begin{matrix} A & T & C & G \end{matrix}
$$

It turns out that there is no real π (i.e., all entries of the vector are real) satisfying

$$
\pi\mathbf{P} = \pi
$$

for this $\mathbf{P}$. In *matrix theory*, we say that $\mathbf{P}$ is *reducible* if by a permutation of rows (and corresponding columns) it can be reduced to the form

$$
\mathbf{P}' = \begin{bmatrix} P_{11} & P_{12} \\ 0 & P_{22} \end{bmatrix},
$$

where P_{11} and P_{22} are square matrices. If this reduction is not possible, then $\mathbf{P}$ is called *irreducible*. The following theorem gives the condition(s) under which a matrix has a real solution.

THEOREM 5.1
(**Generalized Perron-Frobenius theorem**) *A stochastic irreducible matrix* $\mathbf{P}$ *has a real solution (i.e., values of* π *are real) to Equation (5.4).*

The proof of this theorem is outside the scope of this book and we omit it.

It turns out that $\mathbf{P}$ of Equation (5.5) is not irreducible, since the following holds:

$$
\mathbf{P}' = \begin{bmatrix} P_{11} & P_{12} \\ 0 & P_{22} \end{bmatrix} = \left[\begin{array}{cc|cc} 0.3 & 0.4 & 0.3 & 0 \\ 0.25 & 0.25 & 0.25 & 0.25 \\ \hline 0 & 0 & 0.5 & 0.5 \\ 0 & 0 & 0.4 & 0.6 \end{array} \right] \begin{matrix} A \\ C \\ G \\ T \end{matrix}
$$
$$
\begin{matrix} A & C & G & T \end{matrix}
$$

Given a matrix, is there a simple algorithm to check if a matrix is irreducible? We resort to *algebraic graph theory* for a simple algorithm to this problem.

We have already seen that $\mathbf{P}$ is the edge weight matrix of a directed graph with S as the vertex set (also called the FSM earlier in the discussion).

A directed graph is *strongly connected* if there a directed path from x_i to x_j for each x_i and x_j in S. See Figure 5.2 for an example. The following theorem gives the condition under which a matrix is irreducible.

THEOREM 5.2
(**Irreducible matrix theorem**) *The adjacency matrix of a strongly connected graph is irreducible.*

Thus to check if a directed graph is strongly connected, all we need to see is that every vertex is reachable by every other.

A depth first search (DFS) is a linear time,

$$\mathcal{O}(|V| + |E|),$$

recursive algorithm to traverse a graph. It turns out that we can check if the directed graph is strongly connected by doing the following.

1. Fix a vertex $v' \in V$.

2. (a) Carry out a DFS in G from v' and

   (b) check that all vertices were traversed.

   If there exists a vertex that cannot be traveresed then the graph is not strongly connected.

3. Transpose the graph, i.e., reverse the direction of every edge in the graph to obtain G^T.

   (a) Carry out a DFS in G^T from v' and

   (b) check that all vertices were traversed.

   If not, then the graph is not strongly connected.

Thus we can check for strongly connectedness of the graph in just two DFS traversals i.e., in

$$\mathcal{O}(|V| + |E|)$$

time. Why is the algorithm correct?

- The first DFS traversal on the graph G initiated at v ensures that every other vertex is reachable from v.

- The second DFS traversal on the transpose of the graph G^T initiated at v ensures that v is reachable from every other vertex.

- Hence for any two vertices u, w there is at least a path from u to v to w and vice-versa.

This algorithm is a special case of the *Kosaraju Algorithm*.

Algorithm 4 *DFS Traversal Algorithm*

```
DFS(v)
    Mark v
    For each unmarked neighbor u of v
        DFS(u)
```

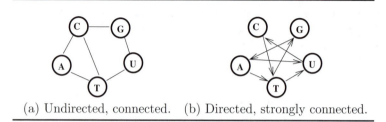

(a) Undirected, connected.   (b) Directed, strongly connected.

FIGURE 5.2:   Examples of connected and strongly connected graphs.

Algorithm 5 *Strongly Connected Check*

> *DFS(G, v)*
> *IF all vertices traversed*
>   *THEN DFS(G^T, v)*
>   *IF all vertices traversed*
>   *THEN StronglyConnected*
>   *ELSE notStronglyConnected*
>   *ELSE notStronglyConnected*

Question 2 (Uniqueness).   Next, it is equally important to check if the solution to Equation (5.4) is unique.

THEOREM 5.3

(Irreducible matrix theorem) *A stochastic, irreducible* **P** *has a unique solution to Equation (5.4).*

The proof of this theorem is outside the scope of this book and we omit it.

Before concluding the discussion, we ask another question: *Does it matter what state we start the Markov process with?* Consider the following transition matrix:

$$\mathbf{P} = \begin{bmatrix} 0 & 1 \\ 1 & 0 \end{bmatrix}$$

Note that **P** is stochastic and irreducible, hence the solution π is unique and is given below:

$$[1/2 \ 1/2] \begin{bmatrix} 0 & 1 \\ 1 & 0 \end{bmatrix} = [1/2 \ 1/2]$$

However,

$$\mathbf{P}^2 = \begin{bmatrix} 0 & 1 \\ 1 & 0 \end{bmatrix}^2 = \begin{bmatrix} 1 & 0 \\ 0 & 1 \end{bmatrix} = Id$$

Thus

$$\mathbf{P}^k = \begin{cases} Id & \text{for } k \text{ even,} \\ \mathbf{P} & \text{for } k \text{ odd.} \end{cases}$$

Since

$$[1/2 \ 1/2]$$

is the unique nonnegative left eigenvector, it is clear that

$$\lim_{k \to \infty} v \cdot \mathbf{P}^k$$

does not exist unless

$$v = [1/2 \ 1/2].$$

Hence we need to have another constraint on $\mathbf{P}$ so that

$$\lim_{k \to \infty} v \cdot \mathbf{P}^k = \pi$$

for every stochastic v.

A sufficient condition for this to happen is that $\mathbf{P}$ is *primitive*, i.e., $\mathbf{P}^k$ is positive for some k. Under these conditions, the *Perron-Frobenius theorem* states that $\mathbf{P}^k$ converges to a rank-one matrix in which each row is the stationary distribution π, that is,

$$\lim_{k \to \infty} \mathbf{P}^k = \mathbf{1}\pi$$

where $\mathbf{1}$ is the column vector with all entries equal to 1. In other words,

> *With large t, the Markov chain forgets its initial distribution and converges to its stationary distribution.*

5.4.2 Computing probabilities

Consider the following stochastic matrix:

$$\mathbf{P} = \begin{bmatrix} 1/4 & 1/4 & 1/4 & 1/4 \\ 1/6 & 1/6 & 1/3 & 1/3 \\ 1/4 & 1/4 & 1/4 & 1/4 \\ 1/6 & 1/6 & 1/3 & 1/3 \end{bmatrix} \quad \begin{matrix} A \\ C \\ G \\ T \end{matrix}$$
$$\phantom{\mathbf{P} = } \begin{matrix} A & \ C & \ G & \ T \end{matrix}$$

We can show that the FSM graph is strongly connected, hence the matrix is irreducible.

Another easy check is that since the matrix has no zero entry, the Markov process is trivially irreducible and has a unique solution to Equation (5.4), given as:

$$\pi = [5/24 \ 5/24 \ 7/24 \ 7/24]$$
$$ \begin{matrix} A & \ C & \ G & \ T \end{matrix} \ .$$

Now, we are ready to compute the probability, $P(x)$, associated with string x, where
$$x = x_0 x_1 x_2 \ldots x_n,$$
for a Markov process specified by $\mathbf{P}$. This is given as follows:
$$P(x) = \pi(x_0) \prod_{i=1}^{n} P(x_{i-1}|x_i)$$
$$= \prod_{i=1}^{n} \mathbf{P}[i-1, i].$$

As a concrete example, let
$$x = \text{CCGA}.$$
Then,
$$P(x) = \pi(\text{C})\ P(\text{C}|\text{C})\ P(\text{C}|\text{G})\ P(\text{G}|\text{A})$$
$$= \pi(\text{C})\ \mathbf{P}[2,2]\quad \mathbf{P}[2,3]\quad \mathbf{P}[3,1]$$
$$= \left(\frac{5}{24}\right) \left(\frac{1}{6}\right)\ \left(\frac{1}{3}\right)\ \left(\frac{1}{4}\right)$$
$$= \left(\frac{5}{1728}\right).$$

5.5   Hidden Markov Model (HMM)

A natural next step in modeling a biological sequence is a *Hidden Markov Model* (or HMM) which has an underlying hidden (unobserved) state process that follows a Markov chain.

For example, it has been found that there is perhaps some correlation between bending propensity or bendability of the DNA segment with promoter regions of eukaryotic genes. Similarly, CG-rich or CG-poor is a compositional property associated with similar DNA functions (see Exercise 44). Observations such as these could influence the process of generating a DNA sequence and they can be captured using 'hidden states'.

A Hidden Markov Model $(\Sigma, S, \mathbf{P}, \mathbf{E})$ is defined as follows:

1. A finite alphabet Σ,

2. a Markov process $(S, \mathbf{P})$ (see Eqn (5.3)) where S is called the set of *hidden states*, and

3. an *emission matrix* $\mathbf{E}$. Each state in S is associated with a probability vector, called the *emission vector*, on Σ. $\mathbf{E}$ is a
$$|S| \times |\Sigma|$$

Hidden Markov Model $(\Sigma, S, \mathbf{P}, \mathbf{E})$
with $\Sigma = \{$A, T$\}$, $S = \{$C,G$\}$

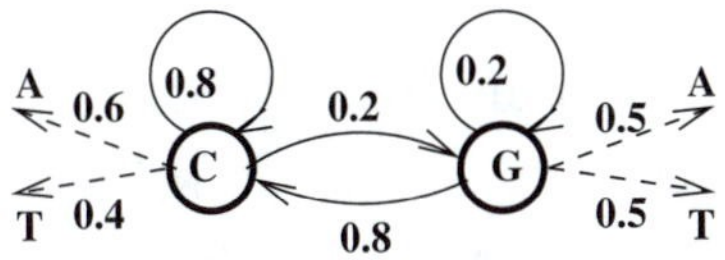

(a)

$$
\begin{array}{c}
\mathbf{P} \\
\begin{array}{cc|c}
0.8 & 0.2 & C \\
0.8 & 0.2 & G \\
\hline
C & G &
\end{array}
\end{array}
$$

(b)

$$
\begin{array}{c}
\mathbf{E} \\
\begin{array}{cc|c}
0.6 & 0.4 & C \\
0.5 & 0.5 & G \\
\hline
A & T &
\end{array}
\end{array}
$$

(c)

(d) A path in the HMM: C G C C G G G G C G C.
(e) An observable from this process: A T T A T T T T A A A.

FIGURE 5.3: (a) Specification of a Hidden Markov process with a finite state machine (or a directed graph) with two states given as S as the states. The dashed edges represent the emission vectors for each state. (b) The transition probability matrix and (c) The emission matrix.

stochastic matrix where each row is an emission vector.

A path in HMM,

$$z = z_0 z_1 z_2 \ldots z_n,$$

is a sequence of states in S and a string,

$$x = x_0 x_1 x_2 \ldots x_n,$$

is a sequence of characters in alphabet Σ. Recall that

$$(S, \mathbf{P})$$

is a Markov process and for an irreducible $\mathbf{P}$ there exists a unique stationary distribution π. Then the probability of x given path z is given as follows:

$$P(x|z) = \pi(z_0) \, P(x_0|z_0) \prod_{i=1}^{n} P(z_i|z_{i-1}) P(x_i|z_i).$$

By our convention,

$$P(x_i|z_i) = \mathbf{E}[z_i, z_i]$$

and

$$P(z_{i+1}|z_i) = \mathbf{P}[z_i, z_{i+1}].$$

Then we get the following:

$$P(x|z) = \pi(z_0) \, \mathbf{E}[z_0, x_0] \prod_{i=1}^{n} \mathbf{P}[z_i, z_{i-1}] \mathbf{E}[z_i, x_i].$$

See Figure 5.3 for an example.

Why use HMMs?   The primary reason for using a Hidden Markov Model is to model an underlying 'hidden' basis for seeing a particular sequence of observation x. This basis, under the HMM model, is the path z.

5.5.1   The decoding problem (Viterbi algorithm)

The output or observation of an HMM is the string x (the sequence of emitted symbols from Σ). Then the possible questions of interest are:

1. What is the most likely *explanation* of x?
    The explanation under this model is a plausible HMM path z. So the question is rephrased as finding an optimal path z^* for a given x such that

$$P(x|z)$$

    is maximized, or,

$$z^* = \arg\left(\max_z P(x|z)\right)$$

2. What is

$$P(x|z^*),$$

    the probability of observing x, given z^*?

3. What is

$$P(x),$$

    the (overall) probability of observing x?

The first and the second question are obviously related and are answered simultaneously by using the *Viterbi Algorithm*. The third question is left as an exercise for the reader (see Exercise 41).

We address the first question here. The algorithm for this problem was originally proposed by Andrew Viterbi as an error-correction scheme for noisy digital communication links [Vit67], hence it is also sometimes called a *decoder*.

LEMMA 5.1
(The decoder optimal subproblem lemma) *Given an HMM,*

$$(\Sigma, S, \mathbf{P}, \mathbf{E}),$$

and a string,

$$x = x_1 x_2 \dots x_n,$$

on Σ, consider

$$x_1 x_2 \dots x_i,$$

the i-length prefix of x. For a state

$$z_j \in S$$

let the path ending at state z_j given as

$$z_{j1} z_{j2} \ldots z_{ji}$$

be optimal, i.e.,

$$z_{j1} z_{j2} \ldots z_{ji} = \arg \left(\max_{z_1 z_2 \ldots z_i} P(x_1 x_2 \ldots x_i | z_1 z_2 \ldots z_i) \right)$$

Let

$$f_{ij}$$

be the probability associated with string

$$x_1 x_2 \ldots x_i,$$

given the optimal path ending at z_j written as:

$$f_{ij} = P(x_1 x_2 \ldots x_i | z_{j1} z_{j2} \ldots z_{ji}).$$

Recall

$$z_j = z_{ji}.$$

Then the following two statements hold:

$$f_{ij} = \max_k \left(f_{(i-1)k} \, P(z_j | z_k) \, P(x_i | z_j) \right) \tag{5.6}$$

and

$$f_{1j} = \pi(z_j) \, P(x_1 | z_j). \tag{5.7}$$

Also, let the optimal path ending at z_j be written as

$$Path(z_j) = z_{j1} z_{j2} \ldots z_{ji}.$$

Further, the i-length prefix of the optimal path is obtained as

$$z_j \, Path(z_{k'}),$$

where

$$k' = \arg \left(\max_k \left(f_{(i-1)k} \, P(z_j | z_k) \, P(x_i | z_j) \right) \right).$$

In spite of its intimidating looks, the lemma is very simple and we leave the proof as an exercise for the reader (Exercise 43). In a nutshell, the lemma states simply that the optimal solution,

$$x_1 x_2 \ldots x_n,$$

to the problem can be built from the optimal solutions to its subproblems

$$x_1 x_2 \ldots x_i.$$

We now describe the algorithm below based on this observation. Let x of length n is the input string. Let F_i be a

$$1 \times |S|$$

matrix where each

$$F_i[j]$$

stores f_{ij} of Lemma (5.1). Similarly,

$$Pth_i[j]$$

stores the corresponding path.

1. In the first phase, array F_1 and Pth_1 are initialized.

2. In the second phase, where the algorithm loops, array F_i and Pth_i are updated based on the observation in the lemma.

3. In the final phase, the algorithm goes over $F_n[k]$ for each k and picks the maximum value.

The algorithm takes

$$\mathcal{O}(|S|\,|\Sigma|^2)$$

time. However, due to the underlying Markov process, F_{i+1} depends only on F_i and not on

$$F_{i'}, \quad i' < i.$$

Thus only two arrays, say F_0 and F_1, are adequate and the arrays can be re-used in the algorithm. Thus the algorithm requires only

$$O(|S|)$$

extra space to run.

Algorithm 6 *Viterbi Algorithm*

Input: $\mathbf{P}, \mathbf{E}, \pi, x$; *Output: opt, opt-path*
                                    // Initialize
For each j
    $F_1[j] \leftarrow \pi(j)\mathbf{E}[j, x_1]$
    $Pth_1[j] \leftarrow z_j$
                                    // Main Loop
For $i = 2, 3, \ldots, n$
    $F_i[j] \leftarrow \max_k \left(F_{i-1}[k]\, \mathbf{P}[k, j]\, \mathbf{E}[x_i, j]\right)$
    $k' \leftarrow \arg\left(\max_k \left(F_{i-1}[k]\, \mathbf{P}[k, j]\, \mathbf{E}[x_i, j]\right)\right)$
    $Pth_i[j] \leftarrow Pth_i[k']z_j$
                                    // Finale
$opt \leftarrow \max_k F_n[k];$ $k' \leftarrow \arg\left(\max_k F_n[k]\right)$
$opt\text{-}path \leftarrow Pth_n[k']$

5.6   Comparison of the Schemes

The Bernoulli scheme is a special case of a Markov chain where the transition probability matrix has identical rows. In other words, state $t+1$ is even independent of state t, in addition to being independent of all the previous states.

Under a Bernoulli scheme, all strings s that have i_j number of states for

$$x_i, \quad i = 1, 2, \ldots N,$$

have the same probability given as

$$P(s) = \prod_i (pr_i)^{i_j}.$$

For example,

$$
\begin{aligned}
P(\text{A A T C C G}) &= P(\text{T C A A C G}) \\
&= P(\text{T G A A C C}) \\
&= P(\text{A C T A C G}) \\
&= P(\text{A C C A T G}) \\
&\;\;\vdots \\
&= (pr_A)^2 pr_T (pr_C)^2 pr_G.
\end{aligned}
$$

However, a Markov chain model induces a probability distribution on the strings that can possibly distinguish each of the cases above. A Hidden Markov Model can even offer a possible explanation for a particular string.

5.7   Conclusion

What scheme best approximates a DNA fragment? What scheme best approximates a protein sequence? These are difficult questions to answer satisfactorily. It is best to understand the problem at hand and use a scheme that is simple enough to be tractable and complex enough to be realistic. Of course, it is unclear whether a more complex scheme is indeed a closer approximation to reality. The design of an appropriate model for a biological sequence continues to be a hot area of research.

5.8   Exercises

Exercise 35 (Constructive model) *Let*

$$pr_{a_1}, pr_{a_2}, \ldots, pr_{a_N}$$

be real values such that

$$\sum_{j=1}^{N} pr_{a_j} = 1,$$

and let

$$r_0 = 0,$$
$$r_j = r_{j-1} + 1000\, pr_{a_j}, \quad \text{for each } 0 < j \leq N.$$

Then show that if

$$0 \leq r < 1000,$$

then there always exists a unique $1 \leq j \leq N$ such that the following holds:

$$r_{j-1} \leq r < r_j.$$

Exercise 36 (Random number generator) *Let a random number generator, RAND(), return a floating point (real) number uniformly distributed between 0 and 1. Modify Equation (5.1) to construct the sequence s.*

Exercise 37 **(Pseudorandom generator)** *There are mainly two methods to generate random numbers.*
The first measures some random physical phenomenon (with some corrections to neutralize measurement biases).
The second uses a computational algorithm that produces long sequences of (apparently) random numbers. The second method is called a pseudo-random number generators and uses an initial seed.

  1. *Design a pseudo-random number generator.*

  2. *Design a pseudo-random permutation generator.*

Hint: Investigate *statistical randomness* and *algorithmic randomness* to use as test for the results. Note that the two problems are topics of current active research.

Exercise 38 (Random permutation)

  1. *Argue that in the second (constructive) model of Section 5.2.4,*

(a) *for a fixed position k the probability of s[k] taking the value σ is*

$$\frac{1}{|\Sigma|},$$

   for each σ ∈ Σ.

   (b) *for a fixed σ ∈ Σ, the probability of position k taking the value σ is*

$$\frac{1}{|\Sigma|},$$

   for each k.

2. *Argue that the two models of a random permutation presented in Section 5.2.3 have identical interpretations.*

Hint: Note that the sum of two random numbers is random and the product of two random numbers is random. Show that a random string/permutaion produced by the second model satisfies the properties of the first model.

Exercise 39 (Random string) *Argue that the two models of a random string presented in Section 5.2.4 have identical interpretations.*

Hint: Show that a random string/permutaion produced by the second model satisfies the properties of the first model.

Exercise 40 *Given an HMM*

$$(\Sigma, S, \mathbf{P}, \mathbf{E}),$$

show that

$$\sum_i P(x_i) = 1,$$

where x_i is any observed string of length of fixed length say k and $P(x_i)$ is the marginal probability of x_i i.e.,

$$P(x_i) = \sum_j P(x_i | z_j),$$

where z_j is an HMM path of length k.

Hint: Assume, for simplicity, $|S| = 4$, $|\Sigma| = 2$ and $k = 4$.

Exercise 41 (Probability space & HMM) *Let a probability space be defined as*

$$(\Omega, 2^{\Omega}, M_P)$$

where the probability distribution M_P is defined by an HMM

$$(\Sigma, S, \mathbf{P}, \mathbf{E}).$$

Let Ω be the space of all strings on Σ of length k. Show that the Kolomogorov's conditions hold for this probability distribution.

Hint: Show that

1. $P(\omega \in \Omega) \geq 0$ and

2. $P(\Omega) = 1$.

Exercise 42 *Let the output of an HMM be the string*

$$x = x_1 x_2 x_3 \ldots x_n$$

(i.e. a sequence of emitted symbols from Σ). What is the (overall) probability of observing x under this model?

Hint: Note that there are many HMM paths of length n that emit x. How many such paths exist?

Exercise 43 *Prove Eqns (5.6) and (5.7) in The Decoder Optimal Subproblem Lemma (5.1).*

Hint: Use proof by contradiction.

Exercise 44 *Design a bi-variate Hidden Markov Model where*

1. *one variable models the CG content (composition, as rich or poor) and*

2. *the other models the bendability of DNA (structure).*

What is the probability space

$$(\Omega, \mathcal{F}, P)?$$

What are the issues with this architecture?

Hint: Note that in practice, bendability is usually a real-valued (continuous) measure.

Exercise 45 *Let D be large collection of long sequences of amino acids. The task is to approximate a Markov Chain process based on D. The twenty states of the Markov process correspond to the twenty amino acids in D,*

$$a_i, \ 1 \le i \le 20.$$

Let

$$\#(a_i a_j)$$

denote the number of times the substring $a_i a_j$ is observed in D. For each i, the following is computed:

$$n_i = \sum_j \#(a_i a_j),$$

$$n = \sum_i n_i.$$

1. *The transition probability matrix $\mathbf{P}$ is computed as follows:*

$$\mathbf{P}[i, j] = \frac{\#(a_i a_j)}{n_i}.$$

2. *The stationary probability distribution is computed as follows:*

$$\pi(a_i) = \frac{n_i}{n}.$$

By this construction, does

$$\pi \mathbf{P} = \pi$$

hold? Why?

Chapter 6

String Pattern Specifications

Keep it simple, but not simpler.
- anonymous

6.1 Introduction

One dimensional data is about the simplest organization of information. It is amazing that deoxyribonucleic acid (DNA) that contains the genetic instructions for the entire development and functioning of living organisms, even as complex as humans, is only linear. The genome, also called the *blueprint* of the organism, encodes the instructions to build other components of the cell, such as proteins and RNA molecules that eventually make up 'life'.

As we have seen in Chapter 5 the biopolymers can be modeled as strings (with a left-to-right direction). In this chapter we explore the problem of specifying string patterns. We begin with a few examples of patterns in biological sequences.

Solid patterns. The ROX1 gene encodes a repressor of the hypoxic functions of the yeast *Saccharomyces cerevisiae* [BLZ93]. The DNA binding motif is recognized as follows:

$$\boxed{\text{CCATTGTTCTC}}$$

This pattern can also be extracted from DNA sequences of the transcriptional factor ROX1.

Rigid patterns (with wild cards). One of the yeast genes required for growth on galactose is called GAL4. Each of the UAS genes contains one or more copies of a related 17-bp sequence called UASGAL,

$$\boxed{\text{G.CAAAA.CCGC.GGCGG.A.T}}$$

Gal4 protein binds to UASGAL sequences and activates transcription from a nearby promoter. Note the presence of wild cards in the pattern.

Extensible patterns (with homologous sets). Fibronectin is a plasma protein that binds cell surfaces and various compounds including collagen, fibrin, heparin, DNA, and actin. The major part of the sequence of fibronectin consists of the repetition of three types of domains, which are called type I, II, and III. Type II domain is approximately forty residues long, contains four conserved cysteines involved in disulfide bonds and is part of the collagen-binding region of fibronectin. In fibronectin the type II domain is duplicated. Type II domains have also been found in various other proteins. The fibronectin type II domain pattern has the form shown below:

C..PF.[FYWI].......C-(8,10)WC....[DNSR][FYW]-(3,5)[FYW].[FYWI]C

Note the presence of homologous characters such as [FYWI] in the motif. The wild cards are represented as '.' characters and the extensible part of the pattern is shown as integer intervals: (8,10) is to be interpreted as 'gap' of length 8 to 10 and the interval (3,5) is similarly interpreted.

Back to introduction. As the pattern specification become less stringent, such as allowing for wild cards, or permitting flexibility in terms of its length, what happens to the size of the output? We explore this in details in this chapter and in the next chapter present a generic discovery algorithm for different classes of patterns.

6.2 Notation

Let the input s be a *string* of length n be over a finite discrete alphabet

$$\Sigma = \{\sigma_1, \sigma_2, \ldots, \sigma_L\},$$

where $|\Sigma| = L$. A substring of s, written as

$$s[i..j], \ 1 \le i \le j \le n,$$

is the string obtained by yanking out the segment from index i to j from s. A character from the alphabet,

$$\sigma \in \Sigma,$$

is called a *solid* character. A wild card or a 'dont care' is denoted as the '.' character. This is to be interpreted as any character from the alphabet at that position in the pattern. The *size* of the pattern is given as

$$|p|,$$

which is simply the number of elements in p. However, sometimes the size may be defined by the area (size) of the segment in the input I that an an occurrence of p covers.

A *nontrivial* pattern p has the following characteristics.

1. The size of p is larger than 1, i.e., $|p| \geq 2$.
   An element of the alphabet,

$$\sigma \in \Sigma,$$

   is a *trivial* (singleton) pattern.

2. The first and last element of p is solid, i.e.,

$$p[1], p[|p|] \in \Sigma.$$

For example, if
$$\Sigma = \{A, C, G, T\},$$
then all the following patterns are *trivial*.

$$p_1 = C\ T\ G\ .\ .$$
$$p_2 = .\ C\ .\ G\ .$$
$$p_3 = .\ .\ .\ .\ .$$
$$p_4 = .\ .\ C\ T\ G$$

Let P denote the set of all nontrivial patterns in s.

The location list, $\mathcal{L}_p$, of a pattern p is the list of positions in s where p occurs. If
$$|\mathcal{L}_p| = k,$$
we also say that the pattern p has a *support* of k in s. For any pattern p,

$$|\mathcal{L}_p| \geq 2.$$

Occurrence & operators $\otimes$ *(meet),* $\preceq$. Let q_1 and q_2 be defined on alphabet

$$\Sigma = \{C, G, T\}.$$

Consider the three following 'alignments' of q_1 with q_2. Considering the elements along each column of this alignment, a disagreement is denoted by a dont care '.' character, in the last row below. The alignment is denoted by (j_1, j_2), which is interpreted as position j_1 of q_1 is aligned withe position j_2 of q_2.

	Alignment I: (1,1)	Alignment II: (2,1)	Alignment III: (3,1)
q_1	G G G T G G G C	G G G T G G G C	G G G T G G G C
q_2	G T T T G C	G T T T G C	G T T T G C
p	G . . T G C . .	. G . T . G . .	. . G T . . G C

This is a brief introduction and the details are to be presented for each class of patterns in their respective sections. For a fixed alignment (j_1, j_2), we use the following terminology.

1. p is usually defined by removing the leading and trailing dont care elements, so that its leftmost and rightmost element is a solid character. Also, let $|p| > 1$. In the rest, we assume p is nontrivial.

2. p is called the *meet* of q_1 and q_2 with alignment (j_1, j_2). Also,

$$p = q_1 \otimes q_2, \text{ for alignment } (j_1, j_2)$$

3. p *occurs* at the location(s) dictated by the alignment in q_1 and q_2. The convention followed depends on the context, but could be taken as the leftmost positions in q_1 and q_2 of the alignment. Also, we say

$$p \preceq q_1 \text{ and } p \preceq q_2.$$

Location list $\mathcal{L}_p$.   The location list, $\mathcal{L}_p$, of a pattern $p \in P$ is the list of positions in the input s where p occurs. If

$$|\mathcal{L}_p| = k,$$

we also say that the pattern p has a *support* of k in s. For any pattern p,

$$|\mathcal{L}_p| \geq 2.$$

6.3   Solid Patterns

A *solid* pattern, p, as the name suggests, has only solid characters. Its *occurrence* is simply defined as the appearance of p in the the input as a substring. In other words, there exist some

$$1 \leq j \leq k \leq n - |p|,$$

such that

$$p = s[j..k].$$

The solid pattern is also known as an *l-mer* in literature, where $|p| = l$.

Consider the example shown in Figure 6.1 which gives P and the location list $\mathcal{L}_p$, for each $p \in P$. It is not difficult to see that the number of nontrivial solid patterns is

$$\mathcal{O}(n^2).$$

We next define *maximal* patterns that helps remove some repetitive patterns in P.

$$s_1 = \text{a b c d a b c d a b c a b.}$$

$$P = \left\{ \begin{array}{llll} \text{a b c d a b c,} & \text{b c d a b c,} & \text{c d a b c,} & \text{d a b c,} \\ \text{a b c d a b,} & \text{b c d a b,} & \text{c d a b,} & \text{d a b,} \\ \text{a b c d a,} & \text{b c d a,} & \text{c d a,} & \text{d a,} \\ \text{a b c d,} & \text{b c d,} & \text{c d,} & \\ \text{a b c,} & \text{b c,} & & \\ \text{a b.} & & & \end{array} \right\}.$$

$$\begin{aligned} \mathcal{L}_{\text{a b c d a b c}} &= \mathcal{L}_{\text{a b c d a b}} \\ &= \mathcal{L}_{\text{a b c d a}} \\ &= \mathcal{L}_{\text{a b c d}} \\ &= \{1, 5\}. \end{aligned}$$

$$\mathcal{L}_{\text{a b c}} = \{1, 5, 9\}.$$

$$\mathcal{L}_{\text{a b}} = \{1, 5, 9, 12\}.$$

$$\begin{aligned} \mathcal{L}_{\text{c d a b c}} &= \mathcal{L}_{\text{c d a b}} \\ &= \mathcal{L}_{\text{c d a}} \\ &= \mathcal{L}_{\text{c d}} \\ &= \{3, 7\}. \end{aligned}$$

$$\begin{aligned} \mathcal{L}_{\text{b c d a b c}} &= \mathcal{L}_{\text{b c d a b}} \\ &= \mathcal{L}_{\text{b c d a}} \\ &= \mathcal{L}_{\text{b c d}} \\ &= \{2, 6\}. \end{aligned}$$

$$\mathcal{L}_{\text{b c}} = \{2, 6, 10\}.$$

$$\begin{aligned} \mathcal{L}_{\text{d a b c}} &= \mathcal{L}_{\text{d a b}} \\ &= \mathcal{L}_{\text{d a}} \\ &= \{4, 8\}. \end{aligned}$$

FIGURE 6.1: P is the set of all nontrivial solid patterns on s_1. $\mathcal{L}_p$ is the location list of p. We follow the convention that, $i \in \mathcal{L}_p$, is the leftmost position (index) in s_1 of an occurrence of p.

6.3.1   Maximality

A maximal solid pattern p is *maximal in length*, if p such cannot be extended to the left or to the right, without decreasing its support, to get a nontrivial

$$p' \neq p.$$

This is indeed the intuitive definition of maximality. It can be shown that this definition is equivalent to the one in Chapter 4 in terms of its location list. We leave this as an exercise for the reader (Exercise 46).

Thus pattern

$$\text{a b c d a b c,}$$

is maximal in s_1 but patterns

$$\text{a b c d a b,} \quad \text{a b c d a,} \quad \text{a b c d,}$$

are not maximal in s_1 since each can be extended to the right with the same support (i.e., $k = 2$). By this definition, what are the maximal patterns? Let

$$P_{maximal}(s_1) \subseteq P(s_1),$$

be the set of all maximal patterns. Then

$$P_{maximal}(s_1) = \{\text{a b c d a b c,} \quad \text{a b c,} \quad \text{a b}\}.$$

How small can $|P_{maximal}(s)|$ be for a given s?

6.3.2   Counting the maximal patterns

We show that the number of maximal solid patterns in a string of size n can be no more than n. This is believed to be a 'folklore' theorem.[1] This section substantiates the folklore using *suffix trees*. Given s, a suffix tree is the compact trie of all suffixes of s. This is explained through a concrete example below.

Given a string s of size n, let

$$\$ \notin \Sigma.$$

We terminate s with $\$$ as

$$s\$,$$

---

[1] An alternative proof is from [BBE+85]: It descends from the observation that for any two substrings p_1 and p_2 of s, if

$$\mathcal{L}_{p_1} \cap \mathcal{L}_{p_2} \neq \emptyset,$$

then p_1 is a prefix of p_2 or vice versa.

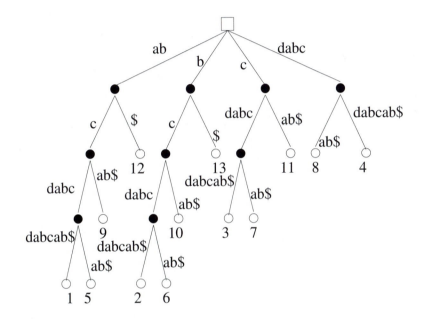

FIGURE 6.2: Suffix tree $T(s_1)$ where $s_1 =$ a b c d a b c d a b c a b $. All the suffixes $suf_i, 1 \leq i \leq 13$ can be read off the labels of the unique path from the root node to the leaf node marked by the integer i.

for reasons that will soon become obvious. This is best explained with a concrete example. A suffix of s, suf_i $(1 \leq i \leq n)$, is defined as

$$suf_i = s[i..n].$$

Continuing the working example, s_1 is written as

$$s_1 = \text{a b c d a b c d a b c a b } \$.$$

Then the suffixes of s_1 are as follows.

$$suf_1 = \text{a b c d a b c d a b c a b } \$$$
$$suf_2 = \text{b c d a b c d a b c a b } \$$$
$$suf_3 = \text{c d a b c d a b c a b } \$$$
$$suf_4 = \text{d a b c d a b c a b } \$$$
$$suf_5 = \text{a b c d a b c a b } \$$$
$$suf_6 = \text{b c d a b c a b } \$$$
$$suf_7 = \text{c d a b c a b } \$$$
$$suf_8 = \text{d a b c a b } \$$$
$$suf_9 = \text{a b c a b } \$$$
$$suf_{10} = \text{b c a b } \$$$
$$suf_{11} = \text{c a b } \$$$
$$suf_{12} = \text{a b } \$$$
$$suf_{13} = \text{b } \$$$

The suffix tree of s_1, written as

$$T(s_1),$$

is a compact trie of all the 13 suffixes and is shown in Figure 6.2. The direction of each edge is downwards. It has three kinds of nodes.

1. The square node is the *root* node. It has no incoming edges.

2. The *internal* nodes are shown as solid circles. Each has a unique incoming edge and multiple outgoing edges.

3. The *leaf* nodes are shown as hollow circles. A leaf node has no outgoing edges.

The edges of $T(s)$ are labeled by nonempty substrings of s and the tree satisfies the following properties.

1. The tree has exactly n leaf nodes are labeled bijectively by the integers

$$1, 2, \ldots, n.$$

2. Each internal node has at least two outgoing edges.

3. No two outgoing edges of a node can be labeled with strings that start with the same character.

4. Suffix suf_i is obtained by traversing the *unique* path from the root node to the leaf node labeled with i and concatenating the edge labels of this path.

$T(s)$ is called the suffix tree of s.

A suffix tree has various elegant properties, but we focus only on those that help prove the folklore.

1. (Property 1): Let r be a label on an incoming edge on note v. Then r appears exactly k times in s where k is the number of leaf nodes reachable from node v.

2. (Property 2): The number of internal nodes, I_s, in $T(s)$ satisfies the following:
$$I_s \leq n.$$

The first property follows from the definition of the suffix tree and we have seen the second property in Chapter 2. We now make the crucial observation about the suffix tree.

THEOREM 6.1
(Suffix tree theorem) *Given s,*

$$p \in P_{maximal}(s)$$

can be obtained by reading off the labels of a path from the root node to an internal node in $T(s)$.

PROOF   Given a node v, let

$$pth(v)$$

denote the string which is the label on the path from the root node to v.

Each maximal pattern is be a prefix of some suffix,

$$suf_i, \ 1 \leq i < n,$$

hence can be read off from the (prefix) of the label of a path from the root node.

Let p be a maximal motif. If

$$p = pth(v),$$

then v cannot be a leaf node since then the motif must end in the terminating symbol $\$$.

Assume the contrary, that there does not exist any internal node v such that

$$p = pth(v).$$

Hence p must end in the middle of an edge label, say r where the edge is incident on some internal node v'. Let

$$r = r_1 r_2,$$

where r_1 is a suffix of the maximal pattern p. Let k be the number of leaf nodes reachable from v'. Then by Property 1 of suffix trees, p appears k times in s. Then the concatenated string

$$p\, r_2$$

must also appear k times in s, contradicting the assumption that p is maximal. Hence for some internal node v,

$$p = pth(v).$$

This concludes the proof.   ⬚

In the example in Figure 6.2, note each

$$p \in P_{maximal}(s_1)$$

is such that there exist node v in $T(s_1)$

However, the converse of this theorem is not true. Let

$$p_1 = \text{b c},$$
$$p_2 = \text{b c d a b c},$$
$$p_3 = \text{c d a b c},$$
$$p_4 = \text{d a b c},$$

and

$$p_1, p_2, p_3, p_4 \notin P_{maximal}(s_1).$$

Notice in Figure 6.2 that for each p_i, $\leq i \leq 4,$, there exists some node v_i in $T(s_1)$ such that

$$pth(v_i) = p_i.$$

Thus the pattern corresponding to

$$p = pth(v),$$

for *any arbitrary* internal node v is not maximal. Since we only need an upper bound on $|P_{maximal}(s)|$, this is not a concern.

THEOREM 6.2
(Maximal solid pattern theorem) *Given a string s of size n, the number of maximal solid patterns in s is no more than n, i.e.,*

$$|P_{maximal}(s)| < n.$$

This directly follows from Property (2) of suffix trees and Theorem (6.1).

6.4 Rigid Patterns

Unlike the solid pattern, a *rigid* pattern, p, has also 'dont care' characters, written as '.'. This is also called a dot character or a wild card. Thus p is defined on

$$\Sigma \cup \{`.'\},$$

also written for convenience as:

$$\Sigma + `.'$$

The pattern is called rigid since the length it covers in all its occurrences is the same (in the next section we study patterns that occupy different lengths).

Next, we define the *occurrence* of a rigid motif. We first define the following. Let

$$\sigma_1, \sigma_2 \in \Sigma.$$

Then we define a partial order relation ($\preceq$) as follows.

$$\text{If } (\sigma_1 = `.') \Leftrightarrow \sigma_1 \prec \sigma_2.$$
$$\text{If } (\sigma_1 \prec \sigma_2) \text{ OR } (\sigma_1 = \sigma_2) \Leftrightarrow \sigma_1 \preceq \sigma_2.$$

A rigid pattern p *occurs* at position j in s if for $1 \leq l \leq |p|$,

$$p[l] \preceq s[j + l - 1]$$

p is said to *cover* the interval

$$[j, j + |p| - 1]$$

on s. Also for two strings with

$$|p| = |q|,$$

we say

$$p \preceq q,$$

if and only if the following holds for $1 \leq l \leq |p|$,

$$p[l] \preceq q[l].$$

Thus the *occurrence* of a rigid motif p in s is simply defined as follows: There exist some

$$1 \leq j \leq k \leq n - |p|,$$

such that

$$p \preceq s[j..k].$$

If p and q are of different lengths, then the smaller string is padded with '.' characters to the right, so the padded string is of the same length as the other, for the comparison. See Exercise 59 for the connection between relation $\preceq$ between patterns and their location lists.

Note that a nontrivial rigid pattern cannot start or end with a '.' character. Consider the example of the last section:

$$s_1 = \text{a b c d a b c d a b c a b}.$$

Let $P(s_1)$ be the set of all nontrivial rigid patterns and we consider a subset, $P'(s_1)$, as shown below.

$$P'(s_1) = \left\{ \begin{array}{l} \text{c d a b c, c d a b, c d a, c d} \\ \text{c . a b c, c . a b, c . a,} \\ \text{c d . b c, c d . b,} \\ \text{c d a . c, c . . b,} \\ \text{c . . b c,} \\ \text{c d . . c,} \\ \text{c . a . c,} \\ \text{c . . . c,} \end{array} \right\} \subset P(s_1).$$

Following the convention that, $i \in \mathcal{L}_p$, is the leftmost position (index) in s_1 of an occurrence of p, for each $p \in P'(s_1)$,

$$\mathcal{L}_p = \{3, 7\}.$$

In fact, if q is a nontrivial pattern in s_1 with

$$\mathcal{L}_q = \{3, 7\},$$

then

$$q \in P'(s_1).$$

6.4.1   Maximal rigid patterns

Since the rigid patterns can have the '.' character in then, maximality for rigid pattern is a little more complicated than that of solid patterns. If the

'.' character in a p can be replaced by a solid character, it is called *saturating* the pattern p. For example

$$p = \text{a . c . e}$$

can be saturated to p_1 or p_2 or p_3 where

$$p_1 = \text{a b c . e,}$$
$$p_2 = \text{a . c d e,}$$
$$p_3 = \text{a b c d e.}$$

A maximal rigid pattern p must satisfy both the conditions below.

1. (*maximal in length*) p is such that it cannot be extended to the left or to the right, without decreasing its support, to get a nontrivial

$$p' \neq p.$$

2. (*maximal in composition*) p is such that there exists no other p' with

$$p \preceq p' \text{ and } \mathcal{L}_p = \mathcal{L}_{p'}.$$

In other words, a maximal p cannot be saturated without decreasing its support.

This is indeed the intuitive definition of maximality for rigid motifs. It can be shown that this definition is equivalent to the one in Chapter 4 in terms of its location list.

Consider $P'(s_1)$ of the running example. Here each pattern can be saturated and the saturated motif can be extended to give

$$\text{c d a b c}$$

as the maximal rigid pattern. Thus it can be shown that the maximal rigid patterns are:

$$P_{maximal}(s_1) = \{\text{a b c d a b c, a b c, a b}\}.$$

It turns out that for s_1 the set of solid maximal patterns is the same as the set of rigid maximal patterns. Consider the following:

$$s_2 = \text{a b d d a c c d a b c a b.}$$

Then the set of maximal rigid patterns is given as

$$P_{maximal}(s_2) = \{\text{a . . d a . c, a b}\}.$$

Following the convention that, $i \in \mathcal{L}_p$, is the leftmost position (index) in s_2 of an occurrence of p,

$$\mathcal{L}_{\text{a . . d a . c}} = \{1, 5\},$$
$$\mathcal{L}_{\text{a b}} = \{1, 9, 12\}.$$

p	size	$\mathcal{L}_p$
$\overset{\longmapsto\quad \ell \quad \longmapsto}{\text{c c c c} \cdots \text{c c c c}}$	ℓ	$\{1, \ell+2\}$
c c c $\cdots$ c c c c	$\ell-1$	$\{1, 2, \ell+2, \ell+3\}$
c c $\cdots$ c c c c	$\ell-2$	$\{1, 2, 3, \ell+2, \ell+3, \ell+4\}$
c $\cdots$ c c c c	$\ell-3$	$\{1, 2, 3, 4, \ell+2, \ell+3, \ell+4, \ell+5\}$
$\vdots$	$\vdots$	$\vdots$
c c c c	4	$\{1, 2, \ldots, \ell-3, \ell+2, \ell+3, \ldots, 2\ell-3\}$
c c c	3	$\{1, 2, \ldots, \ell-2, \ell+2, \ell+3, \ldots, 2\ell-2\}$
c c	2	$\{1, 2, \ldots, \ell-1, \ell+2, \ell+3, \ldots, 2\ell-1\}$

FIGURE 6.3: All nontrivial maximal motifs in s_3 with no '.' elements.

6.4.2　Enumerating maximal rigid patterns

How small can $|P_{maximal}(s)|$ be for a given s?

We have already seen that the number of maximal solid patters is linear in the size of the input. However, we demonstrate here that the number of maximal rigid patterns is very large (possibly exponential in the size of the input).

The discussion in this section is not to be construed as a method for enumeration (a general discovery algorithm is presented in Section 7.2), but merely a means to gain insight into the complexity of interplay of patterns when a dont care character (the '.' character) is permitted in the pattern.

Construct, s_3, an input string of length

$$n = 2\ell + 1,$$

with 2ℓ with a stretch of ℓ c's, followed by one a, followed by ℓ c's as shown below.

$$s_3 = \underset{\longmapsto\quad \ell \quad \longmapsto}{\text{c c c} \cdots \text{c c c}} \, a \, \underset{\longmapsto\quad \ell \quad \longmapsto}{\text{c c c} \cdots \text{c c c}}$$

In the discussion below, we follow the convention that, $i \in \mathcal{L}_p$, is the leftmost position (index) in the input of an occurrence of p.

All nontrivial patterns with no '.' character is shown in Figure 6.3.

Figure 6.4 shows the longest pattern with exactly one '.' character. Each pattern is of length (or size) ℓ and if p_{1i} is a pattern with one '.' character at position i of the pattern, then

$$\mathcal{L}_{p_{1i}} = \{1, \ell - (i-2)\}.$$

The number of such patterns is

$$\ell - 2.$$

p	$\mathcal{L}_p$	i
$\mid\!\!\longleftarrow\quad\ell\quad\longrightarrow\!\!\mid$		
c . c c $\cdots$ c c c c	$\{1,\ell\}$	2
c c . c $\cdots$ c c c c	$\{1,\ell-1\}$	3
c c c . $\cdots$ c c c c	$\{1,\ell-2\}$	4
$\vdots$	$\vdots$	$\vdots$
c c c c $\cdots$. c c c	$\{1,5\}$	$\ell-3$
c c c c $\cdots$ c . c c	$\{1,4\}$	$\ell-2$
c c c c $\cdots$ c c . c	$\{1,3\}$	$\ell-1$

FIGURE 6.4:   The longest patterns with exactly one '.' character at position i in the pattern in s_3. Each is of length ℓ and is maximal in composition but not in length.

It is interesting to note that the longest pattern with exactly one '.' character is not maximal. Each pattern listed above is maximal in composition but not maximal in length. In fact, there is no maximal pattern with exactly one '.' character in s_3.

Figure 6.5 shows the maximal patterns with exactly two '.' elements at positions i and j of the pattern. The maximal pattern can be written as p_{2i}, which has an '.' character at position i and position $\ell+1$ of the pattern. p_{2i} is of length $\ell+i-1$, and

$$\mathcal{L}_{p_{2i}} = \{1, \ell - (i-2)\}.$$

The number of such maximal patterns is $\ell-1$.

Figure 6.7 shows a few examples of patterns with more than two '.' elements and each of these can be constructed from the patterns of Figure 6.5. It can be verified that the number of such maximal patterns with $j+1$ '.' elements is

$$\binom{\ell-1}{j}.$$

Thus including the maximal patterns with no '.' elements and exactly two '.' elements, the total number of maximal patterns is give as:

$$2(\ell-1) + \sum_{j=3}^{\ell-3} \binom{\ell-1}{j}.$$

Note that

$$\ell \approx \frac{n}{2},$$

thus the number of maximal patterns is

$$\mathcal{O}(2^n).$$

p	$\mathcal{L}_p$	i, j
$\|\!\!\longleftarrow \ell + 2 \longrightarrow\!\!\|$ c . c $\cdots$ c . c	$\{1, \ell\}$	$2, \ell + 1$
$\|\!\!\longleftarrow \ \ \ell + 3 \ \ \longrightarrow\!\!\|$ c c . c $\cdots$ c . c c	$\{1, \ell - 1\}$	$3, \ell + 1$
$\|\!\!\longleftarrow \ \ \ \ell + 4 \ \ \ \longrightarrow\!\!\|$ c c . c c $\cdots$ c c . c c	$\{1, \ell - 2\}$	$4, \ell + 1$
$\vdots$	$\vdots$	$\vdots$
$\|\!\!\longleftarrow \ \ \ \ \ 2\ell - 2 \ \ \ \ \ \longrightarrow\!\!\|$ c c c $\cdots$ c . c c c . c $\cdots$c c c	$\{1, 4\}$	$\ell - 2, \ell + 1$
$\|\!\!\longleftarrow \ \ \ \ \ 2\ell - 1 \ \ \ \ \ \longrightarrow\!\!\|$ c c c c $\cdots$ c c . c . c c $\cdots$c c c c	$\{1, 3\}$	$\ell - 1, \ell + 1$
$\|\!\!\longleftarrow \ \ \ \ \ \ 2\ell \ \ \ \ \ \ \longrightarrow\!\!\|$ c c c c $\cdots$ c c c . . c c c $\cdots$c c c c	$\{1, 2\}$	$\ell, \ell + 1$

FIGURE 6.5: All maximal patterns in s_3 with two '.' elements at positions i and j of the pattern. Note that the length of each pattern is $> \ell + 1$.

p	p	$\mathcal{L}_p$
$\longleftarrow \quad \ell + 2 \quad \longrightarrow$	$\longleftarrow \quad \ell + 2 \quad \longrightarrow$	
c $\bullet$ c c c - - - c $\bullet$ c	c $\bullet$ c c c - - - c $\bullet$ c	$1, \ell$
c c $\bullet$ c c - - - c $\bullet$ c c	c c $\bullet$ c c - - - c $\bullet$ c c	$1, \ell$-1
c c c $\bullet$ c - - - c $\bullet$ c c c	c c c $\bullet$ c - - - c $\bullet$ c c c	$1, \ell$-2
$\vdots$	$\vdots$	
c c c - - c $\bullet$ c c $\bullet$ c c - - - c	c c c - - c $\bullet$ c c $\bullet$ c c - - - c	$1, 4$
c c c - - c c $\bullet$ c $\bullet$ c c - - - c c	c c c - - c c $\bullet$ c $\bullet$ c c - - - c c	$1, 3$
c c c - - c c c $\bullet$ $\bullet$ c c - - - c c c	c c c - - c c c $\bullet$ $\bullet$ c c - - - c c c	$1, 2$
$\longleftarrow \ell - 1 \longrightarrow$ 2 $\longleftarrow \ell - 1 \longrightarrow$	$\longleftarrow \ell - 1 \longrightarrow$ 2 $\longleftarrow \ell - 1 \longrightarrow$	
$\longleftarrow \qquad 2\ell \qquad \longrightarrow$	$\longleftarrow \qquad 2\ell \qquad \longrightarrow$	

FIGURE 6.6: The $\ell+1$ collection of maximal patterns with two '.' elements (shown as $\bullet$) on s_3 have been stacked in two different ways (left flushed and right flushed) to reveal the 'pattern' of their arrangement within the maximal patterns.

p	$\mathcal{L}_p$	i,j,k
$\overset{\displaystyle|\!\!\leftarrow\quad \ell+2\quad\rightarrow\!\!|}{\texttt{c . c c}\ \cdots\ \texttt{c . . c}}$	$\{1, \ell-1, \ell\}$	$2, \ell, \ell+1$
$\overset{\displaystyle|\!\!\leftarrow\quad \ell+2\quad\rightarrow\!\!|}{\texttt{c . . c}\ \cdots\ \texttt{c c . c}}$	$\{1, 2, \ell\}$	$2, 3, \ell+1$
$\overset{\displaystyle|\!\!\leftarrow\quad \ell+2\quad\rightarrow\!\!|}{\texttt{c . c c}\ \cdots\ \texttt{. c . c}}$	$\{1, \ell-2, \ell\}$	$2, \ell-1, \ell+1$
$\overset{\displaystyle|\!\!\leftarrow\quad \ell+2\quad\rightarrow\!\!|}{\texttt{c . c .}\ \cdots\ \texttt{c c . c}}$	$\{1, 3, \ell\}$	$2, 4, \ell+1$
$\overset{\displaystyle|\!\!\leftarrow\quad \ell+2\quad\rightarrow\!\!|}{\texttt{c . c c .}\ \cdots\ \texttt{c c . c}}$	$\{1, 4, \ell\}$	$2, 5, \ell+1$
$\overset{\displaystyle|\!\!\leftarrow\quad \ell+3\quad\rightarrow\!\!|}{\texttt{c c . c c}\ \cdots\ \texttt{c . . c c}}$	$\{1, \ell-2, \ell-1\}$	$3, \ell, \ell+1$
$\overset{\displaystyle|\!\!\leftarrow\quad \ell+3\quad\rightarrow\!\!|}{\texttt{c c . c .}\ \cdots\ \texttt{c c . c c}}$	$\{1, 3, \ell-1\}$	$3, 5, \ell+1$
$\overset{\displaystyle|\!\!\leftarrow\quad \ell+3\quad\rightarrow\!\!|}{\texttt{c c . c c .}\ \cdots\ \texttt{c . c c}}$	$\{1, 4, \ell-1\}$	$3, 6, \ell+1$
$\overset{\displaystyle|\!\!\leftarrow\quad \ell+4\quad\rightarrow\!\!|}{\texttt{c c c . c c c .}\ \cdots\ \texttt{. c c c}}$	$\{1, 5, \ell-2\}$	$4, 8, \ell+1$
$\begin{array}{c}|\!\!\leftarrow\qquad 2\ell-3\qquad\rightarrow\!\!|\\ \texttt{c c c c}\cdots\texttt{ . c c . . }\cdots\texttt{c c c c}\\ |\!\!\leftarrow \ell-4 \rightarrow\!\!|\qquad\quad |\!\!\leftarrow \ell-4 \rightarrow\!\!|\end{array}$	$\{1, 4, 5\}$	$\ell-3, \ell, \ell+1$
$\begin{array}{c}|\!\!\leftarrow\qquad 2\ell-3\qquad\rightarrow\!\!|\\ \texttt{c c c c}\cdots\texttt{ . c . c . }\cdots\texttt{c c c c}\\ |\!\!\leftarrow \ell-4 \rightarrow\!\!|\qquad\quad |\!\!\leftarrow \ell-4 \rightarrow\!\!|\end{array}$	$\{1, 3, 5\}$	$\ell-3, \ell-1, \ell+1$
$\begin{array}{c}|\!\!\leftarrow\qquad 2\ell-2\qquad\rightarrow\!\!|\\ \texttt{c c c c c}\cdots\texttt{ . c . . . }\cdots\texttt{c c c c c}\\ |\!\!\leftarrow \ell-3 \rightarrow\!\!|\qquad\quad |\!\!\leftarrow \ell-3 \rightarrow\!\!|\end{array}$	$\{1, 3, 4\}$	$\ell-2, \ell, \ell+1$

FIGURE 6.7: Some maximal patterns in s_3 with three '.' elements each at positions i, j and k of the pattern.

6.4.3   Density-constrained patterns

Is it possible that the number of maximal patterns is high, since the patterns are allowed to have a large stretch of '.' elements? See some of the maximal pattern in s_3 below.

```
c . . . . . . . c
c c . . . . . . c
c . . . . . . c c
c c . . . . . c c
|←——  ℓ  ——→|
```

We define a version of rigid patterns with *density constraints*. This dictates the ratio of the '.' character to the solid characters in the pattern. One way of specifying the density is to impose a bound, d, on the number of consecutive '.' elements in a pattern. This is to be interpreted as follows: In a pattern, no two solid characters can have more than d consecutive '.' elements.

Does the density-constraint help in cutting down the number of maximal (restricted) rigid patterns?

Let $d = 1$, i.e., two solid characters can have at most one '.' character between them. For example, the patterns that meet the density constraints are marked with $\sqrt{}$ (the rest are marked with $\times$) below.

$$p_1 = c \ . \ . \ c \ c \ c \ . \ c \quad \times$$
$$p_2 = c \ . \ c \ . \ c \ . \ . \ c \quad \times$$
$$p_3 = c \ . \ . \ . \ . \ . \ . \ c \quad \times$$
$$p_4 = c \ . \ c \ . \ c \ . \ c \ c \quad \sqrt{}$$
$$p_5 = c \ c \ c \ c \ c \ c \ c \ c \quad \sqrt{}$$

However, we construct an example to show that even under this stringent density requirement, where d takes the smallest possible value of 1, the number of maximal density-constrained rigid patterns can be very large. Recall s_3 of length $2\ell + 1$, defined in the last section:

$$s_3 = c \ c \ c \cdots c \ c \ c \ a \ c \ c \ c \cdots c \ c \ c$$
$$|←—— ℓ ——→| \ |←—— ℓ ——→|$$

Construct s_4 of length $n = 4\ell + 1$,, based on s_3, as follows:

$$s_4 = c \ a \ c \ a \ c \ a \cdots c \ a \ c \ a \ c \ a \ g \ c \ a \ c \ a \ c \ a \cdots c \ a \ c \ a \ c \ a$$
$$|←—— \qquad 2\ell \qquad ——→| \ |←—— \qquad 2\ell \qquad ——→|$$

Note that each

$$c$$

of s_3 is replaced with

$$c \ a$$

in s_4 (and a is replaced with g). Let p be a pattern on s_3 whose alphabet set is:

$$\Sigma_{s_3} = \{a, c\}.$$

p' is constructed from p by replacing each element

$$x \in \Sigma + \text{`.'}$$

with

$$x \, a.$$

For example, p_1 and p_2 are maximal motifs in s_3 below. Notice that p_1' and p_2' meet the density constraint $(d = 1)$.

$$
p_1 = \begin{array}{c} |\!\!\longleftarrow \quad \ell + 2 \quad \longrightarrow| \\ \text{c . . c c } \cdots \text{ c c c . c} \end{array}
$$

$$
p_1' = \begin{array}{c} |\!\!\longleftarrow \qquad\qquad 2\ell + 4 \qquad\qquad \longrightarrow| \\ \text{c a . a . a c a c a } \cdots \text{ c a c a c a . a c a} \end{array}
$$

$$
p_2 = \begin{array}{c} \text{c c c c } \cdots \text{ c c c c . . . . c} \\ |\!\!\longleftarrow \ell - 1 \longrightarrow| |\!\!\longleftarrow \quad \ell \quad \longrightarrow| \, 1 \end{array}
$$

$$
p_2' = \begin{array}{c} \text{c a c a c a } \cdots \text{ c a c a c a . a . a . a } \cdots \text{ . a . a . a c a} \\ |\!\!\longleftarrow \quad 2\ell - 2 \quad \longrightarrow| |\!\!\longleftarrow \quad 2\ell \quad \longrightarrow| \, 2 \end{array}
$$

Next, we make the following claims.

1. If p is pattern in s_3, then p' is a pattern in s_4.

2. If p is maximal in s_3, then p' is maximal in s_4.

The proof of the claim is straightforward and we leave that as an exercise for the reader (Exercise 54). Thus the number of maximal density-constrained $(d = 1)$ rigid patterns in s_4 is also

$$\mathcal{O}(2^n).$$

6.4.4 Quorum-constrained patterns

For any pattern p,

$$|\mathcal{L}_p| \geq 2.$$

Here we define a version of rigid patterns with *quorum constraint*. This dictates that for a specified quorum $k(< n)$, the following must hold:

$$|\mathcal{L}_p| \geq k.$$

Although, this filters out patterns that occur less than k times, it is not sufficient to bring the count of the patterns down. We next show that even for the quorum-constrained rigid patterns, the number of maximal motifs can be very large.

Consider the input string s_3 of Section 6.4.1. The number of quorum constrained maximal motifs can be verified to be at least:

$$(\ell - k + 1) + \sum_{j=k}^{\ell-3} \binom{\ell-1}{j}.$$

The patterns can be constrained to meet both the density and quorum requirements, yet the number of maximal patterns could be very large. This is demonstrated by calculating the number patterns on s_4. We leave this as an exercise for the reader.

6.4.5   Large-$|\Sigma|$ input

Note that the s_3's alphabet set is

$$\{c, a\}$$

and s_4's alphabet is

$$\{c, a, g\}.$$

Both the sets are fairly small. Is it possible that if the alphabet is large, say,

$$\Sigma = \mathcal{O}(n),$$

then the number of maximal patterns is limited? Let

$$\ell = \sqrt{n},$$

and

$$\Sigma = \{e, 0, 1, 2, \dots, \ell-1, \ell\}.$$

For convenience, denote

$$\ell_i = \ell - i.$$

Next we construct s_5 and for convenience display the string in ℓ rows as follows:

$$
\begin{aligned}
s_5 = \; & 0\ 1\ 2\ 3\ \cdots\ \ell_3\ \ell_2\ \ell_1\ \ell \\
& 0\ e\ 2\ 3\ \cdots\ \ell_3\ \ell_2\ \ell_1\ \ell \\
& 0\ 1\ e\ 3\ \cdots\ \ell_3\ \ell_2\ \ell_1\ \ell \\
& 0\ 1\ 2\ e\ \cdots\ \ell_3\ \ell_2\ \ell_1\ \ell \\
& \quad\quad\quad \vdots \\
& 0\ 1\ 2\ 3\ \cdots\ e\ \ell_2\ \ell_1\ \ell \\
& 0\ 1\ 2\ 3\ \cdots\ \ell_3\ e\ \ell_1\ \ell \\
& 0\ 1\ 2\ 3\ \cdots\ \ell_3\ \ell_2\ e\ \ell
\end{aligned}
$$

We make the following claim about s_5.

Let

$$D \subseteq \{1, 2, \ldots, \ell_1\} \text{ and}$$
$$q = 0 \ 1 \ 2 \ 3 \ \cdots \ \ell_3 \ \ell_2 \ \ell_1 \ \ell$$

For each $j \in D$, obtain $q(D)$ of length $\ell + 1$ as follows: replace the element j in p' with the dont care character '.'. For example,

$$D_1 = \{1, 2\}, \quad q(D_1) = 0 \ . \ . \ 3 \ \cdots \ \ell_3 \ \ell_2 \ \ell_1 \ \ell$$
$$D_2 = \{2, \ell_2, \ell_1\}, \quad q(D_1) = 0 \ 1 \ . \ 3 \ \cdots \ \ell_3 \ . \ . \ \ell$$

Next, we then make the following claims:

1. For nonempty sets

$$D_1 \neq D_2,$$

   clearly the following holds:

$$q(D_1) \neq q(D_2).$$

2. For some nonempty D, there exists a unique maximal motif, say p_D, in s_5 which has a prefix $q(D)$. In other words,

$$p_D[1 \ldots 1 + \ell] = q(D).$$

   Following the convention that, $i \in \mathcal{L}_p$, is the leftmost position (index) in s_5 of an occurrence of p, we have

   (a)
$$1 \in \mathcal{L}_{p_D}.$$

   (b) For each $j \in D$, the following position (index) is in $\mathcal{L}_{p_D}$:

$$(\ell + 1)(j - 1) + (j + 1) \in \mathcal{L}_{p_D}.$$

   (c)
$$|\mathcal{L}_{p_D}| = |D| + 1.$$

3. The number of distinct such D's is

$$\mathcal{O}\left(2^{\ell-1}\right).$$

To understand these claims, consider the case where $\ell = 5$. Then s_5' (from s_5) is constructed as below:

$$s_5' = 0 \ 1 \ 2 \ 3 \ 4 \ 5 \ 0 \ e \ 2 \ 3 \ 4 \ 5 \ 0 \ 1 \ e \ 3 \ 4 \ 5 \ 0 \ 1 \ 2 \ e \ 4 \ 5 \ 0 \ 1 \ 2 \ 3 \ e \ 5$$

$$\begin{array}{ccccc} \uparrow & \uparrow & \uparrow & \uparrow & \uparrow \\ i = \ 1 & 7 & 13 & 19 & 25 \end{array}$$

Then all possible D sets and the corresponding $q(D)$ are shown below:

$|D| = 1$		$|D| = 2$		$|D| = 3$		$|D| = 4$	
D	$q(D)$	D	$q(D)$	D	$q(D)$	D	$q(D)$
$\{1\}$	0 . 2 3 4 5	$\{1,2\}$	0 . . 3 4 5	$\{1,2,3\}$	0 . . . 4 5	$\{1,2,$	0 5
$\{2\}$	0 1 . 3 4 5	$\{1,3\}$	0 . 2 . 4 5	$\{1,2,4\}$	0 . . 3 . 5	$3,4\}$	
$\{3\}$	0 1 2 . 4 5	$\{1,4\}$	0 . 2 3 . 5	$\{1,3,4\}$	0 . 2 . . 5		
$\{4\}$	0 1 2 3 . 5	$\{2,3\}$	0 1 . . 4 5	$\{2,3,4\}$	0 1 . . . 5		
		$\{2,4\}$	0 1 . 3 . 5				
		$\{3,4\}$	0 1 2 . . 5				

The maximal pattern p_D and its location list $\mathcal{L}_{p_D}$ for four cases are shown below.

$q(D)$	p_D	$\mathcal{L}_{p_D}$
0 1 2 . 4 5	0 1 2 . 4 5 0 . 2 3 . 5	$\{1, 19\}$
0 1 . . 4 5	0 1 . . 4 5 0 . 2 . . 5	$\{1, 13, 19\}$
0 . . 3 . 5	0 . . 3 . 5	$\{1, 7, 13, 25\}$
0 5	0 5	$\{1, 7, 13, 19, 25\}$

We leave the proof of the claims for a general ℓ as an exercise for the reader (Exercise 55).

Recall that $\ell = \sqrt{n}$. Thus the number of maximal rigid patterns in s_5 is

$$\mathcal{O}\left(2^{n^{\frac{1}{2}}}\right).$$

6.4.6 Irredundant patterns

Given an input s, let $P_{maximal}(s)$ be the set of all maximal patterns in s. A motif $p \in P_{maximal}(s)$ is redundant if

$$p = p_1 \otimes p_2 \otimes \ldots \otimes p_l, \text{ for some alignment } (i_1, i_2, \ldots, i_l)$$

and for

$$p \neq p_i \in P_{maximal}(s), \quad i = 1, 2, \ldots l,$$

and the support of p is obtained from the support of the $p_i's$, i.e.,

$$\mathcal{L}_p = \mathcal{L}_{p_1} \cup \mathcal{L}_{p_2} \cup \ldots \cup \mathcal{L}_{p_l}.$$

In other words, if all the information about a maximal pattern p is contained in some other l maximal patterns, then since p has nothing new to offer, it is a redundant pattern. Also, if need be, p can be deduced (constructed) from these l patterns. Hence the set of irredundant patterns is also called a *basis*. Let $P_{basis}(s)$ be the set of all irredundant rigid patterns in s.

However, there are some details hidden in the notation. We bring these out in the following concrete example. For some input s, let

$$p_0, p_1, p_2, p_3, p_4 \in P_{maximal}(s).$$

and the following holds:

p_1			G	T	T	.	G	A			
p_2	G	G	G	T	G	G	A	C	C	C
p_3			G	T	.	G	A	C	C
p_4			G	T	T	G	A	C
p_0	.	.	.	T	.	G	A	.	.	.

Thus p_0 is obtained by aligning the p_i's as shown. Incorporating the alignment information,

$$p_0 = p_1 \otimes p_2 \otimes p_3 \otimes p_4$$

is annotated as

$$p_0 = p_1(3) \otimes p_2(4) \otimes p_3(2) \otimes p_4(2).$$

Thus $p_i(j)$, is to be interpreted as the jth position of p_i corresponding to the leftmost or position 1 of p_0. Let

$$\mathcal{L} + j = \{i + j \mid \in \mathcal{L}\}.$$

Then, following the convention that, $i \in \mathcal{L}_p$, is the leftmost position (index) in s of an occurrence of p, we have

$$
\begin{aligned}
\mathcal{L}_{p_0} = \ & \mathcal{L}_{p_1} + 3 \\
\bigcup & \mathcal{L}_{p_2} + 4 \\
\bigcup & \mathcal{L}_{p_3} + 2 \\
\bigcup & \mathcal{L}_{p_4} + 2.
\end{aligned}
$$

Next, we state the central theorem of this section [AP04]. We begin with a definition that the theorem uses. Given an input s of length n, for $1 < i < n$, let ar_i be defined as follows:[2]

$$ar_i = s \otimes s \text{ for alignment } (1, i).$$

THEOREM 6.3
(Basis theorem)[AP04] *Given an input s of length n,*

$$P_{basis}(s) = \{ar_i \mid 1 < i < n\}.$$

Thus $|P_{basis}(s)| < n.$

---

[2] ar_i is also called the ith *autocorrelation*.

The proof is straightforward using the vocabulary developed so far and we leave that as an exercise for the reader: see Exercise 61 for the roadmap of the proof. A stronger result (of which this theorem is a corollary) is discussed in [AP04].

For a concrete example, see Exercise 62 for the basis of s_3 of Section 6.4.1. Note that the size of the basis is

$$2\ell - 2,$$

where as the number of maximal patterns was shown to be

$$\mathcal{O}(2^n).$$

Although the size of the basis is linear in the size of the input, it is important to bear in mind that extracting *all* the maximal motifs may still involve more work.

6.4.6.1   Density-constrained patterns

If the patterns are restricted to satisfy the density constraints, then what can be said about the size of the basis for this restricted class of rigid patterns? Let

$$p_1 \preceq p_2.$$

Since p_2 is more specific (has at least as many solid characters as p_1), if p_1 does not meet the density requirement, then p_2 does not meet the density requirements. Hence if p_1 is redundant w.r.t. maximal motifs

$$p_1, p_2, \ldots, p_l,$$

then it is not possible that p_1 meets the density requirement while one or more of p_i, $1 \leq i < l$, does not.

Although, this restriction filters out patterns that do not meet the density requirements, the basis for this restricted class actually gets larger.

THEOREM 6.4
(Density-constrained basis theorem) *Given s of length n, let*

$$P_{basis}^{(d)}(s)$$

denote the basis for the collection of rigid patterns that satisfy some density constraint $d > 0$. Then

$$|P_{basis}^{(d)}(s)| < n^2.$$

An informal argument is as follows. Each ar_i, $1 < i < n$, may get fragmented at regions where the density constraint is not met. For example, let density constraint $d = 2$ and consider

$$ar_i = \text{c a c . . g . . . c c . c . c a t . . . c t.}$$

Then ar_i has three fragments as follows:

$$ar_{i1} = \text{c a c . . g}$$
$$ar_{i2} = \text{c c . c . c a t}$$
$$ar_{i3} = \text{c t}$$

The number of such fragments for each ar_i is no more than n. Hence the size of the basis is

$$\mathcal{O}(n^2).$$

6.4.6.2 Quorum-constrained patterns

If the patterns satisfy the quorum constraint $k \geq 2$, then what can be said about the size of the basis for this restricted class of rigid patterns?

Although, this restriction filters out patterns that occur less than k times, the basis for this restricted class actually gets larger.

THEOREM 6.5

(Quorum-constrained basis theorem) [AP04, PCGS05] *Given s, let*

$$P_{basis}^{(k)}(s)$$

denote the basis for the collection of rigid patterns that satisfy quorum k (> 1). Then :

$$P_{basis}^{(k)}(s) = \left\{ \begin{array}{l} s \otimes s \otimes \ldots \otimes s, \\ \text{for alignment} \\ (1, i_2, i_3, \ldots, i_k) \end{array} \, \middle| \, 1 < i_2 < \ldots < i_k < n \right\}.$$

Thus

$$|P_{basis}^{(k)}(s)| < (n-1)^{k-1}.$$

This follows from the proof of Theorem (6.3). However, to show that such a bound is actually attained, consider the concrete example of s_3 of Section 6.4.1 and quorum $k = 3$. See Figure 6.7 for some maximal patterns p with

$$|\mathcal{L}_p| = 3.$$

It can be verified that the number of such maximal patterns, in s_3, is

$$\binom{\ell - 2}{2}.$$

6.5   Extensible Patterns

Consider two occurrences of a pattern in the following string.

$$s_6 = \boxed{\textbf{C G G T C G}}\; \text{T T} \;\boxed{\textbf{C G C A T A G}}$$

Note that the length of the cover of the two occurrences differ, yet there is some commonality (shown in bold) in the two that is captured as follows:

$$p = \text{C G - T . G}$$

where the dash symbol '-' is used to denote the variable gap. Its two occurrences, with the variable gap replaced by fixed ',' elements are

$$\text{C G . T . G and C G . . T . G}$$

These two rigid patterns are also called *realizations*. Allowing for spacers in a pattern is what makes it extensible. The dash symbol '-' represents the *extensible wild card* and

$$\text{'-'} \notin \Sigma.$$

Thus an extensible pattern is defined on

$$\Sigma + \text{'.'} + \text{'-'}$$

The density constraint d denotes the maximum number of consecutive dots allowed in a string or the maximum size of the gap. We use the *extensible wild card* denoted by the dash symbol '-'.

Given an extensible pattern (string) p, a rigid pattern (string) p' is a *realization* of p if each extensible gap is replaced by the exact number of dots. The collection of all such rigid realizations of p is denoted by

$$R(p).$$

Thus in this example,

$$R(\text{C G - T . G}) = \{\text{C G . T . G},$$
$$\text{C G . . T . G}\}.$$

As discussed earlier, a rigid string p occurs at position i on s if

$$p[j] \preceq s[i + j - 1], \text{ for } 1 \leq j \leq |p|.$$

A extensible string p *occurs* at position l in s if there exists a realization p' of p that occurs at l.

Note than an extensible string p could possibly occur a *multiple number of times* at a location on a sequence s. See Exercise 64 for an illustration. In the rest of the discussion our interest is in only the very first occurrence from the left at a location. This multiplicity of occurrence increases the complexity of the algorithm over that of rigid motifs in the discovery process discussed in Section 7.2.

6.5.1   Maximal extensible patterns

See Exercises 65 and 66 for examples of trivial extensible patterns in the absence of any density requirement. Thus it is very essential for a meaningful extensible pattern to have a density constraint, d. This ensures that the variable gap is no larger than d positions or characters.

A maximal extensible motif p must satisfy all the three conditions below.

1. (*maximal in length*) p is such that it cannot be extended to the left or to the right, without decreasing its support.

2. (*maximal in composition*) p is such that no '.' character can be replaced by a solid character, without decreasing its support.

3. (*maximal in extension*) p is such that no extensible gap character of p can be replaced by a fixed length substring (without extensible gaps), without decreasing its support. In other words, no extensible gap can be replaced by a fixed length substring (including the '.' character) that appears in all the locations in $\mathcal{L}_p$.

An extensible motif that is maximal in length, in composition and in extension is maximal.

Irredundant patterns. Note that the notion of redundancy stems from the co-occurrence of distinct rigid or solid patterns in the same locations (or segments) in the input. Thus if distinct patterns $p_1, p_2, \ldots, p_l$ co-occur, we define the redundant motif

$$p = p_1 \otimes p_2 \otimes \ldots \otimes p_l,$$

for some alignment.

This is straightforward and we leave that as an exercise for the reader (Exercise 67).

6.6   Generalizations

Here we discuss two simple and straightforward generalizations of string patterns: one where an element of the input is replaced by a set of elements (called homologous sets) and the second where the input is a sequence of real values.

6.6.1   Homologous sets

Consider the following input

$$s_6 = [\text{G L T}] \text{ A T L } [\text{G L}] \text{ A T } [\text{A T}] \text{ G}.$$

Note that
$$\Sigma = \{A, L, G, T\}.$$

The input is interpreted as follows.

1. The first position is a either G or L or T; the fifth position is either G or L; similarly the eighth position is A or T.

2. The second position is A; the third position is T and so on.

Thus each element, $s[j]$, is a set for $1 \leq j \leq n$. Usually, only a certain subset of elements can appear at a position and they are called *homologous characters*. For example, some homologous (groups) amino acids are shown below:

$$[L\ I\ V\ M]$$
$$[L\ I]$$
$$[F\ Y\ W]$$
$$[A\ S\ G]$$
$$[A\ T\ D]$$

The partial order, $\preceq$, on sets is defined as follows. For sets x, y,

$$x \preceq y \iff (x \subseteq y),$$

and

$$`.' \preceq x.$$

Following the convention that, $i \in \mathcal{L}_p$, is the leftmost position (index) in s_6 of an occurrence of solid motif

$$p_1 = G\ A\ T,\ \text{and}\ \mathcal{L}_{p_1}\{1, 5\}.$$

$$p_2 = [G\ L]\ A\ T,\ \text{and}\ \mathcal{L}_{p_2}\{1, 5\}.$$

Rigid motifs:
$$p_3 = G\ A\ T\ .\ G,\ \text{and}\ \mathcal{L}_{p_3}\{1, 5\}.$$

$$p_4 = [G\ L]\ A\ T\ .\ G,\ \text{and}\ \mathcal{L}_{p_4}\{1, 5\}.$$

Note that
$$p_1, p_2, p_3 \preceq p_4,$$

with
$$\mathcal{L}_{p_1} = \mathcal{L}_{p_2} = \mathcal{L}_{p_3} = \mathcal{L}_{p_4}.$$

Thus p_i is nonmaximal w.r.t. p_4, for $i = 1, 2, 3$.

6.6.2 Sequence on reals

Consider a sequence of real numbers. Is it possible to rename each distinct real number as an element of a discrete alphabet Σ?

Let s_r be the sequence of reals, and let s_d be the corresponding sequence with each distinct real number renamed as shown below:

$$s_r = 0.7 \quad 3.6 \quad 2.2 \quad 0.75 \quad 2.1 \quad 2.2 \quad 0.80 \quad 6.1 \quad 2.2$$
$$s_d = \ a \quad\ \ b \quad\ \ c \quad\ \ d \quad\ \ e \quad\ \ c \quad\ \ f \quad\ \ g \quad\ \ c$$

It is possible that values 0.75 and 0.80 may be considered fairly close but they have been assigned two different characters d and f. However, if we decide to assign any two values that differ by 0.05 the same character then consider the following scenario:

$$s'_r = 0.75 \quad 0.80 \quad 0.85 \quad 0.90 \quad 0.95 \quad 1.00$$
$$s'_d = \ a \qquad a \qquad a \qquad a \qquad a \qquad a$$

Clearly, 0.75 is not the same as 1.0 but both are assigned a by this scheme. We discuss a systematic method below to convert a sequence of reals to a string on a finite alphabet so that there is a one-to-one correspondence between the patterns in one to the other.

Two real values may be deemed equal if they are sufficiently close to each other. This is usually specified using a parameter $\delta \geq 0$. Two real numbers x and y are equal (written as $x =_r y$)

$$x =_r y \quad \Leftrightarrow \quad |x - y| \leq \delta.$$

To avoid confusion with the decimal point, in this section we denote the dont care character '.' by $\bullet$. The partial order relation ($\preceq$) is defined as follows. For reals x, y,

$$x = y \quad \Leftrightarrow \quad (x =_r y),$$

and

$$\bullet \preceq x.$$

Thus the *occurrence* of a pattern on a real sequence is defined as in Section 6.4.1. Let

$$\delta = 0.12$$

Following the convention that, $i \in \mathcal{L}_p$, is the leftmost position (index) in s_r of an occurrence of p, consider the patterns that have location list

$$\{1, 4, 7\}.$$

$$p_1 = 0.71 \bullet 2.2,$$
$$p_2 = 0.72 \bullet 2.2,$$
$$p_3 = 0.73 \bullet 2.2,$$
$$p_4 = 0.77 \bullet 2.2,$$
$$p_5 = 0.78 \bullet 2.2,$$

$$\vdots$$

In fact there are uncountably infinite patterns with this location list. To circumvent this problem, we allow the patterns to draw their alphabets not from real numbers but from closed real intervals. For example, in this case the first character is replaced by the interval

$$(0.69, 0.81)$$

and the unique pattern corresponding to this location list is:

$$p = (0.69, 0.81) \bullet 2.2, \quad \mathcal{L}_p = \{1, 4, 7\}.$$

The partial order on an interval, (x, y), is defined naturally as follows (x_1, x_2, y_1, y_2 are reals),

$$((x_1, x_2) \preceq (y_1, y_2)) \Leftrightarrow ((x =_r y) \text{ for each } x_1 \leq x \leq x_2 \text{ and } y_1 \leq y \leq y_2).$$

Thus for $i = 1, \ldots, 5$,

$$p_i \preceq p \text{ and } \mathcal{L}_p = \mathcal{L}_{p_i}.$$

In other words each p_i is redundant (or nonmaximal) w.r.t. p.

6.6.2.1   Mapping to a discrete instance

The problem of reporting uncountably infinite patterns on a sequence of reals is solved by an irredundant (or maximal) pattern defined on intervals of reals. In this section, we map this problem to a problem on discrete characters.

This is explained through a concrete example. Consider a sequence of six real numbers

$$7 \; 8 \; 10 \; 4 \; 1 \; 6.$$

They are sorted as

$$1 < 4 < 6 < 7 < 8 < 10 \; .$$

and their position on the real line is shown on the bottom horizontal axis of Figure 6.8. Let

$$\delta = 3,$$

and the size of δ is shown by the length of the (bold) horizontal bar shown for each number. Each unique interval intersection is labeled as

$$a, b, c, d, e, f, g, h$$

as shown. Each real number x can now be written as the set of alphabets that span

$$\left[x - \frac{\delta}{2}, x + \frac{\delta}{2} \right].$$

Thus the given problem has been reduced to an instance of a problem defined on a sequence of multi-sets (homologous sets). The fact that a solution to the former problem can be computed from a solution to the latter is straightforward to see and we leave that as an exercise for the reader (Exercise 57).

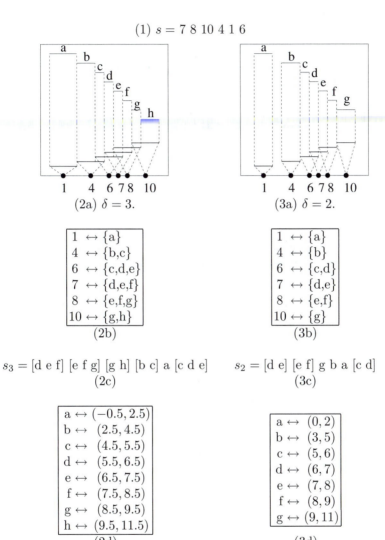

FIGURE 6.8: The input real sequence s is shown in (1). For $\delta = 3$: (2a)-(2b) show the mapping of the reals on the real line to the discrete alphabet. (2c) shows the string s_δ on the discrete alphabet. (2d) shows the mappings of the discrete alphabet to real (closed) intervals. For $\delta = 2$, the same steps are shown in (3a)-(3d).

6.7   Exercises

Exercise 46 *Consider the solid pattern definition of Section 6.3.*

    1. Show that the maximality of a solid pattern as defined in this section is equivalent to the definition of maximality of Chapter 4.

    2. Show that for solid patterns so defined, for a given s,

$$P_{maximal}(s) = P_{basis}(s).$$

Hint: 1. Use proof by contradiction. 2. What is $p = p_1 \oplus p_2$ for solid patterns?

Exercise 47 *Is it possible to have a string s, with*

$$|s| > 1,$$

such that the root node in its suffix tree has exactly one child?

Hint:

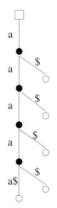

Exercise 48   *1. To construct a suffix tree, s is terminated with a meta symbol $. What purpose does this serve?*

    2. For each node v in the suffix tree $T(s)$, characterize the set $\{pth(v)\}$. Note that

$$P_{maximal}(s) \subset \{pth(v)\}.$$

Hint: 1. What if, $suf_i = s[j..k]$ holds for some $1 \leq j < k \leq n$? 2. Is it the collection of *all* nonmaximal patterns?

Exercise 49 *In Section 6.3, we saw that the number of maximal solid patterns, $P_{maximal}(s)$, in s of length n satisfies*

$$|P_{maximal}(s)| < n.$$

Show that this bound is tight, i.e, construct an example s' where

$$|P_{maximal}(s')| = n - 1.$$

Exercise 50 *Consider the rigid pattern definition of Section 6.4. Show that the maximality of a rigid pattern as defined in this section is equivalent to the definition of maximality of Chapter 4.*

Hint: Use proof by contradiction.

Exercise 51 *Let*

$$s = a\ b\ c\ d\ a\ b\ c\ d\ a\ b\ c\ a\ b.$$

We follow the convention that, $i \in \mathcal{L}_p$, is the leftmost position (index) in s_1 of an occurrence of p. Enumerate all the rigid patterns p in s such that

1. $\mathcal{L}_p = \{1, 5\}$,

2. $\mathcal{L}_p = \{1, 5, 9\}$,

3. $\mathcal{L}_p = \{1, 5, 9, 12\}$,

4. $\mathcal{L}_p = \{2, 6\}$,

5. $\mathcal{L}_p = \{2, 6, 10\}$,

6. $\mathcal{L}_p = \{3, 7\}$,

7. $\mathcal{L}_p = \{4, 8\}$.

Exercise 52 *Consider the following input sequence of length $2\ell + 1$.*

$$s_3 = c\ c\ c\ \cdots\ c\ c\ c\ a\ c\ c\ c\ \cdots\ c\ c\ c$$
$$|\!\!\leftarrow\quad \ell\quad \longrightarrow|\ |\!\!\leftarrow\quad \ell\quad \longrightarrow|$$

1. *Is it true that every nontrivial maximal pattern in s_3 has an occurrence that begins at location 1 in s_3? Why?*

2. *Is it true that every nontrivial maximal pattern in s_3 has an occurrence that ends at location $2\ell + 1$ in s_3? Why?*

3. *Show that there is no nontrivial maximal pattern of size $\ell + 1$.*

4. *Show that there is no nontrivial maximal pattern with only one '.' character.*

5. *Show that every nontrivial maximal pattern with a nonzero number of '.' character is of length $\geq \ell + 1$.*

Exercise 53 *Consider the input string s_3 of Section 6.4.1. Let the patterns satisfy quorum k. Show that the number of quorum constrained maximal motifs is at least:*

$$(\ell - k + 1) + \sum_{j=k}^{\ell-3} \binom{\ell-1}{j}.$$

Hint: The patterns without dont cares and meeting the quorum constraint is $\ell - k + 1$.

Exercise 54 *Consider s_3 of length $2\ell + 1$, and s_4 of length $4\ell + 1$, as follows.*

$$s_3 = c\ c\ c\ \cdots\ c\ c\ c\ a\ c\ c\ c\ \cdots\ c\ c\ c$$
$$\vert\!\longleftarrow\quad \ell\quad \longrightarrow\!\vert\ \vert\!\longleftarrow\quad \ell\quad \longrightarrow\!\vert$$

$$s_4 = c\ a\ c\ a\ c\ a\ \cdots\ c\ a\ c\ a\ c\ a\ g\ c\ a\ c\ a\ c\ a\ \cdots\ c\ a\ c\ a\ c\ a$$
$$\vert\!\longleftarrow\qquad 2\ell\qquad \longrightarrow\!\vert\ \vert\!\longleftarrow\qquad 2\ell\qquad \longrightarrow\!\vert$$

Further, given p, p' is constructed from p by replacing each element

$$x \in \Sigma + \text{'.'}$$

of p with

$$x\ a.$$

Then show that the following statements hold.

1. *If p is pattern in s_3, then p' is a pattern in s_4. Is the converse true? Why?*

2. *p_1 is a maximal pattern in s_3 and p_1' is a maximal pattern in s_4 where*

$$p_1 = \overset{\vert\!\longleftarrow\quad \ell + 2\quad \longrightarrow\!\vert}{c\ .\ .\ c\ c\ \cdots\ c\ c\ c\ .\ c}$$

$$p_1' = \overset{\vert\!\longleftarrow\qquad 2\ell + 4\qquad \longrightarrow\!\vert}{c\ a\ .\ a\ .\ a\ c\ a\ c\ a\ \cdots\ c\ a\ c\ a\ c\ a\ .\ a\ c\ a}$$

3. p_2 is a maximal pattern in s_3 and p_2' is a maximal pattern in s_4 where

$$p_2 = \underbrace{c\ c\ c\ c\ \cdots\ c\ c\ c\ c}_{\ell-1} \ . . . . \underbrace{c}_{1}$$

$$p_2' = \underbrace{c\ a\ c\ a\ c\ a\ \cdots\ c\ a\ c\ a\ c\ a}_{2\ell-2} \ . \underbrace{a\ .\ a\ .\ a\ \cdots\ .\ a\ .\ a\ .\ a\ c\ a}_{2\ell} \ \underbrace{}_{2}$$

4. If p is maximal in s_3, then p' is maximal in s_4.

5. The number of maximal patterns in s_4 is

$$\mathcal{O}(2^n).$$

Also, is the converse of statement 1 true? Why?

Hint: 5. Use the fact that the number of maximal patterns in s_3 is exponential.

Exercise 55 (Large-$|\Sigma|$ example) *See Section 6.4.5 for the construction of input string s_5 and definition of D, $q(D)$ and p_D. Assume in the following*

$$D \neq \emptyset.$$

1. Show that the number of distinct D's is

$$\mathcal{O}\left(2^{\ell-1}\right).$$

2. Show that if $D_1 \neq D_2$, then $q(D_1) \neq q(D_2)$.

3. For each D, there exists a motif p_D in s_5 which has a substring $q(D)$, i.e. for some r,
$$p_D[r \ldots r + \ell] = q(D).$$
Show that p_D is maximal.
Further show that p_D is unique, for each D.

4. Following the convention that, $i \in \mathcal{L}_p$, is the leftmost position (index) in s_5 of an occurrence of p, prove each of the following statements.

   (a)
   $$1 \in \mathcal{L}_{p_D}.$$

   (b) For each $j \in D$,
   $$(\ell+1)(j-1) + (j+1) \in \mathcal{L}_{p_D}.$$

(c)

$$|\mathcal{L}_{p_D}| = |D| + 1.$$

Hint: What are the location lists for each D?

Exercise 56   *1. Let s_8 of length $2\ell + 4$ be defined as:*

$$s_8 = c\,c\,c \cdots\ c\,c\,c\,a\,c\,c\,c\,a\,c\,c\,c \cdots\ c\,c\,c$$
$$\underleftarrow{\qquad} \ell \underrightarrow{\qquad} \qquad\qquad \underleftarrow{\qquad} \ell \underrightarrow{\qquad}$$

Enumerate the maximal patterns in s_8.

2. Let s_9 of length $3\ell + 2$ be defined as:

$$s_9 = c\,c\,c \cdots\ c\,c\,c\,a\,c\,c\,c \cdots\ c\,c\,c\,a\,c\,c\,c \cdots\ c\,c\,c$$
$$\underleftarrow{\quad} \ell \underrightarrow{\quad} \underleftarrow{\quad} \ell \underrightarrow{\quad} \underleftarrow{\quad} \ell \underrightarrow{\quad}$$

Enumerate the maximal patterns in s_9.

Hint: What are patterns with 0,1,2, ... dont care elements? Since the patterns are maximal, consider the autocorrelations of the input sequences.

Exercise 57 (Real sequences) *Consider, s, a sequence on reals. Let*

$$0 \le \delta_1 \le \delta_2.$$

Let s_i be the sequence of multi-sets constructed as described in Section 6.6.2, for δ_i, $i = 1, 2$.

1. Let s_i be defined on Σ_i, $i = 1, 2$. Compare the sizes of Σ_1 and Σ_2.

2. Let n_i be the number of positions in s_i that are singletons, for $i = 1, 2$. How do n_1 and n_2 compare?

3. Show that if

$$[\sigma_1, \sigma_2, \dots, \sigma_m]$$

is a homologous set resulting from the construction with the mappings:

$$\sigma_1 \leftrightarrow (l_1, u_1)$$
$$\sigma_2 \leftrightarrow (l_2, u_2)$$
$$\vdots$$
$$\sigma_m \leftrightarrow (l_m, u_m)$$

Then the following holds:

$$u_1 = l_2,$$
$$u_2 = l_3,$$
$$\vdots$$
$$u_{m-1} = l_m.$$

In other words, the homologous set is mapped to the (closed) interval:

$$[\sigma_1, \sigma_2, \ldots, \sigma_m] \leftrightarrow (l_1, u_m).$$

Hint: See Figure 6.8.

Exercise 58 (Real sequences) *Let* $\delta = 0.5$ *and*

$$s_r = 0.75 \quad 0.80 \quad 0.85 \quad 0.90 \quad 0.95 \quad 1.00$$

Construct s_d *on a finite alphabet set, based on* s_r, *such that patterns on* s_r *correspond to patterns on* s_d.

Hint: Note that the successive intervals on the real intersect at a point.

$$[a\ b]\ [b\ c\ d]\ [d\ c\ e]\ [e\ f\ g]\ [g\ h\ i]\ [i\ j].$$

Exercise 59 (Relation $\preceq$ and location lists $\mathcal{L}_p$) *Let* p_i, $i = 0, 1, 2, \ldots, l$ *be patterns on* s.

1.
$$\text{If } p_1 \hookrightarrow p_2, \text{ then } p_2 \preceq p_1.$$

2.
$$\text{If } \{p_1, p_2, \ldots, p_l\} \hookrightarrow p_0, \text{ then } p_0 \preceq p_i, i = 1, 2, \ldots, l.$$

Prove the following.

(a) Statement (1) for solid patterns.

(b) Statements (1) and (2) for rigid patterns.

(c) Statements (1) and (2) for extensible patterns.

Do the above statements hold for density-constrained and quorum-constrained patterns? Why?

Hint: How are the location lists related?

Exercise 60 (Redundant pattern construction) *For some input s, let p_1, p_2, p_3, p_4 be maximal motifs in s.*

$$p_1 = \text{G T T G G A}$$
$$p_2 = \text{G G G T G G A C C C}$$
$$p_3 = \text{G T . G A C C}$$
$$p_4 = \text{G T T G A C}$$

Four meet operations with the alignments are shown below (see Section 6.4.6):

$$q_1 = p_2 \otimes p_3 \otimes p_4$$

p_2	G G	G T G G A C	C C
p_3		G T . G A C	C
p_4		G T T G A C	
q_1	. .	G T . G A C	. .

$$q_2 = p_2 \otimes p_3$$

p_2	G G	G T G G A C C	C
p_3		G T . G A C C	
q_2	. .	G T . G A C C	.

$$q_3 = p_1 \otimes p_2 \otimes p_3 \otimes p_4$$

p_1		G T	T G G A	
p_2	G G G	T	G G A	C C C
p_3		G T	. G A	C C
p_4		G T	T G A	C
q_3	. . .		T . G A	. . .

$$q_4 = p_1 \otimes p_2 \otimes p_3 \otimes p_4$$

p_1		G T T G	G A	
p_2	G G	G T G G	A C C C	
p_3		G T . G	A C C	
p_4		G T T G	A C	
q_4	. .	G T . G	. . . .	

1. *Are q_1 and q_2 maximal in s? Why?*

2. *Are q_3 and q_4 maximal in s? Why?*

3. *Can $\mathcal{L}_{q_i}$, $1 \leq i \leq 4$, be computed using $\mathcal{L}_{p_i}$, $1 \leq i \leq 4$? Why?*

Hint: 2. & 3. Use proof by contradiction. 4. For q_i, what if there exists some other maximal p_5 such that $q_i \preceq p_5$?

Exercise 61 (Proof of linearity of basis) *Let s be of length n. For $1 < i < n$, let*

$$ar_i = s \otimes s, \text{ for alignment } (1, i).$$

1. *Show that ar_i is a maximal pattern in s, for each i.*

2. *Show that it is possible that*

$$ar_i = ar_j, \quad for \ i \neq j.$$

3. *Let $\mathcal{L}_p$ represent the leftmost position of the occurrence of p in s.*

   (a) *Let*
   $$\mathcal{L}_p = \{i_1 < i_2\}.$$

   Then show the following.

   i. *p is a substring of $ar_{i_2-i_1+1}$.*
   ii. *If p is maximal then $p = ar_{i_2-i_1+1}$.*

   (b) *Let*
   $$\mathcal{L}_p = \{i_1 < i_2 < \ldots < i_k\}.$$

   If p is maximal then show that

   $$p = ar_{i_2} \otimes ar_{i_3} \otimes \ldots \otimes ar_{i_k}, \quad for \ alignment \ (1, 1, \ldots, 1).$$

4. *Show that*
   $$P_{basis} = \{ar_i \mid 1 < i < n\}.$$

Hint: 1. Use proof by contradiction. 2. See Exercise 62. 3. Follows from the definition of maximality and the meet operator $\otimes$. Show the following.

$$p = ar_{i_2-i_1+1}$$
$$= s \otimes s, \quad for \ alignment \ (1, i_2 - i_1 + 1).$$

4. Show the following:

$$p = s \otimes s \otimes \ldots \otimes s,$$
$$for \ alignment \ (1, i_2 - i_1 + 1, i_3 - i_1 + 1, \ldots, i_k - i_1 + 1)$$
$$= ar_{i_2} \otimes ar_{i_3} \otimes \ldots \otimes ar_{i_k},$$
$$for \ alignment \ (1, 1, \ldots, 1).$$

Exercise 62 *Let s_3 of length $2\ell + 1$ be defined as*

$$s_3 = c \ c \ c \cdots \ c \ c \ c \ a \ c \ c \ c \cdots \ c \ c \ c$$
$$\underbrace{\hspace{2.2cm}}_{\ell} \quad \underbrace{\hspace{2.2cm}}_{\ell}$$

Then show the following:

1. *$ar_{\ell+1} = ar_{\ell+2}$.*

2. $|P_{basis}| = 2\ell - 2$.

Hint: s_3's basis is:

```
C C
C C C
C C C C

    ⋮

C C C - - C C
C C C - - C C C
C C C - - C C C C
C . C C C - - - C . C
C C . C C - - - - C . C C
C C C . C - - - - C . C C C

          ⋮

C C C - - C . C C . C C - - - C
C C C - - C C . C . C C - - - C C
C C C - - C C C . . C C - - - C C C
```
$$\longleftarrow \ell - 1 \longrightarrow \quad 2 \quad \longleftarrow \ell - 1 \longrightarrow$$

Exercise 63 (Basis construction algorithm)

1. *Design an algorithm to compute the basis for a string, based on Theorem (6.3).*

2. *What is the time complexity of the algorithm?*

Hint: Each ar_i can be constructed in $\mathcal{O}(n)$ time, even under density constraint d. Thus the basis can be constructed in time

$$\mathcal{O}(n^2)$$

See [AP04] for an efficient incremental algorithm. See also [PCGS05] for a nice exposition.

Exercise 64 (Multiple occurrences at i) *Let*

$$s = \text{A T C G A T A}.$$

What are the occurrences, with the left most position $i = 1$ in s, of

$$p = \text{A . C} - \text{A ?}$$

Hint: The two occurrences of the extensible pattern at position 1 in s are:

$$\boxed{\text{A T C G A}}\ \text{T A}$$
$$\boxed{\text{A T C G A T A}}$$

Exercise 65 (Trivial extensible patterns) *Given an input s, let p be a rigid pattern with*

$$|\mathcal{L}_p| > 2.$$

1. *Construct an extensible pattern p' that must occur in s with*

$$|\mathcal{L}_{p'}| \geq 2.$$

2. *How many extensible patterns can be constructed based on p?*

Hint: Let

$$\mathcal{L}_p = \{i_1, i_2, \ldots, i_{k-1}, i_k\}.$$

1. Let

$$p' = p-p$$

Then

$$\mathcal{L}_{p'} = \{i_1, i_2, \ldots, i_{k-1}\}.$$

2. Let

$$p' = p-p-p$$

Then

$$\mathcal{L}_{p'} = \{i_1, i_2, \ldots, i_{k-2}\}.$$

And, so on.

Exercise 66 (Nontrivial extensible patterns) *Consider the following extensible pattern*

$$p = p_1-p_2-\ldots-p_l$$

where each of p_i, $1 \leq i \leq k$ is rigid. Under what conditions is p nontrivial, i.e., it cannot be simply constructed from p_i and $\mathcal{L}_{p_i}$, $1 \leq i \leq l$?

Hint: Let the quorum be $k(\geq 2)$. If some p_i is such that it occurs less than k times and this p_i is used in more than one occurrence of p.

Exercise 67 *Given two extensible patterns p_1 and p_2, that co-occur at a position i in s, devise a method to compute*

$$p = p_1 \otimes p_2, \text{ for alignment } (i_1, i_2).$$

Hint:

$p_1 =$	A G A	–	C T A A	–	A . G	–	A
$p_2 =$	G	– – –	T G A A A	– –	A A C . G		
$p =$	G	– – –	T . A	–	A	– – –	A

Exercise 68 (Invariance under reversal)   *The reverse of a string s is denoted as $\bar{s}$. For example, if*

$$s = A\ C\ G\ G\ T\ T\ C,$$

then

$$\bar{s} = C\ T\ T\ G\ G\ C\ A.$$

1. *If p occurs at j in s, let the center of occurrence, j_c, be given as*

$$j_c = j + \frac{|p| - 1}{2}.$$

   Note that j_c may not always be an integer, but that does not matter. We follow the convention that, $i \in \mathcal{L}_p$, is the center of the occurrence of p in s. If p is a pattern in s, then show the following:

   (a) *$\bar{p}$ is a pattern in $\bar{s}$, and*

   (b) *$\mathcal{L}_p = \mathcal{L}_{\bar{p}}$.*

2. *If p is maximal (nonmaximal) in s, then show that $\bar{p}$ is maximal (nonmaximal) in $\bar{s}$.*

3. *If p is irredundant (redundant) in s, then show that $\bar{p}$ is irredundant (redundant) in $\bar{s}$.*

4. *Show the following:*

$$|P(s)| = |P(\bar{s})|,$$
$$|P_{maximal}(s)| = |P_{maximal}(\bar{s})|,$$
$$|P_{basis}(s)| = |P_{basis}(\bar{s})|.$$

Comments

String patterns is about the simplest idea, in terms of its definition, in bioinformatics. It is humbling to realize how complicated the implications of simplicity can be. This area has gained a lot from a vibrant field of research in computer science, called *stringology* (not to be confused with *string theory*, from high energy physics).

Chapter 7

Algorithms & Pattern Statistics

7.1 Introduction

In the previous chapter, we described a whole array of possible characterizations of patterns, starting from the simple l-mer (solid patterns) to rigid patterns with dont care characters to extensible patterns with variable length gaps. Further, the element of a pattern could be drawn from homologous sets (multi-sets). In this chapter we take these intuitive definitions to fruition by designing practical discovery algorithms and devising measures to evaluate the significance of the results.

7.2 Discovery Algorithm

We describe an algorithm [ACP05] for a very general form of pattern:

extensible pattern on multi-sets (homologous sets)
with density constraint d and quorum constraint k.

Input: The input is a string s of size n and two positive integers, density constraint d and quorum $k > 1$.

Output: The density (or extensibility) parameter d is interpreted as the maximum size of the gap between two consecutive solid characters in a pattern. The output is all maximal extensible patterns that occur at least k times in s.

The algorithm can be adapted to extract rigid motifs as a special case. For this, is suffices to interpret d as the maximum number of dot characters between two consecutive solid characters.

Algorithm: The algorithm works by converting the input into a sequence of possibly overlapping *cells*. A maximal extensible pattern is a sequence of (overlapping) cells. Given a pattern p defined on

$$\Sigma + \text{`.'} + \text{`-'}$$

a substring $\hat{p}$, on p is a *cell*, also denoted by a triplet,

$$\langle \sigma_1, \sigma_2, \ell \rangle,$$

is defined as follows:

1. $\hat{p}$ begins in σ_1 and ends in σ_2 where

$$\sigma_1, \sigma_2 \subseteq \Sigma.$$

   Note that σ_i, $i = 1, 2$, is a set of solid characters (homologous set), possibly singleton.

2. $\hat{p}$ has only nonsolid intervening characters.
   ℓ is the number of intervening dot characters `.'.
   If the intervening character is the extensible character, `-', then ℓ takes a value of -1.

$C(p)$ is the collection of all cells of p. For example, if

$$p = \text{A G .. C} - \text{T . [G C]}$$

then the cells of p, $C(p)$ are:

$\hat{p}$	$\langle \sigma_1, \sigma_2, \ell \rangle$
A G	$\langle \text{A, G}, 0 \rangle$
G .. C	$\langle \text{G, C}, 2 \rangle$
C − T	$\langle \text{C, T}, -1 \rangle$
T . [G C]	$\langle \text{T, [G C]}, 1 \rangle$

This will be also used later in the probability computations of the patterns.

Initialization Phase: The cell is the smallest *extensible* component of a maximal pattern and the string can be viewed as a sequence of overlapping cells. The initialization phase has the following steps.

Step 1: Construct patterns that have exactly two solid characters in them and separated by no more than d spaces or `.' characters. This is done by scanning the string s from left to right. Further, for each location, the start and end position of the cell are also stored.

Step 2: The extensible cells are constructed by combining all the cells with at least one dot character and the same start and end solid characters. The location list is updated to reflect the start and end position of each occurrence.

If B is the collection of all such cells in an input s, then it can be verified that

$$|B| \leq (2 + d) |\Sigma|^2.$$

Define the following order relation, where $\sigma \in \Sigma$,

$$\text{`--'} \prec \text{`.'} \prec \sigma$$

The cells in B are arranged in descending lexicographic order. For two solid characters, we can arbitrarily pick any order without affecting the results. See the concrete example below for an illustration. The idea is that cells are processed in the order of 'saturation', which roughly means maximal in composition and extensibility (but not necessarily in length). Thus, for example

$$\begin{aligned}
A\,C &\succ A\,.\,T \\
&\succ A\,.\,.\,G \\
&\succ A\,.\,.\,.\,C \\
&\succ A - C.
\end{aligned}$$

Similarly,

$$\begin{aligned}
C\,A &\succ A\,.\,A \\
&\succ G\,.\,.\,A \\
&\succ C\,.\,.\,.\,A \\
&\succ C - A.
\end{aligned}$$

Iteration Phase: The algorithm works by starting with a pattern in B, and iteratively using compatible cells in B to generate an extended pattern. The process is repeated until this pattern can no longer be extended. This 'maximally' extended pattern is emitted and then we move on to the next cell in B.

However, once a pattern p is emitted, it cannot be withdrawn. Hence it is required to make sure that p is maximal and there can be no other p' that can be generated later to make p nonmaximal w.r.t. p'. How can this condition be ensured?

This is done by processing the cells in a decreasing order of saturation. For

$$p_1, p_2 \in B,$$

p_2 can possibly be added to the right of p_1 if the last character of p_1 is the same as the first character of p_2. We say that p_1 is *compatible* with p_2. Further,

$$p = p_1 \oplus p_2, \text{ when } p_1 \text{ is compatible with } p_2,$$

where p is the concatenation of p_1 and p_2 with an overlap at the common end and start character. For example, if

$$\begin{aligned}
p_1 &= C\,.\,G \\
p_2 &= G\,.\,.\,T
\end{aligned}$$

then p_1 is compatible with p_2 and

$$p = p_1 \oplus p_2$$
$$= \text{C. G . . T}$$

Note that p_2 is not compatible with p_1. Also, the locations list of p is appropriately generated as follows:

$$\mathcal{L}_p = \{(i,j) \mid (i,l) \in \mathcal{L}_{p_1}, (l,j) \in \mathcal{L}_{p_2}\}.$$

The algorithm is straightforward and is explained through a concrete example. The two points to bear in mind are as follows.

1. Amongst all possible candidate cells, always pick the most saturated one (at each step). This ensures that the patterns are generated in the desirable order: If p' is nonmaximal w.r.t. a maximal pattern p, then p is always generated (emitted) before p'.

2. A pattern under construction is first extended to the right, till it can be right-extended no more. Then it is extended to the left till it can be left-extended no more. Then it is ready to be emitted. Before emitting, it is checked against the emitted patterns to see if it is nonmaximal.

The overall algorithm could either be simply iterative or recursive (to take advantage of partial computations). We describe a recursive version below (note that the 'backtrack' in the discussion can be implicitly captured by recursive calls).

The details of the algorithm are left as an exercise for the reader (Exercise 70) which can be gleaned from the concrete example discussed below. The reader is also directed to [ACP05] for other details.

Concrete example. For example, let

$$\text{density constraint } d = 2 \text{ and quorum } k = 2,$$

with input string

$$s = \quad \text{C A G C A G T C T C.}$$

In Steps 1 and 2 of the initialization, the set of cells B' is generated. B'_{left} shows the elements of B' sorted by the starting element of the cell and and each column follows the $\preceq$ ordering of the cells. Similarly, B'_{right} shows the elements of B' sorted by the last elements of the cell and the elements of each column is similarly ordered. Thus each column shows the 'saturation' ordering of the cells and the cells are processed in the order displayed here.

$$
B'_{left} = \left\{
\begin{array}{llll}
A\,G, & C\,A, & G\,C, & T\,C, \\
A\,.\,C, & C\,T, & G\,T, & T\,.\,T, \\
A\,.\,T, & C\,.\,G, & G\,.\,A, & T\,.\,.\,T, \\
A\,.\,.\,A, & C\,.\,C, & G\,.\,C, & T-T. \\
A\,.\,.\,C, & C\,.\,.\,C, & G\,.\,.\,G, & \\
A-C, & C\,.\,.\,T, & G\,.\,.\,T, & \\
& C-C, & G-C, & \\
& C-T, & G-T, &
\end{array}
\right\},
$$

$$
B'_{right} = \left\{
\begin{array}{llll}
C\,A, & G\,C, & A\,G, & G\,T, \\
G\,.\,A, & T\,C, & C\,.\,G, & C\,T, \\
A\,.\,.\,A, & A\,.\,C, & G\,.\,.\,G, & A\,.\,T, \\
& C\,.\,C, & & T\,.\,T, \\
& G\,.\,C, & & C\,.\,.\,T, \\
& C\,.\,.\,C, & & G\,.\,.\,T, \\
& A\,.\,.\,C, & & C-T, \\
& T\,.\,.\,C, & & G-T, \\
& A-C, & & T-T. \\
& C-C, & & \\
& G-C, & &
\end{array}
\right\}.
$$

To avoid clutter, the cells that do not meet the quorum constraints have been removed to produce B_{left} and B_{right}. See Exercise 69 (1) for a mild warning about this step.

$$
B_{left} = \left\{
\begin{array}{llll}
A\,G, & C\,A, & G-C, & T\,C. \\
A-C, & C\,.\,G, & G-T, & \\
& C-C, & & \\
& C-T, & &
\end{array}
\right\},
$$

$$
B_{right} = \left\{
\begin{array}{llll}
C\,A, & T\,C, & A\,G, & C-T, \\
& A-C, & C\,.\,G. & G-T. \\
& C-C, & & \\
& G-C, & &
\end{array}
\right\}.
$$

Note that

$$
(i, j) \in \mathcal{L}
$$

denotes that the cell begins at position i and ends at position j in the input. Again, to avoid clutter, we do not enumerate all the location lists of the cells.

We show only a few examples of cells with their location lists below.

$$\mathcal{L}_{C\ A} = \{(1,2),(4,5)\},$$
$$\mathcal{L}_{A\ G} = \{(2,3),(5,6)\},$$
$$\mathcal{L}_{A\ -\ C} = \{(2,4),(5,8)\},$$

We now show the steps involved in constructing the maximal extensible patterns.

1. **(Pick cell in order of saturation)** Consider the cells that start with A and have at least two occurrences. Then we have the following:

$$p_1 = A\ G$$
$$p_2 = A\ -\ C$$

What should the first choice be? Between the two, pattern p_1 is more 'saturated' than p_2 and p_1 is picked first.

$$p_1 = \boxed{A\ G}$$

2. **(Explore right)** Next, we pick cells that are compatible with p_1. We look in B_{left} for cells that start with G.

$$p_3 = G\ -\ C,$$
$$p_4 = G\ -\ T,$$

We pick p_3.

$$q_1 = p_1 \oplus p_3$$
$$= \boxed{A\ G\ -\ C}, \quad \text{and}$$
$$\mathcal{L}_{q_1} = \{(2,4),(5,8)\}.$$

3. **(Explore right)** We continue to explore the right and search for cells, p, such that q_1 is compatible with p. We look in B_{left} for cells that start with C.

$$p_5 = C\ A$$
$$p_6 = C\ .\ G$$
$$p_7 = C\ -\ C$$
$$p_8 = C\ -\ T$$

Adding p_5, p_6, p_7 does not meet the quorum requirements. The only option is p_8.

$$q_2 = q_1 \oplus p_8$$
$$= \boxed{A\ G\ -\ C\ -\ T}, \quad \text{and}$$
$$\mathcal{L}_{q_2} = \{(2,7),(5,9)\}.$$

4. **(Explore right)** We continue to explore the right and search for cells, p, such that q_2 is compatible with p. We look in B_{left} for cells that start with T.

$$p_9 = \text{T C}$$

$$
\begin{aligned}
q_3 &= q_2 \oplus p_9 \\
&= \boxed{\text{A G} - \text{C} - \text{T C}}, \quad \text{and} \\
\mathcal{L}_{q_3} &= \{(2, 8), (5, 10)\}.
\end{aligned}
$$

5. **(Explore left)** No more cells can be added to the right in q_3. Now we try to add cells to the left of q_3 and look for cells p such that p and q_3 are compatible. We look in B_{right} for cells that end with A.

$$p_5 = \text{C A}$$

Thus

$$
\begin{aligned}
q_4 &= p_5 \oplus q_3 \\
&= \boxed{\text{C A G} - \text{C} - \text{T C}}, \quad \text{and} \\
\mathcal{L}_{q_4} &= \{(1, 8), (4, 10)\}.
\end{aligned}
$$

6. **(Emit maximal pattern)** However, no more cells can be added to the right or to the left of q_4, hence emitted.

$$q_4 = \boxed{\boxed{\text{C A G} - \text{C} - \text{T C}}}$$

7. **(Backtrack)** We can backtrack to utilize some partial computations (using the recursive call mechanism). As we backtrack, step 5 had only one option; step 4 had only one option; the very last option was used in step 3. However, in step 2, another option can still be explored.

8. **(Explore right)** We are back in the state of step 2 and at this stage we wish to extend

$$p_1 = \boxed{\text{A G}}$$

to the right with

$$p_4 = \text{G} - \text{T}.$$

But this does not meet the quorum constraint, so we explore the left now.

9. **(Explore left)** We look in B_{right} for cells that end in A.

$$p_5 = \text{C A}$$

Thus

$$q_5 = p_5 \oplus p_1$$
$$= \boxed{\text{C A G}}, \text{ and}$$
$$\mathcal{L}_{q_5} = \{(1,3),(4,6)\}.$$

10. **(Explore left)** We explore extending q_5 to the left without success.

11. **(NO emit)** Before emitting q_5, it is checked for maximality against the emitted pattern q_4 and it turns out that q_5 is nonmaximal w.r.t. q_4, hence it cannot be emitted.

12. **(Backtrack)** We must backtrack to step 1. In step 1, we can now use the second choice and use cell p_2:

$$p_2 = \boxed{\text{A} - \text{C}}$$

13. **(Explore right)** We explore the right and search for cells, p, such that p_2 is compatible with p :

$$p_5 = \text{C A}$$
$$p_6 = \text{C . G}$$
$$p_7 = \text{C} - \text{C}$$
$$p_8 = \text{C} - \text{T}$$

Adding p_5, p_6, p_7 does not meet the quorum requirements. The only option is p_8.

$$q_6 = p_2 \oplus p_8$$
$$= \boxed{\text{A} - \text{C} - \text{T}}, \quad \text{and}$$
$$\mathcal{L}_{q_6} = \{(2,7),(5,9)\}.$$

14. **(Explore right)** We continue to explore the right and search for cells, p, such that q_6 is compatible with p:

$$p_9 = \text{T C}$$

Adding p_9 does not meet the quorum requirements. No more cells can be added to the right in q_6.

15. **(Explore left)** Now we try to add cells to the left of q_6 and look for cells p such that p and q_6 are compatible:

$$p_5 = \text{C A}$$

Thus

$$q_7 = p_5 \oplus q_6$$
$$= \boxed{\text{C A} - \text{C} - \text{T}}, \quad \text{and}$$
$$\mathcal{L}_{q_7} = \{(1,7),(4,9)\}.$$

However, no more cells can be added to the right or to the left of q_7.

16. **(Emit maximal pattern)** q_7 is checked against q_4 which was emitted before. q_7 is not nonmaximal w.r.t. q_4, hence emitted.

$$q_7 = \boxed{\boxed{\text{C A} - \text{C} - \text{T}}}$$

17. **(Repeat iteration)** In fact, we should repeat the whole process with cells starting with C, G and T as well. But we skip those details here.

See Exercises 70 and 71 for other details on the algorithm.

7.3 Pattern Statistics

We have seen in the last sections that the specification of a pattern can be very flexible, resulting in a large number of legitimate patterns in an input. Can we assign a level of statistical significance to these patterns?

In Chapter 5 we have seen that biopolymers can be modeled in various ways. In the following sections, we discuss the probabilities of the occurrence of the various classes of string patterns under these different models (i.i.d. and the Markov model).

7.4 Rigid Patterns

The discussion here can be easily adapted for solid patterns as a special case.

We begin our treatment by deriving some simple expressions for the probability, pr_q, of a rigid pattern q over

$$\Sigma + \text{`.'}$$

Let q be a rigid pattern generated by an i.i.d. source (see Chapter 5) which emits $\sigma \in \Sigma$ with probability

$$pr_\sigma.$$

Note

$$\sum_{\sigma} pr_{\sigma} = 1.$$

Let the number of times σ appears in q be given by

$$k_{\sigma}.$$

Then probability of occurrence of q, pr_q, is given as

$$pr_q = \prod_{\sigma \in \Sigma} (pr_{\sigma})^{k_{\sigma}}. \tag{7.1}$$

Thus, the dot character implicitly has a probability of 1. This fact alone can raise some debate about the model, but we postpone this discussion to Section 7.5.3.

Markov Chain

Next, we obtain the form of pr_q for a pattern when input q is assumed to be generated by a Markov chain (see Chapter 5). For the derivation below, we assume the Markov chain has order 1. Let

$$pr^{(k)}_{\sigma_1,\sigma_2}$$

denote the probability of moving from σ_1 to σ_2 in k steps.

Let s be a stationary, irreducible, aperiodic Markov chain of order 1 with state space Σ ($|\Sigma| < \infty$). Further,

$$\pi_{\sigma}$$

is the equilibrium (stationary) probability of $\sigma \in \Sigma$ and the

$$(|\Sigma| \times |\Sigma|)$$

transition probability matrix is as follows:

$$P[i, j] = pr^{(1)}_{\sigma_i, \sigma_j}.$$

Recall the definition of *cell* from Section 7.2. For a rigid pattern q, for each cell

$$\langle \sigma_1, \sigma_2, \ell \rangle \in C(q),$$

is such that $\ell \geq 0$. It is easy to see that when $\ell \geq 0$, the cell represents the $(\ell+1)$-step transition probability given by $P^{\ell+1}$, i.e.,

$$pr_{\sigma_1(\cdot)\ell\sigma_2} = P^{\ell}[\sigma_1, \sigma_2].$$

Thus for a rigid pattern q',

$$pr_{q'} = \pi_{q'[1]} \left(\prod_{\langle \sigma_1, \sigma_2, \ell \rangle \in C(q')} P^{\ell}[\sigma_1, \sigma_2] \right). \tag{7.2}$$

7.5 Extensible Patterns

Let q be an extensible pattern with density constraint d, i.e., the extensible gap character '-' is to be interpreted as upto d dot '.' characters. $R(q)$ is the set of all possible realizations of q and each realization is a rigid pattern. For example if

$$q = A . C - G,$$

and $d = 4$, the realizations of q are:

$$R(q) = \{q_0, q_1, q_2, q_3, q_4\},$$

where

$$q_0 = A . C G$$
$$q_1 = A . C . G$$
$$q_2 = A . C . . G$$
$$q_3 = A . C . . . G$$
$$q_4 = A . C . . . . G$$

Extensible patterns display various characteristics that makes the probability computation nontrivial. See Exercise 81 for an illustration. Let

$$R(q) = \{q_1, q_2, \ldots, q_r\}.$$

Note that

$$pr_q = pr_{R(q)}$$
$$= pr_{q_1 + q_2 + \ldots + q_r}.$$

where $pr_{q_1 + q_2 + \ldots + q_r}$ is the probability of occurrence of q_1 or q_2 or $\ldots$ q_r.

If no two q_i's in $R(q)$ co-occur, then we can simply add the individual probabilities of q_i's. But, unfortunately, it is possible that they can co-occur. In other words, it is possible for an extensible pattern to occur multiple times at the same location i. Continuing the example, q occurs three times (as q_1, q_2 and q_3) at position 1 of s as shown below:

$$s = A \ G \ C \ A \ G \ G \ G$$

q_1 on s	$\boxed{A \ G \ C \ A \ G}$ G G
q_2 on s	$\boxed{A \ G \ C \ A \ G \ G}$ G
q_3 on s	$\boxed{A \ G \ C \ A \ G \ G \ G}$

So, we consider two kinds of extensible patterns:

1. (nondegenerate) Extensible patterns that can occur only once at a position in the input.

2. (degenerate) Extensible patterns that may occur multiple times at a position in the input.

7.5.1 Nondegenerate extensible patterns

Here we discuss nondegenerate extensible patterns or the ones that occur only once at a site. Then

$$pr_q = \sum_{q_j \in R(q)} pr_{q_j}.$$

Hence we need to compute pr_{q_j} where q_j is a rigid pattern. Assume q_j is a rigid pattern with no dot characters. From Equation (7.1),

$$pr_{q_j} = \prod_{\sigma \in \Sigma} (pr_\sigma)^{k_\sigma},$$

where σ appear k_σ times in q_j. In other words, only the solid characters contribute nontrivially to the computation of pr_{q_j}. Hence, if q is not rigid,

$$pr_q = \sum_{q_j \in R(q)} pr_{q_j}$$

$$= \sum_{q_j \in R(q)} \prod_{\sigma \in \Sigma} (pr_\sigma)^{k_\sigma}$$

$$= |R(q)| \prod_{\sigma \in \Sigma} (pr_\sigma)^{k_\sigma}$$

Annotated gaps. Sometimes, the extensibility of an extensible pattern may be represented, not by d but by a special annotation, say α, for a gap character. Note that d always represents the following possible dot characters

$$\{1, 2, \ldots, d\}.$$

It is possible to represent an arbitrary collection of gaps, say such as

$$\alpha = \{2, 4, 5, 7\}.$$

Let the number of extensible gaps be e each with annotation (set)

$$\alpha_i, \quad 1 \le i \le e.$$

For example an extensible pattern q could be written as

$$q = A \; .^{\{1,2\}} \; C \; .^{\{2,3\}} \; G,$$

where the annotation sets are

$$\alpha_1 = \{1, 2\},$$
$$\alpha_2 = \{2, 3\}.$$

All the rigid realizations of q are:

$$q_1 = \text{A . C . . G}$$
$$q_2 = \text{A . C . . . G}$$
$$q_3 = \text{A . . C . . G}$$
$$q_4 = \text{A . . C . . . G}$$

Thus the number of possible rigid realizations of q is all possible combinations

$$(l_1, l_2, \ldots, l_e),$$

where each $l_i \in \alpha_i$ and the total number of rigid realizations is given by:

$$\prod_{i=1}^{e} |\alpha_i|.$$

Then

$$pr_q = \prod_{\sigma \in \Sigma} (pr_\sigma)^{k_\sigma} \prod_{i=1}^{e} |\alpha_i|. \qquad (7.3)$$

But

$$|R(q)| = \prod_{i=1}^{e} |\alpha_i|.$$

Hence,

$$pr_q = |R(q)| \prod_{\sigma \in \Sigma} (pr_\sigma)^{k_\sigma}. \qquad (7.4)$$

Multi-sets (homologous sets).   Consider the case where a solid character, $q[i]$, of the pattern is a set of homologous characters. Then, since only one of the homologous characters may appear at an occurrence, the probability of occurrence, $pr_{q[i]}$, of $q[i]$ is given by

$$pr_{q[i]} = \sum_{\sigma \in q[i]} pr_\sigma. \qquad (7.5)$$

Thus, if q is a nondegenerate extensible pattern on homologous sets, using Equations (7.4) and (7.5), its probability of occurrence is given by

$$pr_q = |R(q)| \prod_{q[i] \neq \text{'.','-'}} \left(\sum_{\sigma \in q[i]} pr_\sigma \right). \qquad (7.6)$$

Markov chain

If q is a nondegenerate extensible pattern then,

$$pr_q = \sum_{q' \in R(q')} pr_{q'}. \tag{7.7}$$

Using Equations (7.2) and (7.7), for a nondegenerate extensible pattern q, using the Markov chain model, we have

$$pr_q = \pi_{q[1]} \left(\sum_{q' \in R(q)} \left(\prod_{\langle \sigma_1, \sigma_2, \ell \rangle \in C(q')} P^\ell[\sigma_1, \sigma_2] \right) \right). \tag{7.8}$$

When sets of characters or homologous sets are used in patterns, the *cell* is appropriately defined so that σ_1 and σ_2 are sets of homologous characters, possibly singletons. Then the following holds.

$$pr_q = \left(\sum_{\sigma \in q[1]} \pi_\sigma \right) \left(\sum_{q' \in R(q)} \left(\prod_{\substack{\langle \sigma_1, \sigma_2, \ell \rangle \\ \in C(q')}} \left(\sum_{\substack{\sigma_a \in \sigma_1, \\ \sigma_b \in \sigma_2}} P^\ell[\sigma_a, \sigma_b] \right) \right) \right) \tag{7.9}$$

7.5.2 Degenerate extensible patterns

Next we consider the scenario where the patterns possibly occur multiple times at a position i in the input. The following discussion holds for both the i.i.d. and the Markov model.

Let $M^l(q)$ denote a set of strings that has only the solid characters of at least l occurrences of the pattern q at position one of the string. Here we display every other character (not solid for pattern q) as $\square$. For example, consider the pattern

$$q = \text{C} - \text{G}.$$

with

$$R(q) = \{\text{C . G,   C . . G,   C . . . G}\}.$$

q occurs once on each $q' \in M^1(q)$, where,

$$M^1(q) = \{ \text{C} \square \text{G},$$
$$\text{C} \square \square \text{G},$$
$$\text{C} \square \square \square \text{G}\}.$$

q occurs twice on each $q' \in M^2(q)$, where

$$M^2(q) = \{ \text{C} \square \text{G G},$$
$$\text{C} \square \square \text{G G},$$
$$\text{C} \square \text{G} \square \text{G}\}.$$

q occurs three times on $q' \in M^3(q)$, where

$$M^3(q) = \{ \text{ C } \square \text{ G G G}\}.$$

We will compute the probability of $q' \in M^l$, for each l, and q' is treated like a rigid pattern and the probability of occurrence of the $\square$ character is assumed to be 1 (like the dot character). The probability of the occurrence of the set $M^l(q)$, $pr(M^l(q))$, is given by

$$pr(M^l(q)) = \sum_{q' \in M^k(q)} pr_{q'}.$$

Let q be a degenerate (possibly with multiple occurrences at a site) extensible pattern and

$$|R(q)| = r.$$

Then, using the inclusion-exclusion principle (see Theorem (3.2) in Chapter 3),

$$pr_q = pr\left(M^1(q)\right) - pr\left(M^2(q)\right) + \ldots + (-1)^{r+1}pr\left(M^r(q)\right) \quad (7.10)$$

$$= \sum_{k=0}^{r-1}(-1)^k pr\left(M^{k+1}(q)\right). \quad (7.11)$$

Approximating the probability. Notice that for a degenerate pattern, Equation (7.6) is the zeroth order approximation of Equation (7.11). The first order approximation is

$$pr_q \approx pr\left(M^1(q)\right) - pr\left(M^2(q)\right),$$

and the second order approximation is

$$pr_q \approx pr\left(M^1(q)\right) - pr\left(M^2(q)\right) + pr\left(M^3(q)\right),$$

and so on.

Using Bonferroni's inequalities (see Chapter 3), if k is odd, then a kth order approximation of pr_q is an overestimate of pr_q.

7.5.3 Correction factor for the dot character

When two patterns q_1 and q_2 both get a significant score (as defined in Equation (7.13)), and are very close to each other, it becomes desirable to discriminate one from the other. This calls for a more careful evaluation of the score [ACP07].

In the previous definitions, we used the assumption that a pattern is generated by a single stationary source. This model undergoes the restriction that also a mismatch is produced by such source, whereas in reality it is the concatenation of a series of events that generate this mismatch. We revisit the earlier model and refine the treatment of the wild card under the i.i.d. assumption. The dot character is treated as 'any' character emitted by the source and thus its probability is assigned to be 1. However, in computing the probability of the leftmost occurrence of a pattern the dot character actually corresponds to a mismatch. A mismatch occurs when in comparing two input sequences at particular positions the two characters differ. This probability as the complement of having two independent extractions from an urn return the same character, hence:

$$pr_{\text{dot}} = 1 - \sum_{\sigma \in \Sigma} pr_\sigma^2.$$

Expression (7.3) is corrected as:

$$pr_q = \prod_{\sigma \in \Sigma} (pr_\sigma)^{k_\sigma} \prod_{i=1}^{e} \left(|\alpha_i| \, pr_{\text{dot}}^{\alpha_i} \right) \tag{7.12}$$

Using

$$pr_{\text{dot}} < 1,$$

instead of

$$pr_{\text{dot}} = 1,$$

could be interpreted as a probabilistic way to to include a "gap penalty" in the previous formulation.

7.6   Measure of Surprise

Irrespective of the particular model or representation chosen, the tenet of pattern discovery equates overrepresentation (or underrepresentation) of a motif with surprise, hence with interest. Thus, any motif discovery algorithm must ultimately weigh motifs against some threshold, based on a score that compares empirical and expected frequency, perhaps with some normalization. The departure of a pattern q from expectation is commonly measured the by so-called z-scores ([LMS96]), which have the form

$$z(q) = \frac{f(q) - E(q)}{N(q)},$$

where

1. $f(q) > 0$ represents the observed frequency,

2. $E(q) > 0$ an expectation and

3. $N(q) > 0$ is a normalization function.

For a given z-score function, set of patterns P, and real positive threshold α (see Section 3.3.4 for a discussion on threshold α), patterns such that

$$z(q) > \alpha$$

or

$$z(q) < -\alpha$$

are respectively *overrepresented* or *underrepresented*, or simply *surprising*.

7.6.1   z-score

Let pr_q be the probability of the pattern q occurring at any location i on the input string s with

$$n = |s|$$

and let k_q be the observed number of times it occurs on s. Assuming that the occurrence of a pattern p at a site is an i.i.d. process, ([Wat95], Chapter 12), for large n and $k_q \ll n$,

$$\frac{k_q - npr_q}{\sqrt{npr_q(1 - pr_q)}} \to Normal(0, 1).$$

See Chapter 3 for properties of normal distributions. Thus the z-score for a pattern q is given as

$$z(q) = \frac{k_q - npr_q}{\sqrt{npr_q(1 - pr_q)}}. \tag{7.13}$$

7.6.2   χ-square ratio

The z-score in not the only way to measure events that occur with unexpected frequency. In applications related to classification and clustering, such as, e.g., with protein families, a pattern q is considered to be overrepresented if a surprisingly large number of sequences from an ensemble contain each *at least* one occurrence of q. In this context, a large total number of occurrences of q in any particular sequence is immaterial and may be misleading as a measure, since the relevant fact is that the motif is shared across multiple sequences.

Let pr_q be the probability assigned to motif q, computed according to any of the models above. Assuming t sequences

$$s_1, s_2, ..., s_t,$$

to be given, the expected number of occurrences of the pattern in s_i is approximately

$$\mu_i = pr_q|s_i|.$$

By the law of rare events (Poisson distribution), the probability of finding q at least once in s_i is

$$pr^{(i)} = 1 - e^{-\mu_i}.$$

See Chapter 3 for a discussion on Poisson distributions. Then the expected number of sequences containing q at least once is

$$k_e = \sum_{i=1}^{t} pr^{(i)}$$

$$= \sum_{i=1}^{t} \left(1 - e^{-\mu_i}\right)$$

$$= t - \sum_{i=1}^{t} e^{-\mu_i}$$

$$= t - \sum_{i=1}^{t} e^{-pr_q|s_i|}.$$

Let k_q be the observed number of sequences that contain q. Then the statistical significance of a given discrepancy between the observed and the estimated is assessed by taking the χ-square ratio as follows:

$$\chi(q) = \frac{(k_q - k_e)^2}{k_e}.$$

7.6.3   Interplay of combinatorics & statistics

Pattern discovery methods use combinatorial checks such as maximality (and redundancies) and statistical ones such as pattern probabilities as mechanisms to trim the collection of patterns. So do these diverse criteria corroborate or contradict each other?

Note that if pattern q_1 is nonmaximal w.r.t. q_2, then

$$pr_{q_1} > pr_{q_2},$$

and

$$k_{q_1} = k_{q_2},$$

where k_{q_1} is the observed frequency of q_1 and k_{q_2} is that of q_2. Further let,

$$pr_{q_1}, pr_{q_2} < \frac{1}{2}.$$

Then the z-scores of the two patterns satisfy the following [ACP05]:

$$z(q_1) \geq z(q_2).$$

Thus, it is reassuring to learn that the observed frequency of occurrence being equal, a maximal motif always has a smaller probability than the non-maximal version, hence its degree of surprise (overrepresentation) is only stronger [ACP05, ACP07].

Thus, roughly speaking, a pattern that is combinatorially 'uninteresting' is so statistically as well.

Rank	z-score	Motif
1	**7,60E+07**	**R.A.T[LV].C.P-(2,3)G.HP....AC[ATD].L....[ASG]**
2	21416,8	A..[LV].C.P-(2,3)G.HP-(1,2,4)[ASG].[ATD]
3	8105,33	A-(1,4)T....P-(2,3)G.HP....[ATD]-(3)L....[ASG]
4	5841,85	[ATD].T....P-(1,2,3)G.HP-(1,2,4)A.[ATD]
5	4707,62	P.[ASG]-(2,3,4)P....AC[ATD].L....[ASG]
6	4409,21	A..[LV]...P-(2,3)G.HP-(1,2,4)A.[ATD]
7	3086,17	P-(1,2,3)[ASG]..P-(4)AC[ATD].L....[ASG]
8	3068,18	R..[ATD]....P-(2,3)G.HP-(1,2,4)[ASG].[ATD]
9	2615,98	[ASG][ATD]-(1,3,4)P....AC[ATD].L....[ASG]
10	2569,66	[ASG]-(1,2,3,4)P....AC[ATD].L....[ASG]
11	2145,6	G-(2,3)P....AC[ATD].L....[ASG]

FIGURE 7.1: The functionally relevant motif is shown in bold for Streptomyces subtilisin-type inhibitors signature (id PS00999). Here 20 sequences of about 2500 bases were analyzed.

7.7 Applications

We conclude the chapter by showing some results on protein and DNA sequences obtained by using the ideas in the chapter. The experiments [1] involve automatic extraction of significant extensible patterns from some suitable collection of sequences. The interested reader is directed to [ACP07] for further details.

---

[1] The experiments use a system called Varun which is accessible at: www.research.ibm.com/computationalgenomics.

Rank	z-score	Motif
1	**295840**	**[LIM]-(1,2,3,4)[STA][FY]DPC[LIM][ASG]C[ASG].H**
2	**2,86E+05**	**[LIM]-(1,2,3,4)[ASG][FY]DPC[LIM][ASG]C[ASG].H**
3	**155736**	**R-(1,4)[FY]DPC[LIM][ASG]C[ASG].H**
4	78829	[LIM]-(1,2,3,4)[STA].DPC[LIM][ASG]C[ASG].H
5	76101,9	[LIM]-(1,2,3,4)[ASG].DPC[LIM][ASG]C[ASG].H
6	34205,6	[STA]-(1,4)DPC[LIM][ASG]C[ASG].H
7	30325,1	[LIM]-(1,2,3,4)[STA][FY]D.C[LIM][ASG]C..H
8	29276	[LIM]-(1,2,3,4)[ASG][FY]D.C[LIM][ASG]C..H
9	20527,3	[ASG]-(1,4)DPC[LIM][ASG]C[ASG].H
10	17503,4	[LIM]-(1,2,3,4)[ASG]..PC[LIM][ASG]C[ASG].H

FIGURE 7.2:   The functionally relevant motifs are shown in bold for Nickel-dependent hydrogenases (id PS00508). Here 22 sequences of about 23,000 bases were analyzed.

On protein data.   *Streptomyces subtilisin-type inhibitors (*id PS00999*):* Bacteria of the Streptomyces family produce a family of proteinase inhibitors characterized by their strong activity toward subtilisin. They are collectively known as SSIs: Streptomyces Subtilisin Inhibitors. Some SSIs also inhibit trypsin or chymotrypsin. In their mature secreted form, SSIs are proteins of about 110 residues [TKT+94]. The functionally significant motif is discovered as the top ranking one out of 470 extensible motifs (Figure 7.1).

*Nickel-dependent hydrogenases (*id PS00508*):* These are enzymes that catalyze the reversible activation of hydrogen and is further involved in the binding of nickel and which occur widely in prokaryotes as well as in some eukaryotes. There are various types of hydrogenases, but all of them seem to contain at least one iron-sulfur cluster. They can be broadly divided into two groups: hydrogenases containing nickel and, in some cases, also selenium (the [NiFe] and [NiFeSe] hydrogenases) and those lacking nickel (the [Fe] hydrogenases). The [NiFe] and [NiFeSe] hydrogenases are heterodimer that consist of a small subunit that contains a signal peptide and a large subunit [VCP+95]. All the known large subunits seem to be evolutionary related; they contain two cysteine motifs; one at their N-terminal end; the other at their C-terminal end. These four cysteines are involved in the binding of nickel. In the [NiFeSe] hydrogenases the first cysteine of the C-terminal motif is a selenocysteine which has experimentally been shown to be a nickel ligand. Again, this functionally significant motif is detected in the top three out of 4150 extensible motifs (Figure 7.2).

Rank	z-score	Motif
1	**24,3356**	**TTTGCTCA**
2	16,1829	AAAAATGT
3	16,1829	AACTTAAA
4	16,1829	AAATCATG
5	**16,0438**	**TTTGCTC**
6	11,9715	ATAAAAA
7	11,9715	AAAAATG
8	11,9715	ACTTAAA

FIGURE 7.3: Motifs extracted from DNA sequences of the transcriptional factor: CuRE.

DNA sequences   The system automatically discovers the published motifs for CuRE and UASGABA data in the top positions as shown in Figures 7.3 and 7.4.

7.8   Exercises

Exercise 69 (Cells) *Consider the discovery method discussed in Section 7.2. Let d be the density parameter and the alphabet is Σ.*

1. *Construct an example s to show that it is possible that a cell p occurs k times in s but a maximal pattern p′ where p is a substring of p′ may occur more than k times in s.*

2. *If B is the collection of all cells in the input at the initialization phase, show that*
$$|B| \leq (2 + d)\,|\Sigma|^2.$$

3. *Prove that if the cells are processed in the order of saturation, the maximal motifs are emitted before their nonmaximal versions.*

Hint: 1. Let $d = 3$, then for each pair, say, A, C $\in \Sigma$, the possible $d + 2$ cells are:
$$\text{A C}, \quad \text{A . C}, \quad \text{A . . C}, \quad \text{A . . . C}, \quad \text{A} - \text{C}.$$

Rank	z-score	Motif
1	**8469,49**	**G.CAAAA.CCGC.GGCGG.A.T**
2	1056,48	A.CGC.GCTT.G.AC.G.AA
3	528,79	GG.A.TC.T.T.G.TA.T.GC
4	527,143	TT.GA.ATG.TTT.T.TC
5	263,566	GT.CG.T.AT.G.ATA.G
6	263,293	TT.TC.T.C.CC.AAAA
7	263,293	GAT.ATA.AA.A.AG.A
8	263,293	CA.A.TA.TCA.TT.CT
9	263,293	T.TA.G.T.TTT.CTTC
10	263,022	T.ATA.T.TATTAT.A
11	131,499	ATA.A.AA.AG.A.AA
12	131,499	T.TTT.CTT.T.CC.A
13	131,364	G.TGT.AT.AT.TAA
14	131,229	C.T.AATAA.AAAT
15	131,229	TAT.G.TAATC.CT

FIGURE 7.4:   Motifs extracted from DNA sequences of the transcriptional factor: UASGABA.

3. Let density constraint $d = 2$. The extensible pattern p and its two occurrences. Cell, C T, occurs only once in the input s.

$$s = \text{A G A G C T}$$
$$p = \text{A G} - \text{C T}$$

A G $\boxed{\text{A G C T}}$

$\boxed{\text{A G}}$ A G $\boxed{\text{C T}}$

Exercise 70 (Discovery algorithm) *Consider the discovery method discussed in Section 7.2.*

1. *What is the running time complexity of the algorithm? Suggest some heuristics to improve the running time.*

2. *Generalize the method to handle sequences on multi-sets.*

Hint: 1. If in the processing of a cell, all its locations have been used, can it be removed from the B set?

Exercise 71 (Generalized suffix tree) *How is the 'suffix' tree shown below related to the algorithm discussed in Section 7.2. Note that the labels in the edges of the tree use the '.' and '-' symbols. The density parameter $d = 2$*

and the quorum is $k = 2$.

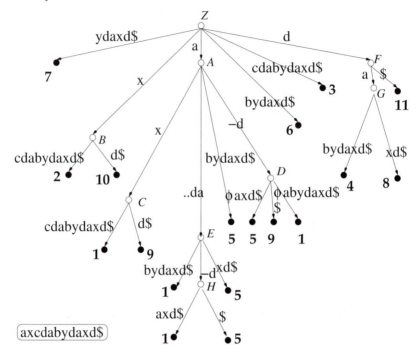

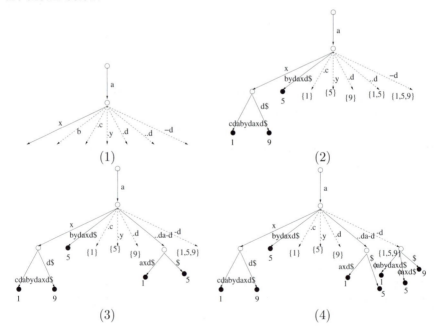

Hint: This is taken from [CP04]. Some steps in the construction of this tree are shown below.

Exercise 72 (Density constraints (d, D)) *The density constraint d denotes the maximum number of gaps (dot characters) between any two consecutive solid characters in the pattern. Sometimes another constraint, D, is used which gives the minimum length of each rigid segment (having both solid and dot characters) in the pattern. Thus a segment between two consecutive extensible gap ('−') characters is constrained to be of length D.*
Modify the algorithm of Section 7.2 to include this D constraint.

1. *Incorporate the change at the iteration phase.*

2. *Incorporate the change at the initialization phase.*

Hint: 2. Is it possible to first build only the rigid segments (using rigid cells) thus enlarging the set of cells B. Note that there can be no pattern of the form, say,

$$A \ C \ . \ - \ G,$$

where the dot character is adjacent to the '−' character.

Exercise 73 (Maximality check) *In the algorithm of Section 7.2, a pattern p_1 is checked for maximality against, say some p_2, before it is emitted. Let*

$$|\mathcal{L}_{p_1}| = |\mathcal{L}_{p_2}|.$$

Devise a method to check if extensible pattern p_1 is nonmaximal w.r.t. p_2, without checking the location lists $\mathcal{L}_{p_1}$ and $\mathcal{L}_{p_2}$.

Hint: For rigid patterns p_1 and p_2, p_1 needs to be aligned at some position j from the left. For extensible motifs, check for solid elements that are in one but not in the other.

For the problems below, see Section 5.2.4 for a definition of *random strings*.

Exercise 74 *Let*

$$p = G \ A \ A \ T \ T \ C.$$

1. *What is the odds of seeing p at least 10 times in a 600 base long random DNA fragment?*

2. *What is the expected copy number of p in a random DNA fragment of length n (i.e., how many times do you expect to see p in the DNA fragment)?*

Exercise 75 *Assume that the chance of occurrence of purines (adenine or guanine) and pyrimidines (thymine or cytosine) in a selected strand of DNA is the same. Then, is the chance of seeing*

at least 7 purines in a random DNA strand of 10 bases

the same as seeing

at least 14 in a random strand of 20 bases?

Why?

Hint: Yes, it is a trick question.

Exercise 76 *Assume a random DNA sequence s of length n. Let pr_z denote the probability of occurrence nucleotide*

$$z = \{A, C, G, T\}$$

in s. Let

$$pr_A = pr_C = pr_G = pr_T = 0.25.$$

1. *What is the probability of seeing a pattern p of length l?*

2. *How do the odds change if p must occur at least $k > 1$ times?*

3. *What can you conclude if $n \gg l$?*

What are the answers to the three questions under the assumption:

$$pr_A = pr_T = 0.2$$
$$pr_C = pr_G = 0.3.$$

Exercise 77 *In a random string on*

$$\{A, C, G, T\}$$

(note each character occurs with equal probability), what is the longest stretch of A's you expect to see?

Exercise 78 (Restriction enzyme) *A position specific restriction enzyme cleaves double stranded DNA at restriction sites characterized solely by the base composition. Thus a DNA sequence can be broken up into fragments. Assuming the restriction site to be the pattern*

$$GAATTC,$$

what is the average length of a fragment? Assume that each base occurs at a position with equal probability and independently.

Exercise 79 *What is the probability of occurrence of a 10-mer consisting of at least*

1. *two A's,*

2. *at least one C,*

3. *at least one G and*

4. *at least two T's*

in a random DNA sequence?

Hint: See Section 11.2.2.

Exercise 80 (Restriction site) *Let the probability p of an occurrence of a mutation (or a restriction site) be*

$$\frac{1}{1024}.$$

What is the average distance between two mutation (or restriction sites)? What is the standard deviation?

Hint: The probability distribution of the number k of Bernoulli trials needed to get one success, defined on the domain

$$\Omega = \{1, 2, 3, \ldots\}$$

is called a *geometric distribution*. The probability mass function is given by

$$pr(k) = (1 - p)^{k-1}p.$$

A random variable following this distribution is usually denoted as

$$X \sim Exponential(k; p)$$

Show that

$$E[X] = \frac{1}{p},$$
$$V[X] = \frac{1 - p}{p^2}.$$

Exercise 81 (Pitfall of extensibility) *Let q be an extensible pattern on some input with the set of realizations,*

$$R(q) = \{q_1, q_2, \ldots, q_l\},$$

where $l > 1$. Since at any location in the input one of $R(q)$ occurs, the probability of occurrences of q is the sum of the probability of occurrence of the rigid motif

$$q_j \in R(q), 1 \leq j \leq l.$$

Thus the probability of occurrence of q, pr_q, is given as

$$pr_q = \sum_{q_j \in R(q)} pr_{q_j}.$$

What is wrong with this computation of pr_q?

Hint: Consider the following example. Let $\Sigma = \{A, C\}$, with

$$pr_A = pr_C = 0.5$$

Let extensible pattern

$$q = A - C$$

and the realizations of q be

$$R(q) = \{A\ C,\quad A\ .\ C,\quad A\ .\ .\ C,\quad A\ .\ .\ .\ C,\quad A\ .\ .\ .\ .\ C\}.$$

Then

$$pr_{A\ C} = pr_{A\ .\ C} = pr_{A\ .\ .\ C}$$
$$= pr_{A\ .\ .\ .\ C} = pr_{A\ .\ .\ .\ .\ C} = 0.25$$

It follows

$$pr_q = 1.25 > 1.$$

Exercise 82 *Let*

$$\Sigma = \{A,\ C,\ F,\ G,\ L,\ V\}$$

and the probability of occurrence of $\sigma \in \Sigma$ is pr_σ with

$$\sum_{\sigma \in \Sigma} pr_\sigma = 1.$$

Let

$$q_1 = A\ C\ .\ .\ L$$

1. *Compute the probability of occurrence of q_1 or q_2 in a random string where*

$$q2 = G \; V \; L \; F \; F$$

2. *Compute the probability of occurrence of q_1 or q_2 in a random string where*

$$q2 = A \; . \; L \; F$$

Hint: 2. q_1 and q_2 may co-occur; use Inclusion-Exclusion Principle.

Exercise 83   *1. Let pattern q_1 be nonmaximal w.r.t. q_2. Show that under i.i.d. distribution,*

$$pr_{q_2} < pr_{q_1}.$$

2. *Let pattern q_1 be redundant w.r.t. q_2. Show that under i.i.d. distribution,*

$$pr_{q_2} < pr_{q_1}.$$

Do the same results hold under a Markov model?

Exercise 84 (Alternative definition of maximality) *Recall the maximality definition discussed in the chapter: Given s, q_1 is nonmaximal w.r.t. q_2 if and only if,*

$$q_1 \prec q_2 \;\; and$$
$$\mathcal{L}_{q_1} = \mathcal{L}_{q_2}.$$

Consider an alternative definition of maximality: Given s, pattern q_1 is non-maximal w.r.t. q_2 if and only if

$$q_1 \prec q_2.$$

1. *Let*

$$s_1 = A \; C \; G \; T \; A \; C \; G \; T \; C \; G \; T \; G \; T.$$

   Enumerate the maximal solid patterns in s_1 for each of the maximal definitions.

2. *Let*

$$s_2 = A \; C \; G \; T \; A \; C \; G \; T.$$

   Enumerate the maximal solid patterns in s_2 for each of the maximal definitions.

3. *Then is it true that the z-scores satisfy $z(q_1) \geq z(q_2)$?*

4. *Compare the two definitions of maximality.*

Hint: Definition 1:

$$P_{maximal}(s_1) = \{\text{A C G T}, \ \text{C G T}, \ \text{G T}\},$$
$$P_{maximal}(s_2) = \{\text{A C G T}\}.$$

Definition 2:

$$P_{maximal}(s_1) = P_{maximal}(s_2)$$
$$= \{\text{A C G T}\}.$$

3. Note that, under this definition, it is possible that

$$|\mathcal{L}_{q_1}| > |\mathcal{L}_{q_2}|.$$

4. How do the two z-scores compare in each of the scheme?

Chapter 8

Motif Learning

The science of learning
is indeed a fine art.
- anonymous

8.1 Introduction: Local Multiple Alignment

We ask a basic question: What is a *motif* in a biological sequence? One possible meaningful definition is to look for structural or functional implications of a segment and if this can be (unambiguously) associated with the segment, then the segment qualifies to be a motif. To put it another way, a segment shared by multiple protein or nucleic acid sequences is a motif as it could possibly tell us about evolution, structure or even function.

Using this premise, in the absence of supporting information such as three-dimensional structures or details of chemical interactions of residues or effects of mutations or even function, is it possible to identify segments as 'potential' motifs? The recognition of these segments then relies on patterns shared by multiple protein or nucleic acid sequences. To elicit these shared regions, the most appropriate approach is to align the sequences. The alignment could be *global*, where the task is to align the entire sequences as best as possible or it could be *local* where the focus is on shorter segments of the sequences. These short segments qualify as motifs. In other words motifs can also be viewed as the consensus derived from a *local multiple alignment* of sequences. This process is also called *learning motifs* from the sequences.

This is complicated by the fact that the target motif (or sometimes also called a *signal*) may vary greatly among different sequences (say proteins). The challenge is to discover these subtle motifs. An example of such a motif is shown in Figure 8.1 which is taken from [LAB+93]. It shows repeating motifs in prenyltransferase subunits: *Ram1* (Swiss-Prot accession number P22007) and *FT-β* (Swiss-Prot Q02293) are the β subunits of farnesyltransferase from the yeast *Saccharomyces cerevisiae* and rat brain respectively. *Cdc43* (Swiss-Prot P18898) is the β subunits of type II geranylgeranyltransferase from the

	129	---	GPFGGGPGQLSH	LA-
	181	---	GGFKTCLEVGEV	DTR
Ram1	230	---	GGFGSCPHVDEA	HGG
	279	---	RGFCGRSNKLVD	GC-
	331	---	PGLRDKPQAHSD	FY-
			SSYSCTPNDSPH	

	122	---	GGFGGGPGQYPH	LA
	173	---	GSFLMHVGGEVD	VR
FT-β	221	---	GGIGGVPGMEAH	GG
	269	---	GGFQGRCNKLVD	GC
	331	---	GGLLDKPGKSRD	FY

	191	---	GAFGAHNEPHSG	--
Cdc43	240	SDD	GGFQGRENKFAD	TC
	309	---	GGFSKNDEEDAD	LY

FIGURE 8.1:   An example of sequences aligned by the motif that is shown boxed. Notice the extent of dissimilarity of the motif across its occurrences.

rat brain.[1]

8.2   Probabilistic Model: Motif Profile

Recall that a motif is simply a sequence of characters from the alphabet possible interspersed with dont care characters. For example a motif

G K K . D D

is interpreted as a segment whose

1. first element is always G (glycine),

2. second and third elements are always K (lysine),

3. fifth and sixth elements are always D (asparatic acid) and

4. the fourth element could potentially be any residue.

---

[1]See the cited paper for any further details on this example.

The emphasis is here on 'always'. Clearly, the fourth element violated the 'always' criterion giving a 'not always' condition, hence delegated down to a dont care. [2] So, can we find a middle ground between 'always' and dont care?

A probabilistic model of a motif associates a real number between zero and one (probability) with each element (residue) may occur in a sequence. This is defined formally below.

Consider an alphabet of size $L = |\Sigma|$ as follows.

$$\Sigma = \{\sigma_1, \sigma_2, \ldots, \sigma_L\}.$$

A motif of size l is then defined by an $|\Sigma| \times l$ matrix ρ

$$\rho = \begin{bmatrix} \rho_{11} & \rho_{12} & \cdots & \rho_{1l} \\ \rho_{21} & \rho_{22} & \cdots & \rho_{2l} \\ \rho_{31} & \rho_{32} & \cdots & \rho_{3l} \\ \vdots & \vdots & & \vdots \\ \rho_{r1} & \rho_{r2} & \cdots & \rho_{rl} \end{bmatrix}. \tag{8.1}$$

The rough interpretation of ρ_{rc} is that it is the probability of position c in the motif to be σ_r. Thus, for each $1 \leq j \leq l$,

$$\sum_{1=1} \rho_{ij} = 1.$$

In other words, each column of matrix ρ adds up to 1. This matrix ρ is also called the *probability matrix*.

For ease of exposition, in the rest of the chapter, the subscript r in ρ_{rc} may also be written as

$$\rho_{\sigma_r c}.$$

8.3 The Learning Problem

The central task is defined as follows: *Given t sequences s_i defined on some finite alphabet Σ of length n_i each and a motif length l, the task is to find a motif of length l that is shared in all the t sequences.*

In the most general scenario:

1. the motif may or may nor occur in a sequence and

2. when it does, the motif may occur multiple times in a sequence.

---

[2]In fact this discontinuity gets in the way of elegant formalization under some combinatorial models.

On occurrences.   If the motif were to occur at most once in each sequence, it is usually called *unsupervised* learning. This means that a motif may or may may not occur in each sequence.

If the motif occurs at least once in each sequence, the discovery of motif under such condition is usually called *supervised learning.*

We will consider the very special case when the motif occurs *exactly once* in each sequence in the following discussion.

The million dollar question is: when does ρ deserve the dignity of a *motif?* In principle, any alignment of a segment of length l of the t sequences can potentially produce a probability matrix ρ. We address this issue by insisting that the motif discovered is the *best* amongst all such l length motifs. This has two implications:

1. A quantitative measure, F, must be defined

$$F : R \to \mathbb{R},$$

   where R is the set of all motif profiles ρ (of the same length l). In other words, for a given collection of input sequences,

$$F(\rho) = v(\in \mathbb{R}).$$

   We discuss possible forms of F in the following section.

2. The problem will produce only one result for a fixed l.

8.4   Importance Measure

Given two motif profiles on an input data set, how can they be compared? In the absence of any other supporting information, we discuss below two ways of comparing motifs. The first uses a log likelihood measure and the second uses information content.

8.4.1   Statistical significance

The task is to assign a numerical value that can be associated with the (statistical) significance of the motif. The

$$\log(\text{likelihood})$$

is such a measure of the profile. The higher the value, usually the more (statistically) significant the motif profile.

The positions in the input data is divided into

1. motif positions and

2. nonmotif (or background) positions.

Further, each motif position has an associated column c depending on what position in the motif covers this position in the input. Thus each position is annotated as follows

$$C_{ij} = \begin{cases} 0 & s_{ij} \text{ is a nonmotif position,} \\ 1 & s_{ij} \text{ is position 1 of the motif,} \\ 2 & s_{ij} \text{ is position 2 of the motif,} \\ \vdots & \\ c & s_{ij} \text{ is position } c \text{ of the motif,} \\ \vdots & \\ l & s_{ij} \text{ is position } l \text{ of the motif.} \end{cases}$$

Define ρ_{r0} as the probability of character

$$\sigma_r \in \Sigma$$

in all nonmotif positions. Thus column 0 of the matrix describes the 'background'.

For ease of exposition, the row r in the matrix ρ will be replaced by the character it represents, i.e., σ_r. Also, we will switch between notation

$$\rho_{ij}$$

and

$$\rho[i, j],$$

depending on the context, for ease of understanding. Note that we use column 0 in the ρ matrix to denote the nonmotif or background probabilities of each character in the alphabet. Then, given the input and the matrix ρ for rows

$$1 \le r \le |\Sigma|$$

and columns

$$0 \le c \le l,$$

the log of the likelihood is given as

$$F_1 = \log \left(\prod_{i=1}^{t} \left(\prod_{j=1}^{n_i} \rho[s_{ij}, C_{ij}] \right) \right)$$

$$= \sum_{i=1}^{t} \left(\sum_{j=1}^{n_i} \log(\rho[s_{ij}, C_{ij}]) \right).$$

Let f be a $|\Sigma| \times (l+1)$ matrix and each entry $f_{\sigma c}$ denotes the number of positions (given by i and j) in the input with annotation c, i.e.,

$$C_{ij} = c,$$

and value σ, i.e.,

$$s_{ij} = \sigma.$$

Then

$$F_1 = \sum_{\sigma \in \Sigma} \sum_{c=0}^{l} f_{\sigma c} \log(\rho_{\sigma c}).$$

Yet another effective measure is by taking a ratio with the background probabilities as follows:

$$F_2 = \sum_{\sigma \in \Sigma} \sum_{c=1}^{l} f_{\sigma c} \log \left(\frac{\rho_{\sigma c}}{\rho_{\sigma 0}} \right).$$

Thus F_1 (and also F_2) can be computed given

1. the $|\Sigma| \times (l+1)$ probability matrix ρ and

2. the $|\Sigma| \times (l+1)$ frequency matrix f.

Consider a concrete example with the solution (alignment) shown below.

Alignment

$$
\begin{array}{rcl}
s_1 & = & A\,\fbox{$A\ C\ C$}\,T\ A \\
s_2 & = & A\ T\ \fbox{$G\ T$}\ A\ G\ G \\
s_3 & = & A\ T\,\fbox{$A\ C\ T$}\,A \\
\text{consensus} & & \fbox{$A\ C\ G$}
\end{array}
$$

Note that this is a very small example and statistical methods work well for larger data sets. However we use this to explain the formula. Let the probability matrix ρ be given as follows. Then using the alignment and matrix ρ, the frequency matrix f can be constructed as follows.

$$
\rho =
\begin{bmatrix}
0.4 & 0.9 & 0.1 & 0.09 \\
0.01 & 0.04 & 0.8 & 0.1 \\
0.4 & 0.04 & 0.09 & 0.8 \\
0.19 & 0.02 & 0.01 & 0.01 \\
\hline
0 & 1 & 2 & 3
\end{bmatrix},
\quad
f =
\begin{bmatrix}
5 & 3 & 0 & 0 \\
0 & 0 & 2 & 1 \\
2 & 0 & 0 & 1 \\
3 & 0 & 1 & 1 \\
\hline
0 & 1 & 2 & 3
\end{bmatrix}
\quad
\begin{matrix}
A \\
C \\
G \\
T \\
\leftarrow c
\end{matrix}.
$$

Then the two measures are:

$$F_1 = \sum_{\sigma \in \Sigma} \sum_{c=0}^{l} f_{\sigma c} \log(\rho_{\sigma c})$$

$$= \quad 5 \log 0.4 + 2 \log 0.4 + 3 \log 0.19$$
$$+ \ 3 \log 0.9$$
$$+ \ 2 \log 0.8 + \log 0.01$$
$$+ \ \log 0.1 + \log 0.8 + \log 0.01$$
$$= -10.3773$$

$$F_2 = \sum_{\sigma \in \Sigma} \sum_{c=1}^{l} f_{\sigma c} \log \left(\frac{\rho_{\sigma c}}{\rho_{\sigma 0}} \right)$$

$$= \quad 3 \log \left(\frac{0.9}{0.4} \right)$$
$$+ \ 2 \log \left(\frac{0.8}{0.01} \right) + \log \left(\frac{0.01}{0.19} \right)$$
$$+ \ \log \left(\frac{0.1}{0.01} \right) + \log \left(\frac{0.8}{0.4} \right) + \log \left(\frac{0.01}{0.19} \right)$$
$$= 3.60625$$

Note that F_1 is almost independent of the number of sequences t and the length of each sequence n_i. Also, this value always improves (increases) with increasing motif length, as well as the input size. This is the well-recognized issue of *model selection* in statistics. One of the effective way of dealing with this is using the following 'normalization' for F_1:

$$\sum_{i=1}^{t} \left(\log \frac{1}{n_i - l + 1} + \sum_{\sigma \in \Sigma} \sum_{c=1}^{l} f_{\sigma c} \log(\rho_{\sigma c}) + (n_i - l) \sum_{\sigma \in \Sigma} f_{\sigma 0} \log(\rho_{\sigma 0}) \right).$$

8.4.2 Information content

The relationship between the motif profile and information content of the motif (with respect to the input) was made by Stormo [Sto88]. For a quick introduction to information theory, we direct the reader to Exercise 21 in Chapter 3.

Let s denote the input (which is a collection of t sequences). The information content of each position c in the motif profile is defined as

$$I_c(s) = \sum_{\sigma \in \Sigma} \rho_{\sigma c} \log \left(\frac{\rho_{\sigma c}}{f_\sigma} \right),$$

where f_σ is the number of times σ appears in input s. Thus the overall information content of the profile is

$$I_\rho(s) = \sum_{c=1}^{l} I_c(s)$$

$$= \sum_{\sigma \in \Sigma} \sum_{c=1}^{l} \rho_{\sigma c} \log \left(\frac{\rho_{\sigma c}}{f_\sigma} \right).$$

The information content is yet another measure to compare different motif profiles. The higher the value of $I_\rho(s)$, usually the more significant is the motif profile.

8.5   Algorithms to Learn a Motif Profile

Recall that in this chapter we discuss the learning problem where the motif occurs *exactly once* in *every* input sequence. Let ρ be the motif profile sought.

The motif represented by the profile ρ occurs in all the t sequences suggesting an alignment of l positions in each sequence. The remaining

$$n_i - l$$

positions in each sequence is often called the nonmotif region or the background. Let Z denote this occurrence (or alignment) information. The detail of what Z represents will depend on the particular method in use.

The motif length l and the t input sequences are given. We make the following assumption.

1. Given ρ, the occurrence/alignment information Z can be estimated.

2. Given Z, motif profile ρ can be estimated.

Then, intuitively, there are two possible iterative schemes to solve the learning problem as shown in Figure 8.2. We begin with an initial estimate and improve the result over the iterations. The iterations terminate when the difference across the iterations is below some δ, that is considered insignificant. In fact, the actual value of δ depends on the problem domain.

It is not too difficult to see that the method is correct. Does the solution improve during the course of the iterations? We leave this as an exercise for the reader (see Exercise 88).

Running time complexity.   Let the size of the input be given as

$$N_I = \sum_{i=1}^{t} n_i.$$

Method 1:	Method 2:
1. Initialize ρ_0	1. Initialize Z_0
2. Repeat	2. Repeat
(a) Re-estimate Z from ρ	(a) Re-estimate ρ from Z
(b) Re-estimate ρ from Z	(b) Re-estimate Z from ρ
3. Until change in ρ is small	3. Until change in Z is small

FIGURE 8.2: Given the input sequences and the motif length, two possible learning methods to learn a motif profile.

In each of the methods, each iteration (Steps 2-3) takes time

$$\mathcal{O}\left(l|\Sigma|N_I\right).$$

Assuming l and $|\Sigma|$ to be constants, each iteration takes only linear time. The number of iterations depend on the data and in practice this is believed to be small.

Issues with a learning method. It is important to note some of the issues with a learning method such as the one described above.

1. (pattern length) The length l of the motif (profile) is a fixed parameter. Thus to find motifs of different lengths, the algorithm must be run multiple times.

2. (one at a time) The method finds only one motif at a time. Once the motif is found, the input must be modified to remove the occurrence of this motif, for subsequent searches for other motifs.

3. What is the best initial estimate?

   The solution offered by the method depends on the initial estimate. Ideally, all possible initial estimates should be tried, but this is not a practical option.

4. (phase shift) If $p[1..l]$ is the true signal, a learning method may converge to a subpattern of p, i.e., starting from say the second or third position. Hence in practice, after the pattern has been found, some further investigation is done to the left of the pattern to make sure that a subpattern is not being reported as the pattern.

If such nontrivial issues are associated with discovering profiles, then why use profiles at all?

One major appealing feature is that it allows some flexibility in the pattern description. Yet another attractive feature is that it provides a (statistical) significance value inherently associated with a motif.

8.6   An Expectation Maximization Framework

In this section we place Method 1, described above, in an *expectation maximization* framework [BE94]. In fact Lawrence and Reilly [LR90] introduced the expectation maximization approach as a means for extracting a motif profile from a given data set.

Under this model, we let Z be represented by z and z is the matrix of offset probabilities where z_{ij} is the probability that the shared motif starts in position j of sequence i.

In the following sections, we will first describe methods to

1. estimate z, given ρ and

2. estimate ρ, given z.

We begin by discussing the initial estimate used in the algorithm.

8.6.1   The initial estimate ρ_0

The method iteratively converges to a profile ρ. However, this depends on the initial estimate ρ_0. Thus different inital estimates, for the same data set and the same length l of the motif could discover different motif profiles.

In practice, the initial estimate is taken from a substring in the data. For example, let the estimate be taken from position $j = 4$ for a motif profile of length $l = 3$ from s_1 where

$$s_1 = \text{A G G} \boxed{\text{C T T}} \text{A G C T G}.$$

The most obvious profile model for this appears to be

$$\rho_0 = \begin{bmatrix} 0.0 & 0.0 & 0.0 \\ 0.1 & 0.0 & 0.0 \\ 0.0 & 0.0 & 0.0 \\ 0.0 & 1.0 & 1.0 \end{bmatrix} \begin{matrix} A \\ C \\ G \\ T \end{matrix}. \tag{8.2}$$

However this motif profile will never change over the iterations. The proof is straightforward and we leave that as an exercise for the reader (Exercise 87).

Thus, in practice, no entry of the ρ_0 matrix is set to 0.0. However, since one entry in the column is biased towards one character, the following is a good initial estimate for the example above.

$$\rho_0 = \begin{bmatrix} 0.15 & 0.15 & 0.15 \\ 0.55 & 0.15 & 0.15 \\ 0.15 & 0.15 & 0.15 \\ 0.15 & 0.55 & 0.55 \end{bmatrix} \begin{matrix} A \\ C \\ G \\ T \end{matrix}.$$

Thus the following is a good rule of thumb to initialize a column c to some $x \in \Sigma$. Let $L = |\Sigma|$. Then column c is initialized to the following.

$$
\begin{bmatrix}
0.5/L\text{-}1 \\
0.5/L\text{-}1 \\
\vdots \\
0.5 \\
\vdots \\
0.5/L\text{-}1
\end{bmatrix}
\begin{matrix}
\sigma_1 \\
\sigma_2 \\
\cdot \\
x \\
\cdot \\
\sigma_L
\end{matrix}
$$

Also note that for this column c

$$
\sum_{r=1}^{L} \rho_{rc} = (L-1)\frac{0.5}{L-1} + 0.5
$$
$$
= 1.0
$$

8.6.2   Estimating z given ρ

Under a combinatorial model, we have the occurrence at position j of sequence i to be either a 'yes' or a 'no'. Under a motif profile model, the occurrence is the probability z_{ij}, thus

$$
0 \le z_{ij} \le 1,
$$

for i and j. This is best explained through a concrete example.

Let s_{ij} denote the jth character of the ith input sequence. For ease of exposition, we let the row r of the matrix ρ to directly denote the character it represents (A, C, G or T for instance). The probability of occurrence of a given ρ at position j in sequence s_i is given by

$$
z'_{ij} = \prod_{c=1}^{l} \rho_{\sigma_c c},
$$

where

$$
s_{i(j+c-1)} = \sigma_c.
$$

In other words the sequence s_i has the character σ_c at position $j+c-1$. This is a straightforward interpretation of the motif profile. For example, let the motif profile where $l = 5$ be given as

$$
\rho =
\begin{bmatrix}
0.6 & 0.3 & 0.05 & 0.7 & 0.1 \\
0.2 & 0.1 & 0.05 & 0.1 & 0.6 \\
0.1 & 0.5 & 0.1 & 0.1 & 0.1 \\
0.1 & 0.1 & 0.8 & 0.1 & 0.2
\end{bmatrix}
\begin{matrix}
A \\
C \\
G \\
T
\end{matrix}
\cdot
$$

Let $j = 4$ and the sequence s_1 be

$$s_1 = A\ G\ G\ \boxed{C\ T\ T\ A\ G}\ C\ T\ G.$$

Then for position $j = 4$ of sequence s_1,

$$z'_{14} = \prod_{c=1}^{5} \rho_{\sigma_c c}$$
$$= \rho[C, 1]\, \rho[T, 2]\, \rho[T, 3]\, \rho[A, 4]\, \rho[G, 5]$$
$$= (0.2)\,(0.5)\,(0.1)\,(0.7)\,(0.1)$$
$$= 0.0007$$

Since this is a probability, we normalize this by summing over all the values in sequence s_i. Thus, for each i,

$$z_{ij} = \frac{z'_{ij}}{\sum_{j=1}^{n_i - l + 1} z'_{ij}}. \tag{8.3}$$

8.6.3 Estimating ρ given z

Consider the scenario where the occurrence probabilities, z_{ij}, for all possible values of i and j are given. Note that j is a function of i and should ideally be written as j_i, but we omit the subscript to avoid clutter. i and j take the following values:

$$1 \leq i \leq t \text{ and}$$
$$1 \leq j \leq (n_i - l + 1), \text{ for each } i.$$

Also, since z_{ij} are probabilities, under the assumption that a motif occurs exactly once per sequence,[3] we assume the following for each i.

$$\sum_{j=1}^{n_i - l + 1} z_{ij} = 1. \tag{8.4}$$

For each sequence s_i, a position x_i, $1 \leq x_i \leq n_i - l + 1$, is chosen with probability

$$\frac{z_{ix}}{\sum_{j=1}^{n_i - l + 1} z_{ij}}$$

---

[3]Under a more general assumption,

$$\sum_{i=1}^{t} \left(\sum_{j=1}^{n_i - l + 1} z_{ij} \right) = 1.$$

Thus the following t positions are chosen randomly:

$$x_1, x_2, \ldots, x_t$$

This suggests an l-wide alignment of the t sequences. A method for estimating ρ from this alignment is discussed in Section 8.7.1.

Back to the method. We will now place Method 1 in the framework of expectation maximization.

We use the following observation [DH70]: *The likelihood of the profile given the training data is the probability of the data given the profile.* The training data here refers to sequences that are aligned by the motif. Of course, we need to find an alignment of the sequences.

Recall Bayes' theorem (Theorem (3.1)):

$$P(E_1|E_2) = \frac{P(E_2|E_1)}{P(E_2)} P(E_1),$$

where events E_1 and E_2 are in the same probability space and $P(E_2) > 0$.

We begin by defining a convenient variable which keeps track of the occurrence of the motif (profile) at a position in the input. This position, for each i is given as

$$1 \leq j \leq n_i - l + 1.$$

Recall that n_i is the length of the sequence s_i and l is the length of the motif. A motif may or may not occur at this position j. Let X_{ij} be an indicator variable where

$$X_{ij} = \begin{cases} 1 & \text{if motif starts at position } j \text{ in } s_i, \\ 0 & \text{otherwise.} \end{cases}$$

Further, let $\rho^{(q)}$ denote the estimate of ρ and $z^{(q)}$ denote the estimate of z after q iterations. Given $\rho^{(q)}$, the probability of sequence s_i, given the start position of the motif (profile), is

$$P\left(s_i | X_{ij} = 1, \rho^{(q)}\right) = \prod_{c=1}^{l} \rho_{x_c c}^{(q)},$$

where

$$s_{i(j+c-1)} = x_c.$$

Note that using our notation,

$$z_{ij} = P\left(s_i | X_{ij} = 1, \rho^{(q)}\right).$$

Using Bayes' theorem,

$$P\left(X_{ij} = 1 | \rho^{(q)}, s_i\right) = \frac{P\left(s_i | X_{ij} = 1, \rho^{(q)}\right) P^0(X_{ij} = 1)}{\sum_{c=1}^{n_i - l + 1} P\left(s_i | X_{ic} = 1, \rho^{(q)}\right) P^0(X_{ic} = 1)}, \quad (8.5)$$

where $P^0(X_{ij} = 1)$ is the prior probability that the motif starts at position j in sequence s_i. Since no information is available about the occurrence of the motif, P^0 is assumed to be uniform. Thus, for each $1 \le i \le t$,

$$P^0(X_{ij} = 1) = \frac{1}{n_i - l + 1}, \text{ for } 1 \le j \le n_i - l + 1.$$

Then the denominator in Equation (8.5) simplifies to

$$\sum_{c=1}^{n_i-l+1} P\left(s_i | X_{ic} = 1, \rho^{(q)}\right) P^0(X_{ic} = 1)$$

$$= \sum_{c=1}^{n_i-l+1} P\left(s_i | X_{ic} = 1, \rho^{(q)}\right) \left(\frac{1}{n_i - l + 1}\right)$$

$$= \sum_{c=1}^{n_i-l+1} P\left(s_i | X_{ic} = 1, \rho^{(q)}\right).$$

Then Equation (8.5) is rewritten as

$$z_{ij} = P\left(X_{ij} = 1 | \rho^{(q)}, s_i\right)$$

$$= \left(\frac{1}{n_i - l + 1}\right) \frac{P\left(s_i | X_{ij} = 1, \rho^{(q)}\right)}{\sum_{c=1}^{n_i-l+1} P\left(s_i | X_{ic} = 1, \rho^{(q)}\right)}.$$

For each i, the constant above is fixed, and

$$\sum_{j=1}^{n_i-l+1} z_{ij} = \frac{1}{n_i - l + 1},$$

hence there no loss by using the following simplification:

$$z_{ij} = \frac{P\left(s_i | X_{ij} = 1, \rho^{(q)}\right)}{\sum_{c=1}^{n_i-l+1} P\left(s_i | X_{ic} = 1, \rho^{(q)}\right)}. \tag{8.6}$$

Under this simplification,

$$\sum_{j=1}^{n_i-l+1} z_{ij} = 1.$$

Notice that Equation (8.6) has the same form as Equation (8.3) of Section 8.6.2.

Thus we have shown that Method 1 can be viewed as an expectation maximization strategy.

There has been a flurry of activity around this problem [EP02, KP02b]. For instance, Improbizer [AGK+04] also uses expectation maximization to determine weight matrixes of DNA motifs that occur improbably often in the input data.

8.7 A Gibbs Sampling Strategy

In this section we use Method 2 of Section 8.5 through a Gibbs sampling strategy [LAB+93].

Gibbs sampling is a strategy to generate a sequence of samples from the joint probability distribution of two or more random variables. This is applicable when the joint distribution is not known explicitly, but the conditional distribution of each variable is known. The sampling method is to generate an instance from the distribution of each variable in turn, conditional on the fixed values of the other variables.

We first define the alignment vector Z of Figure 8.2. Let Z be a t-size vector where each entry Z_i denotes the first position of the motif in s_i. Note that here we do not care about the alignment of the nonmotif or background region of the input.

The initial estimate Z_0 is computed as follows. For each sequence s_i, a position from 1 to n_i is chosen randomly under a uniform distribution.

8.7.1 Estimating ρ given an alignment

This is a straightforward estimation defined as follows. For convenience, we use the alternate notation for s_{ij} as

$$s[i, j].$$

For a fixed column c $(1 \leq c \leq l)$, define a t-sized vector of characters defined as

$$V_c[i] = s[i, Z_i + c - 1],$$

where $1 \leq i \leq t$. Then

$$\rho_{\sigma c} = \frac{\text{Number of times } \sigma \text{ appears in array } V_c}{t}.$$

Pseudocounts & ρ corrections. However, a value of zero in an entry of the ρ matrix is not desirable since this value never changes over the iterations (see Exercise 87). We make the corrections as follows

$$\rho'_{\sigma c} = \frac{b_\sigma + \text{Number of times } \sigma \text{ appears in array } V_c}{B + t}.$$

where b_σ is a σ_r dependent pseudocount and B is the sum of the pseudocounts. This also implies that

$$\sum_{\sigma \in \Sigma} \rho'_{\sigma c} = 1,$$

for each c.

How do pseudocounts fit into the framework? In Bayesian analysis prior probabilities are used (which are usually subjective) for the values of the estimated parameters. A common choice for such priors is the Dirichlet distribution which amounts to an addition of a pseudocount to the actual counts as shown above. They are to be interpreted as a priori expectations of the character occurrences (even possibly different in different positions, i.e., values of c, though we have treated them as the same for all values of c above) in the input.

The Gibbs sampling algorithm. We presented above a very general method of estimating ρ from the alignment. However, under the Gibbs sampling strategy, a sequence, s_y, is picked at random and removed from the collection and the motif probabilities are computed exactly as above with the remaining $t - 1$ sequences.

This step is also called the 'predictive update step'.

8.7.2 Estimating background probabilities given Z

The background or the nonmotif region is computed in the same manner as ρ is. Note that each sequence s_i has a l wide motif region and the remaining is background. Thus each position in the sequence s_i is

$$\begin{cases} \text{motif region} & \text{if } Z_i \leq j \leq Z_i + l - 1, \\ \text{background} & \text{otherwise.} \end{cases}$$

The background probabilities for the input are estimated for each $\sigma \in \Sigma$ as

$$\rho'_{\sigma 0} = \frac{b_\sigma + \text{Number of times } \sigma \text{ appears in the background}}{B + \text{Size of the background}}.$$

The background probability can also be computed for each sequence s_i by considering the background of only s_i.

8.7.3 Estimating Z given ρ

This step is also called the 'sampling step'. For each sequence s_i and each position

$$1 \leq j \leq n_i - l + 1$$

in this sequence, we compute the pattern probability Q_j and the background probability P_j and take the ratio

$$A_j = \frac{Q_j}{P_j}.$$

Q_j is computed as described in Section 8.6.2. In sequence s_i, the position j is chosen with probability

$$\frac{A_j}{\sum_j A_j},$$

and Z_j is set to this picked value.

8.8 Interpreting the Motif Profile in Terms of p

A motif is usually identified by its description p, its occurrence and the number of times, K, it occurs in the input. In our case we have $K' = t$. However, consider a general case when

$$K' \neq t.$$

Recall that the task is to discover or recover motif profiles, for some fixed motif length l given t sequences. what qualifies as a motif profile?

1. The motif has at least K, called quorum, occurrences in the input. In other words
$$K' \geq K.$$

2. An approximation of the probabilities ρ_{ij} is given by:
$$\rho_{ij} \approx \frac{\mu_{ij}}{K'},$$

   where μ_{ij} is the number of occurrences where σ_i occurs in the jth position. Notice that by this,

$$\rho_{ij} \leq 1.0, \text{ and}$$
$$\sum_{i=1}^{r} \rho_{ij} = 1.0,$$

for each j.

Next, a single character must dominate significantly in a position, say j, to specify a 'solid' character in the motif. One way of defining this is as follows: Given some fixed

$$0 < \delta < 1$$

there must be some $1 \leq i' \leq r$ such that

$$\rho_{i'j} - \rho_{ij} > \delta, \text{ for all } i \neq i'.$$

Then the motif at position j takes the value $\rho_{i'j}$. If this does not hold, then that position is defined to be a dont-care.

3. How is p defined? For example, consider the following profile.

$$\rho = \begin{bmatrix} 0.6 & 0.3 & 0.05 & 0.7 & 0.1 \\ 0.2 & 0.1 & 0.05 & 0.1 & 0.6 \\ 0.1 & 0.5 & 0.1 & 0.1 & 0.1 \\ 0.1 & 0.1 & 0.8 & 0.1 & 0.2 \end{bmatrix} \begin{matrix} A \\ C \\ G \\ T \end{matrix},$$

This is interpreted as a motif of length 5 defined as:

$$AGTAC$$

Notice that the definition of the *occurrence* of a motif is intricately associated with the definition of the motif profile.

8.9   Exercises

Exercise 85 (Statistical measures) *What is the relationship between measure F_2 and the information content I, for an input set of sequences and a motif profile ρ of length l, shown below:*

$$F_2 = \sum_{\sigma \in \Sigma} \sum_{c=1}^{l} f_{\sigma c} \log \left(\frac{\rho_{\sigma c}}{\rho_{\sigma 0}} \right),$$

$$I = \sum_{\sigma \in \Sigma} \sum_{c=1}^{l} \rho_{\sigma c} \log \left(\frac{\rho_{\sigma c}}{f_\sigma} \right).$$

Recall that column 0 of ρ stores the background probabilities of the characters; $f_{\sigma c}$, $c \geq 1$, is the number of times σ appears in position c of the motif occurrence in the input sequences; f_σ is the number of times σ appears in the input.

Exercise 86 (Estimating ρ) *See Section 8.5 for definition of the terms used here.*
Given z, consider the following scheme for estimating ρ using sequence s_i. For each $1 \leq c \leq l$, define

$$\rho_{\sigma c} = \sum_{s_i[j+c-1]=\sigma} z_{i(j+c-1)}.$$

1. *Show that the for each c,*

$$\sum_{\sigma \in \Sigma} \rho_{\sigma c} = 1.$$

2. *Is it possible to have an entry of zero in the profile matrix? Why?*

3. *What is the inadequacy of this scheme?*

Hint: 1. Use Equation (8.4). 3. Consider the following.

$$s_1 = \quad \text{A} \quad \text{C} \quad \text{G} \quad \text{A} \quad \text{A} \quad \text{C} \quad \text{G} \quad \text{G} \quad \text{A} \quad \text{A}$$
$$z_{1j} = \quad 0.05 \;\; 0.2 \;\; 0.1 \;\; 0.1 \;\; 0.05 \;\; 0.2 \;\; 0.05 \;\; 0.1 \;\; 0.1 \;\; 0.05$$

Let the subscript r of ρ denote the character σ_r. By this scheme,

$$\rho_{A1} = 0.05 + 0.1 + 0.05 + 0.1 + 0.05 = 0.35$$
$$\rho_{C1} = 0.2 + 0.2 = 0.4$$
$$\rho_{G1} = 0.1 + 0.05 + 0.1 = 0.25$$

What is ρ_{A2}, ρ_{A3}?

Exercise 87 (Initial estimate of ρ)

1. *For some r and some c, let*

$$\rho_{rc} = 0.0.$$

Then argue that at all subsequent iterations in the algorithm (both Methods 1 and 2) ρ_{rc} is likely to remain 0.0.

2. *For some r and some c, let*

$$\rho_{rc} = 1.0.$$

Then argue that at all subsequent iterations in the algorithm (both Methods 1 and 2) ρ_{rc} is likely to remain 1.0.
In other words the motif is likely to have σ_r at position c in all iterations.

3. *Let the initial estimate ρ_0 in the Expectation Maximization approach be as follows.*

$$\rho_0 = \begin{bmatrix} 0.0 & 0.0 & 0.0 \\ 0.1 & 0.0 & 0.0 \\ 0.0 & 0.0 & 0.0 \\ 0.0 & 1.0 & 1.0 \end{bmatrix} \begin{matrix} A \\ C \\ G \\ T \end{matrix}$$

Then argue that at all iterations q $\rho^{(q)}$ is likely to be ρ_0. In other words, at each iteration the motif is likely to be

$$C \; T \; T.$$

Hint: 1. See the update procedures. 2. Note that all the other entries in that column must be zero. 3. From 1 & 2.

Exercise 88 (Proof of convergence) *In the iterative procedure (both Methods 1 and 2), can you argue that the solution improves over the iterations? In other words,*

$$F(\rho^{(q)}) > F(\rho^{(q')}),$$

is likely to hold for a measure F and iteration q > q'.
Let F be defined to be F_1, F_2 or information content I of Section 8.4.

Hint: How is ρ or the alignment Z estimated? How likely is it that the F value decreases over an iteration?

Exercise 89 (Σ size) *Discuss the effect of alphabet size $|\Sigma|$ on the learning algorithm.*

Hint: The number of nucleic acids is 4 and the number of amino acids is 20. So can a system that discovers transcription factors in DNA sequences also discover protein domain motifs? Why? What parameters need to change? What about a binary sequence?

Exercise 90 (Multiple occurrences) *Discuss the main issues involved in allowing multiple occurrences of a motif in a sequence in the input data.*

Hint: For $l = 3$, see occurrences below.

$$s_1 = A\,\boxed{A\ C\ C}\,T\ A$$

$$s_2 = \boxed{A\ T\ G}\,T\,\boxed{A\ G\ G}$$

$$s_3 = A\ T\,\boxed{A\ C\ T}\,A$$

This gives two possible alignments:

	Alignment 1		Alignment 2
s_1	$A\,\boxed{A\ C\ C}\,T\ A$	s_1	$A\,\boxed{A\ C\ C}\,T\ A$
s_2	$\boxed{A\ T\ G}\,T\ A\ G\ G$	s_2	$A\ T\ G\,\boxed{T\ A\ G\ G}$
s_3	$A\ T\,\boxed{A\ C\ T}\,A$	s_3	$A\ T\,\boxed{A\ C\ T}\,A$
motif?	$A\ C\ G$	motif?	$A\ C\ G$

If s_3 also had 2 occurrences, how many alignments could there be? In the worst case, how many alignments are possible? How is the multiplicity of this kind incorporated in the probability computations?

Exercise 91 (Generalizations)

1. *Discuss how Expectation Maximization (Method 1) presented in this chapter can be generalized to handle multiple occurrences in a single sequence of the input.*

2. *Discuss how the Gibbs Sampling approach (Method 2) can be generalized to handle multiple occurrences in a single sequence of the input.*

3. *Can the methods be extended to incorporate unsupervised learning? Why?*

Hint: 1. How should z be updated? 2. What should Z_0 be ? How is Z_0 updated? 3. One of the major difficulties is in guessing in which of the sequences the motif is absent while estimating ρ or Z.

Exercise 92 (Substitution, scoring, Dayhoff, PAM, BLOSUM matrices) *A substitution matrix or a scoring matrix M is such that, value $M[i,j]$ is used to score the alignment of residue i with residue j in an alignment of two protein sequences. Two such matrices are discussed below.*

1. *Margaret Dayhoff and colleagues developed the PAM (Percent Accepted Mutation) series of matrices. In fact Margaret pioneered protein comparisons and data basing and developed the model of protein evolution encapsulated in a substitution matrix. This matrix is also called the* Dayhoff matrix.

   (a) *The values are derived from global alignments of closely related sequences.*

   (b) *Matrices for greater evolutionary distances can be extrapolated from those of smaller ones.*

   (c) *The number with the matrix, such as PAM60, refers to the evolutionary distance; the larger the number, the greater the distance.*

2. *Steve Henikoff and colleagues developed the BLOSUM (BLOcks SUbstitution Matrix) series of matrices.*

   (a) *The values are derived from local (ungapped) alignments of distantly related protein sequences.*

   (b) *All matrices are directly calculated, no extrapolation is used.*

   (c) *The number with the matrix, such as BLOSUM60, refers to the minimum percent identity of the blocks used to construct the matrix; the larger the number, the smaller the distance.*

What is the relationship between substitution matrix M and the probability matrix ρ of Equation (8.1), if any? Is M stochastic? Is M symmetric? Why?

Chapter 9

The Subtle Motif

While there is algorithms,
there is hope.
- anonymous

9.1 Introduction: Consensus Motif

As we saw in Chapter 8 the problem of detecting common patterns across biopolymers such as DNA sequences for locating interesting segments such as regulatory sites, transcription binding factors or even drug target binding sites, is indeed a challenging problem.

In this chapter we discuss a combinatorial modeling of the same problem: the signal is viewed as a *consensus* segment of the different sequences. The inadequacy, in a sense, of the learning methods and a rationale for the use of combinatorial model by the community, is discussed in the next section.

However, even in the combinatorial framework, not surprisingly the problem continues to be challenging. The main difficulty is that these motifs have subtle variations at each occurrence and also the location of the variation may differ at each occurrence. Nevertheless there is enough commonality to qualify this segment as a motif. This commonality is often known as a *subtle motif.*

One of the advantages of using this model over a learning model is that a combinatorial model is more suitable for dealing with the signal occurring over different length segments in the input (due to say insertion or deletion of bases). See an example in Figure 9.1 taken from [CP07] of a transcription binding factor. The different values in the *pos* column show that the alignment moves the sequences around.[1] Only the segments that contain the signal is shown in the figure. Observe how the signal differs at each occurrence. The '-' in the alignment corresponds to a gap.

---

[1] See the cited paper for any further details on this example.

No	pos	Predictions	M	I
0	−101	T G A C G T C A		1
1	−299	T G C − G T C A	1	
2	−71	T G A C A T C A	1	1
3	−69	A T G A − G T C A G		2
4	−527	T G C G A T G A	2	1
6	−173	T G A − C T A A	2	
7	−1595	T G A − A T G A	2	
8	−221	T G G − G T C T	2	
9	−69	T G A − C T G C	3	
10	−105	T G A − A T C A	1	
12	−780	T G C − G T C A	1	
14	−1654	A T G A − A T C A	1	1
15	−69	A T G A − G T C A A		2
16	−97	T G A − G T A A	1	
17	−1936	A T G A − A T C A	1	1
signal		T G A G T C A		

FIGURE 9.1:   An example of a subtle motif (signal) as a transcription binding factor in human DNA. Notice that at each occurrence the motif is some 'edit distance' away from the consensus signal. The edit operations are mutation (M) and insertion (I): see Section 9.3 for details on edit distance.

9.2   Combinatorial Model: Subtle Motif

The subtle motif problem has been of interest to both biologists and computer scientists. A satisfactory practical solution has been elusive although the problem is defined very precisely:

Problem 9 (The consensus motif problem):   *Given t sequences s_i on an alphabet Σ, a length $l > 0$ and a distance $d \geq 0$, the task is to find all patterns p, of length l that occur in each s_i such that each occurrence p'_i on s_i has at most d mismatches with p.*

The problem in this form made its first appearance in 1984 [WAG84]. In this discussion, the alphabet Σ is

$$\{A, C, G, T\}$$

and the problem is made difficult by the fact that each occurrence of the pattern p may differ in some d positions and the occurrence of the consensus pattern p may not have

$$d = 0$$

in any of the sequences.

In the seminal paper [WAG84], Waterman and coauthors provide exact solutions to this problem by enumerating *neighborhood* patterns, i.e., patterns that are at most d Hamming distance from a candidate pattern. Sagot gives a good summary of the (computational) efforts in [Sag98] and offers a solution that improves the time complexity of the earlier algorithms by the use of generalized suffix trees. These clever enumeration schemes, though exact, have a drawback that they run in time exponential in the pattern length. Some simple enumeration schemes are discussed in Section 9.6.

How do the methods discussed in Chapter 8 fit in? As we have seen this problem of detecting common subtle patterns across sequences is of great interest. Various statistical and machine learning approaches, which are inexact but more efficient, have been proposed in literature [LR90, LAB+93, BE94, HS99]. One of the questions that can be asked to compare and test the efficacy of such motif discovery systems is:

> *Given a set of sequences that harbor (with mutations) k motifs, what percentage of the k motifs does the system recover?*

When k is large, the learning methods of Chapter 8 recover a large percentage of the k embedded motifs. Yet another question to ask is:

> *Given a set of sequences that harbor (with mutations) ONE motif p, does the system recover p?*

This is a rather difficult criterion to meet since the learning algorithms use some form of local search based on Gibbs sampling or expectation maximization. Hence it is not surprising that these methods may miss p.

However, a question of this form is a biological reality. Consider the following, somewhat contrived, variation of Problem 9 which is an attempt at simplifying the computational problem.

Problem 10 (The planted (l, d)-motif problem): *Given t sequences $s_i i$ on Σ, a pattern p of length l is embedded in s'_i, with exactly d errors (mutations), to obtain the sequence s_i of length n, for each $1 \le i \le t$. The task is to recover p, given s_i, $1 \le i \le t$, and the two numbers l and d.*

Pevzner and Sze [PS00] made the question more precise and provided a benchmark for the methods, by fixing the following parameters:

1. $n = 600$ (length of each input sequence is 600),

2. $t = 20$ (the number of sequences is 20),

3. $l = 15$ (the length of the signal or motif is 15) and

4. $d = 4$ (the number of mutations is exactly 4 in each embedding).

A solution to this apparently simplified problem was so difficult, that this was dubbed the *challenge problem.*

This chapter discusses methods that solves problems of this flavor. This formalization, in a sense, is the combinatorial version of the problem discussed in Chapter 8. A further generalization, along with a method to tackle it, is presented in the concluding section of the chapter.

We first clarify the different 'motifs' used in this chapter. The central goal is to detect the *consensus* or the *embedded* or the *planted* motif in the given data sets which is also sometimes referred to as the *signal* in the data or the *subtle signal.* When a motif is not qualified with these terms, it refers to a substring that appears in multiple sequences, with possible wild cards.

9.3   Distance between Motifs

We begin by going through some very basic definitions of distance between motifs. Let Σ be the alphabet on which the input string s, as well as any pattern p is defined.

Hamming distance.   Given two patterns (or strings) p_1 and p_2 of length l each, the *Hamming distance* is defined as the number of positions $1 \leq j \leq l$ such that

$$p_1[j] \neq p_2[j].$$

For example, consider

$$
\begin{array}{llllll}
p_1 = & A & C & A & T & G \\
p_2 = & A & T & A & T & A \\
\hline
 & & X & & & X
\end{array}
$$

The Hamming distance between p_1 and p_2 is given as

$$Hamming(p_1, p_2) = 2,$$

since

$$p_1[1] = p_2[1],\ p_1[3] = p_2[3],\ \text{and } p_1[4] \neq p_2[4]$$

but

$$p_1[2] \neq p_2[2] \text{ and } p_1[5] \neq p_2[5]$$

marked as 'X' in the alignment.

Edit distance.   Given a pattern (or string) p_1 various edit operations can be performed on p_1 to produce p_2. Here we describe three edit operations that are effective on a position j on p_1 as follows.[2] The edit distance between p_1 and p_2 is written as:

$$eDis(p_1, p_2).$$

1. Mutation (M): $p_1[j]$ is changed to some

$$\sigma(\neq p_1[j]) \in \Sigma.$$

   For example if $p_1 = ACC$ then mutation at $j = 2$ can give any one of the following as p_2

$$p_2 = AGC \text{ or } ATC \text{ or } AAC.$$

   In this case, we write

$$eDis(p_1, p_2) = 1.$$

2. Deletion (X): $p_1[j]$ is deleted. For example if $p_1 = ACC$ then deletion at $j = 2$ gives p_2 as

$$p_2 = AC.$$

   In this case, we write

$$eDis(p_1, p_2) = 1.$$

3. Insertion (I): A $\sigma \in \Sigma$ is inserted at position j. For example if $p_1 = ACC$ then insertion at $j = 2$ can give any one of the following as p_2

$$p_2 = AACC \text{ or } ACCC \text{ or } AGCC \text{ or } ATCC.$$

   In this case, we write

$$eDis(p_1, p_2) = 1.$$

Note that a deletion or insertion results in a different length of the pattern, i.e.,

$$|p_1| \neq |p_2|.$$

To summarize, we define the distance between two substrings (or motifs) as either Hamming or an edit distance. Note that the latter is more general, since the Hamming distance usually assumes that p_1 and p_2 are of the same length. This is written as

$$dis(p_1, p_2),$$

and the context will define whether this distance is Hamming or an edit distance.

---

[2]It is possible to have edit operations defined on a segment of the string instead of a single location j. For example *inversion* on position 2-4 can transform

$$p_1 = G\underline{ACT}C \text{ to } p_2 = G\underline{TCA}C.$$

Let the occurrence of a motif in the input be o. For a given length l and d $(< l)$, a motif p is a subtle motif if at each occurrence, o, in the input

$$dis(p, o) \leq d.$$

For example let $l = 5$ and $d = 2$ and the three occurrences in the sequences are shown below.

$$s_1 = \boxed{\text{T A T C C}}\ \text{T}$$

$$s_2 = \boxed{\text{A C T C A}}\ \text{C}$$

$$s_3 = \text{C}\ \boxed{\text{T C C A A}}$$

$$\text{subtle motif } p = \quad \text{T C T C A}$$

Notice that in sequence 1, the motif differs at positions 2 and 5; in sequence 2 at position 1; and at positions 3 and 4 in sequence 3.

Let the occurrences in the three sequences (shown boxed above) be termed o_1, o_2 and o_3. At each occurrence the motif differs in at most d $(=2)$ positions. This is also written as:

$$dis(p, o_1) = 2 \leq d,$$
$$dis(p, o_2) = 1 \leq d,$$
$$dis(p, o_3) = 2 \leq d.$$

Thus one must look at all the occurrences to infer what the motif must be.

9.4   Statistics of Subtle Motifs

A subtle motif is defined by its length parameter l and the edit distance d with which it occurs in the given t sequences. For different values of these parameters, does it leave some detectable clues behind? We explore this by studying the statistics of subtle motifs in a very general setting.

We recall some basic definitions here. Given t sequences of length l each, a pattern satisfies *quorum* K if it occurs in

$$K' \geq K$$

of the given t sequences. Further it is of *maximal* size h, if in each of the K' occurrences, the size cannot be increased without decreasing the number of occurrences K'.

For simplicity, the sequences are the same length l and all the t sequences are aligned and we will further assume that a pattern occurs at most once in each sequence.

Given a motif, let the embedded signal in each sequence be constructed with some d edit operations. Given one of these edit operations, we assume

1. mutation (M), with probability of mutation given as q_M,

2. deletion (X), with probability of deletion given as q_X and

3. insertion (I), with probability of insertion given as q_I.

Since the only permissible edit operations are these three,

$$q_M + q_X + q_I = 1.$$

The model. We consider the following simplified model. Given a fixed pattern (or signal),

$$p_{signal},$$

of length l, we construct t sequences from p_{signal}. To construct each sequence, d positions in the pattern p_{signal} are picked at random and an edit operation (mutation with probability q_M, deletion with probability q_X and insertion with probability q_I) is applied to produce the sequence. Then we study these t sequences.

In other words, given t sequences we assume that they are aligned. For example, the table below on the left shows exactly one edit applied to the signal motif and the table on the right shows the alignment of these embedded motifs.

Edits	signal = ACGTAC	Alignment
M	A C G T C C	A C C – T c C
X	A G T A C	A – G – T A C
I	A C G A T A C	A C G a T A C
M	A C C T A C	A C c – T A C
M	G C G T A C	g C G – T A C

Assume that d out of the l positions are picked at random on the embedded motif for exactly one of the edit operations, insertion, deletion or mutation. l can be viewed as the size of the motif. Recall that we assume that the sequences are correctly aligned. Then if a position in the aligned sequence is a mismatch, then either it is due to

1. a mutation (whose probability is q_M) or

2. an insertion (whose probability is q_I).

Then the probability of this position to be a dot character (mismatch) is

$$\frac{d}{l} \left(q_M + q_I \right).$$

Next, the probability q of a position to be a solid character in a motif is:

$$q = 1 - \frac{d}{l} \left(q_M + q_I \right). \tag{9.1}$$

q for three scenarios is shown below.

		q_M	q_X	q_I	q
1)	Exactly d mutations	1	0	0	$1 - d/l$
2)	Exactly d edits	1/3	1/3	1/3	$1 - 2d/3l$
3)	Exactly d edits with, equiprobable indel and mutation	1/2	1/4	1/4	$1 - 3d/4l$

When *no more than d' edit operations* are carried out on the embedded motif, it is usually interpreted as each collection of

$$0, 1, 2, \dots, d'$$

positions being picked with equal probability, and thus

$$d = \frac{d'}{2}$$

for Equation (9.1).

Estimating the probability of occurrence of a motif.   Recall that q is the probability of a position (character) in the input data to match a character in the pattern (signal). Let H be the number of solid characters and let the motif appear in at least K sequences.

For instance in the following alignment, for the first 4 rows, i.e., $k = 4$, the pattern has $H = 3$ solid characters, namely A, T and C, shown in bold at the bottom row. In other words, in these four rows, the solid characters appear in *each* row of the aligned sequences.

Alignment	Pattern	
A C G − T c C	A C G − T c C	←
A − G − T A C	A − G − T A C	←
A C G a T A C	A C G a T A C	←
A C c − T A C	A C c − T A C	←
g C G − T A C	g C G − T A C	
	A **T** **C** k	

For a pattern p with some H solid characters, let p occur in some k sequences (and not in the remaining $(t − k)$ sequences). Then

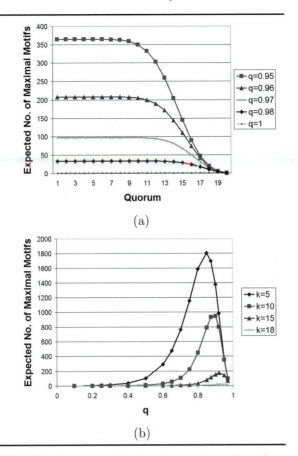

FIGURE 9.2: For $t = 20$, $l = 20$, the expected number of maximal motifs $E[Z_{K,q}]$, is plotted against (a) quorum K shown along the X-axis (for different values of q which are close to 1.0), and, (b) against q shown along the X-axis (for different values of quorum K).

1. the probability of matches in the (aligned) H solid characters in the k rows is

$$q^{H k},$$

   and

2. the probability of at least one mismatch in the (aligned) H positions in the remaining $(t - k)$ sequences, is

$$\left(1 - q^H\right)^{t-k}.$$

Thus the probability of occurrence of pattern p is given by

$$\left(1 - q^H\right)^{t-k} q^{H k}. \tag{9.2}$$

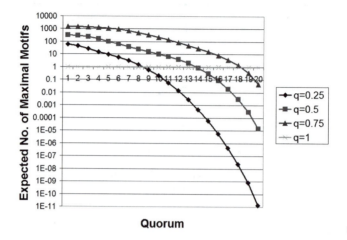

FIGURE 9.3: For $t = 20$, $l = 20$, the expected number of maximal motifs $E[Z_{K,q}]$, is plotted against quorum K shown along the X-axis, for different values of q, in a logarithmic scale. Unlike the plot in Figure 9.2, the value of q here varies from 0.25 to 1.0. Notice that when $q = 1$, the curve is a horizontal line at $y = 1$. Note that for DNA sequences, $q = 0.25$ corresponds to the random input case.

If E_p denotes the event that p occurs in some fixed k sequences then for any two distinct events, i.e.,

$$p_1 \neq p_2,$$

E_{p_1} and E_{p_2} are not necessarily mutually exclusive. However, if the pattern is *maximal*, i.e., H is the maximum number of solid characters seen in the k sequences, then for a fixed set of k sequences, there is at most one maximal pattern that occurs in these k sequences and not in the remaining $t - k$ sequences. Further, when the pattern is maximal there is a guarantee of mismatch in the remaining $(l - H)$ positions in all the k rows and the probability of this mismatch is given as

$$(1 - q^k)^{l-H}. \tag{9.3}$$

Thus to summarize, the probability of occurrence of some pattern with exactly H solid characters in exactly k sequences is given by

$$\left(1 - q^H\right)^{t-k} q^{H\,k} (1 - q^k)^{l-H}. \tag{9.4}$$

Thus if

$$P_{maximal}(K, H, q)$$

is the probability that some maximal pattern with H solid characters and quorum K occurs in the input data, then using equation (9.4),

$$P_{maximal}(K, H, q) = \sum_{k=K}^{t} \binom{t}{k} \left(1 - q^H\right)^{t-k} q^{Hk} (1 - q^k)^{l-H}. \qquad (9.5)$$

Let

$$Z_{K,q}$$

be a random variable denoting the number of maximal motifs with quorum K and q as defined above, and,

$$E[Z_{K,q}]$$

denotes the expectation of $Z_{K,q}$. Using linearity of expectations (for a fixed t and l),

$$E[Z_{K,q}] = \sum_{h=1}^{l} \binom{l}{h} P_{maximal}(K, h, q)$$

$$= \sum_{h=1}^{l} \binom{l}{h} \left(\sum_{k=K}^{t} \binom{t}{k} \left(1 - q^h\right)^{t-k} q^{hk} (1 - q^k)^{l-h} \right).$$

Now, it is rather straightforward to estimate $E[Z_{K,q}]$ given different values of q corresponding to different scenarios. Figures 9.2 and 9.3 show some examples.

9.5 Performance Score

Next, we define measures to evaluate the predictions of the subtle signal. We describe two simple measures that are commonly used [TLB+05, CP07]. Note that unlike in the motif learning process, these measures are not designed to drive the algorithm but to give a quantitative measure of how good the predictions are postmortem. This is used on data (possibly benchmark) where the correct solution is known either by other means or simply through the knowledge of the data construction process.

1. Let P be the set of all positions covered by the prediction and S be the same set for the embedded motif. The score of the prediction P, with respect to the embedded motif, can be given as:

$$score = \frac{|P \cap S|}{|P \cup S|}.$$

The score is 1 if the prediction is 100% correct. However, even for values much smaller than one, the embedded motif may be computed correctly. However, this measure is rather stringent.

2. The *solution coverage* (SC) is defined as the number of sequences that contains at least one occurrence of the predicted motif whose distance from the prediction is within the problem constraint, i.e., bounded by d. Again if the coverage is equal to the total number of sequences t, then the prediction can be considered 100% correct.

9.6   Enumeration Schemes

We first study some enumeration schemes to detect the subtle signal. Given t input sequences and a motif length l with distance d, the task is to detect the subtle signal in the input.

We next discuss the estimation of the set of potential signals C_{signal}. It is tempting to use an exact algorithm when the alphabet size is not too large.

9.6.1   Neighbor enumeration (exact)

An obvious method is to generate a set of potential signals, C_{signal}, which is the set of all possible l-mers. It is easy to see that, in this case,

$$|C_{signal}| = |\Sigma|^l.$$

Next each $p \in C_{signal}$ is checked against the input sequences.

Note that in this naive scheme, C_{signal} is independent of the input sequences. An easy improvement over such a blatant enumeration scheme is to generate a more restricted version of C_{signal} that depends on the input. Here we describe such an enumeration scheme.

Step 1 (Computing $C_1, C_2, \ldots, C_t$).   For each input sequence s_i

$$C_i = \{p \mid p \text{ is a substring of length } l \text{ in } s_i\}.$$

This can be obtained by a single scan of s_i from left to right, and at each location j, extracting a pattern p as

$$p = s[j \ldots (j + l - 1)].$$

and then removing the duplicates. An auxiliary information, j, the location of p on the sequence s_i, is associated with p.

Next, it is easy to see that, for each sequence s_i,

$$|C_i| \le |s_i| - l + 1. \tag{9.6}$$

Step 2 (Computing $C'_1, C'_2, \ldots, C'_t$). For each $p \in C_i$ construct the 'neighborhood' patterns as follows:

$$Nebor(p, d) = \{p' \mid Hamming(p, p') = d\}. \tag{9.7}$$

Next for each C_i, construct C'_i:

$$C'_i = \{p' \mid p' \in Nebor(p, d) \text{ where } p \in C_i\}.$$

The auxiliary information associated with each p' is

$$\{p_1, p_2, \ldots, p_r\} \subset C_i,$$

which is the set of r patterns such that p' is at distance d from each of these patterns.

What is the size of each C'_i? The number of positions that are mutated in the pattern p is d, thus the number of distinct patterns with some mutations is these positions is no more than

$$\binom{l}{d}.$$

Further, if the original value at one of the positions is σ, then it can take any value from the set

$$\Sigma \setminus \{\sigma\}.$$

Thus the total number of distinct patterns at a distance d from a pattern is no more than

$$\binom{l}{d} (|\Sigma| - 1)^d. \tag{9.8}$$

Using Equations (9.6) and (9.8),

$$|C'_i| \leq |C_i| \binom{l}{d} (|\Sigma| - 1)^d$$

$$\leq (|s_i| - l + 1) \binom{l}{d} (|\Sigma| - 1)^d$$

$$= \mathcal{O}\left(|s_i| \binom{l}{d} |\Sigma|^d\right).$$

Step 3 (Computing C_{signal}). It is easy to see that the possible embedded patterns (signals) are given by the set C_{signal}:

$$C_{signal} = C'_1 \cap C'_2 \cap \ldots \cap C'_{t-1} \cap C'_t.$$

For each

$$p'' \in C_{signal},$$

there exists some

$$p_1 \in C_1, p_2 \in C_2, \ldots, p_t \in C_t,$$

such that p'' is at distance d from each of the t patterns. In turn, each of these t patterns, corresponds to some location(s) j_i on each sequence s_i

$$j_1, j_2, \ldots, j_t.$$

Thus it can be said that signal p'' is embedded at location j_1 in s_1, j_2 in s_2, $\ldots, j_t$ in s_t.

We use the following example to illustrate the method.

Example 2 *The input is defined as follows:*

$$
\begin{aligned}
s_1 &= TATCC, & \Sigma &= \{A, T, C\}, \\
s_2 &= ACTCA, & t &= 3, \\
s_3 &= CTTTC, & l &= 4 \ and \ d = 1.
\end{aligned}
$$

Step 1 on the example.   The C_i sets are computed as follows:

$$
\begin{aligned}
C_1 &= \{TATC, ATCC\}, \\
C_2 &= \{ACTC, CTCA\}, \\
C_3 &= \{CTTT, TTTC\}.
\end{aligned}
$$

Step 2 on the example.   The C_i' sets are computed as follows:

$$
\begin{aligned}
C_1' &= \{Nebor(TATC, 1), Nebor(ATCC, 1)\} \\
Nebor(TATC, 1) &= \{AATC, CATC, TTTC, TCTC, \\
&\qquad TAAC, TACC, TATA, TATT\}, \\
Nebor(ATCC, 1) &= \{ATCA, ATCT, ATAC, ATTC, \\
&\qquad AACC, ACCC, CTCC, TTCC\}.
\end{aligned}
$$

$$
\begin{aligned}
C_2' &= \{Nebor(ACTC, 1), Nebor(CTCA, 1)\}, \\
Nebor(ACTC, 1) &= \{ACTA, ACTT, ACAC, ACCC, \\
&\qquad AATC, ATTC, TCTC, CCTC\}, \\
Nebor(CTCA, 1) &= \{CTCT, CTCC, CTAA, CTTA, \\
&\qquad CACA, CCCA, ATCA, TTCA\}.
\end{aligned}
$$

$$C'_3 = \{Nebor(CTTT, 1), Nebor(TTTC, 1)\}$$
$$Nebor(CTTT, 1) = \{CTTA, CTTC, CTAT, CTCT,$$
$$CATT, CCTT, ATTT, TTTT\},$$
$$Nebor(TTTC, 1) = \{TTTA, TTTT, TTAC, TTCC,$$
$$TATC, TCTC, ATTC, CTTC\}.$$

Step 3 on the example. The set of potential signals is estimated as follows:

$$C_{signal} - C'_1 \cap C'_2 \cap C'_3$$
$$= \{TCTC, ATTC\}.$$

Since

$$TCTC \in \begin{cases} Nebor(TATC, 1), \\ Nebor(ACTC, 1), \quad \text{and} \quad ATTC \in \begin{cases} Nebor(ATCC, 1), \\ Nebor(ACTC, 1), \\ Nebor(TTTC, 1), \end{cases} \\ Nebor(TTTC, 1), \end{cases}$$

this gives two alignments of the input s_i's as shown below.

		Using $TCTC$:			Using $ATTC$:
s_1	=	$T\ A\ T\ C$ C	s_1	=	T $A\ T\ C\ C$
s_2	=	$A\ C\ T\ C$ A	s_2	=	$A\ C\ T\ C$ A
s_3	=	C $T\ T\ T\ C$	s_3	=	C $T\ T\ T\ C$
consensus		$T\ C\ T\ C$	consensus		$A\ T\ T\ C$

This shows that there are two embedded signals,

$$TCTC \text{ and } ATTC,$$

that satisfy the $l = 4$, $d = 1$ constraints.

This concludes the discussion on exact neighbor enumeration. See [RBH05] for more details on this approach and some practical implementations of the exact approach.

9.6.2 Submotif enumeration (inexact)

The central observation used in this approach is as follows. *The submotif of the embedded signal (motif) occur in the input more often than a randomly selected submotif.* The method is inexact since some solutions can be missed as we will see in the concrete example below.

The method works in the following steps. Given l, fix some k $(< l)$.

1. Randomly pick k out of l positions to create a *mask*.

2. A mask is used to pick up k-mers from the input sequences.

3. Certain masks occur in multiple sequences, suggesting a local multiple alignment. By the observation that a random submotif is not likely to occur too many times, this can be used to extract the signal (subtle motif).

Consider a concrete example where $l = 4$. We discuss cases $k = 2$ and $k = 3$ for a sample input below.

Case $k = 2$. *Step 1.* If the k position picked at random are 2 and 4, then the mask takes a value 1 in positions 2, 4 and 'dont-care' in the remaining positions, encoded as

$$. 1 . 1$$

The exhaustive list of masks for parameters $l = 4$, $k = 2$ is shown below:

$$mask_1 = 1\ 1\ .\ .$$
$$mask_2 = 1\ .\ 1\ .$$
$$mask_3 = 1\ .\ .\ 1$$
$$mask_4 = .\ 1\ 1\ .$$
$$mask_5 = .\ 1\ .\ 1$$
$$mask_6 = .\ .\ 1\ 1$$

Step 2. A mask is used to pick up k-mers from the input sequences. An l-length mask can be used on a n-length sequence

$$n - (l + 1)$$

times. For example, the 2-mers picked for $mask_1$ and $mask_2$ are shown below.

s_1	$T\ A\ T\ C\ C$	$T\ A\ T\ C\ C$	s_1	$T\ A\ T\ C\ C$	$T\ A\ T\ C\ C$
$mask_1$	$1\ 1\ .\ .$	$1\ 1\ .\ .$	$mask_2$	$1\ .\ 1\ .$	$1\ .\ 1\ .$
2-mers	$T\ A$	$A\ T$	2-mers	$T\quad T$	$A\quad C$

The complete list of 2-mers picked by the masks are listed below.

	s_1 $TATCC$	s_2 $ACTCA$	s_3 $CTTTC$
$mask_1$	$TA..;\ AT..$	$AC..;\ CT..$	$CT..;\ TT..$
$mask_2$	$T.T.;\ A.C.$	$A.T.;\ C.C.$	$C.T..\ T.T.$
$mask_3$	$T..C;\ A..C$	$A..C;\ C..A$	$C..T;\ T..C$
$mask_4$	$.AT.;\ .TC.$	$.CT.;\ .TC.$	$.TT.;\ .TT.$
$mask_5$	$.A.C;\ .T.C$	$.C.C;\ .T.A$	$.T.T;\ .T.C$
$mask_6$	$..TC;\ ..CC$	$..TC;\ ..CA$	$..TT;\ ..TC$

Step 3. The local alignments suggested by some of the masks are shown below.

l-mers & alignment
no. of support

$$\left.\begin{array}{ccc} s_1 & s_2 & s_3 \\ T.T. & 1 & +0 & +1 \\ T..C & 1 & +0 & +1 \end{array}\right\} \quad \begin{array}{l} s_1 = \\ s_3 = C \end{array} \boxed{\begin{array}{c} T\,A\,T\,C\,|C \\ T\,T\,T\,C \end{array}} \quad \text{(I)}$$

$$\left.\begin{array}{ccc} s_1 & s_2 & s_3 \\ .T.C & 1 & +0 & +1 \end{array}\right\} \quad \begin{array}{l} s_1 = T \\ s_3 = C \end{array} \boxed{\begin{array}{c} A\,T\,C\,C \\ T\,T\,T\,C \end{array}} \quad \text{(II)}$$

$$\left.\begin{array}{ccc} s_1 & s_2 & s_3 \\ .TC. & 1 & +1 & +0 \end{array}\right\} \quad \begin{array}{l} s_1 = T \\ s_2 = A \end{array} \boxed{\begin{array}{c} A\,T\,C\,C \\ C\,T\,C\,A \end{array}} \quad \text{(III)}$$

$$\left.\begin{array}{ccc} s_1 & s_2 & s_3 \\ ..TC & 1 & +1 & +1 \end{array}\right\} \quad \begin{array}{l} s_1 = \\ s_2 = \\ s_3 = C \end{array} \boxed{\begin{array}{c} T\,A\,T\,C\,|C \\ A\,C\,T\,C\,|A \\ T\,T\,T\,C \end{array}} \quad \text{(IV)}$$

$$\left.\begin{array}{ccc} s_1 & s_2 & s_3 \\ C.C. & 0 & +2 & +0 \end{array}\right\} \quad - \quad \text{(V)}$$

Consensus alignment of the three sequences give the signal as shown below.

Using alignments (I), (III), (IV):
$$\begin{array}{ll} s_1(v_{11}) = & \boxed{\begin{array}{c} T\,A\,T\,C\,|C \end{array}} \\ s_2(v_{21}) = & A\,C\,T\,C\,|A \\ s_3(v_{32}) = C & T\,T\,T\,C \\ \text{consensus} & T\,C\,T\,C \end{array}$$

Case $k = 3$. Next, consider the same example with $k = 3$.

Step 1. The exhaustive list of masks for parameters $l = 4$, $k = 3$ is shown below:

$$\begin{array}{l} mask_1 = 1\,1\,1\,. \\ mask_2 = 1\,1\,.\,1 \\ mask_3 = 1\,.\,1\,1 \\ mask_4 = .\,1\,1\,1 \end{array}$$

Step 2.

	s_1 $TATCC$	s_2 $ACTCA$	s_3 $CTTTC$
$mask_1$	$TAT.;\ ATC.$	$ACT.;\ CTC.$	$CTT.;\ TTT.$
$mask_2$	$TA.C;\ AT.C$	$AC.C;\ CT.A$	$CT.T;\ TT.C$
$mask_3$	$T.TC;\ A.CC$	$A.TC;\ C.CA$	$C.TT;\ T.TC$
$mask_4$	$.ATC;\ .TCC$	$.CTC;\ .TCA$	$.TTT;\ .TTC$

Step 3.

$$l\text{-mers \&}\qquad\qquad\text{alignment}$$

no. of support

$$\left.\begin{array}{ccc} s_1 & s_2 & s_3 \\ T.TC & 1 & +0 & +1 \end{array}\right\}\ \begin{array}{l} s_1 = \\ s_3 = C \end{array}\ \boxed{\begin{array}{cccc} T & A & T & C \end{array}}C \\ \boxed{\begin{array}{cccc} T & T & T & C \end{array}}$$

Notice that this does not extract the l length signal. It just extracts the following

$$\text{T . T C}$$

This example illustrates the fact that this enumeration is inexact, since one of the solutions is missed in Case 2. It is also quite possible that a solution is missed in all possible values of k.

9.7   A Combinatorial Algorithm

The enumeration scheme, though exact, is very compute-intensive. For most real data, it is prohibitively time consuming to use such an exact method. Here we describe a combinatorial approach [PS00] (*Winnower*) that reduces the given problem to finding *cliques* in a t-partite graph.

A graph $G(V, E)$ is t-partite, if the vertex set can be partitioned as

$$V = V_1 \cup V_2 \cup \ldots \cup V_t,$$

where for $i \neq j$, $1 \leq i, j \leq t$,

$$V_i \cap V_j = \emptyset,$$

and for

each pair $v_{i_1}, v_{i_2} \in V_i$, $v_{i_1} v_{i_2} \notin E$ holds.

In other words, the vertex set can be partitioned into t (nonintersecting) sets such that the edges go across the sets but not within each set.

Given a graph $G(V, E)$, a subgraph

$$G(V' \subset V, E' \subset E),$$

is a *clique* if for

each pair $v_1, v_2 \in V'$, $v_1 v_2 \in E'$ holds.

In other words, a clique is a subgraph where every two vertices is connected by an edge.

It is important to note that the problem of finding a t-sized clique in a t-partite graph is NP-complete. Then why reduce the given problem to yet another difficult problem? The clique finding problem is well studied and various heuristics have been designed to effectively solve the problem and we wish to exploit these insights to solve our problem at hand. However, to avoid digression we do not discuss the details of solving the clique problem here.

Back to Problem 1. We begin with the following observation: If

$$d = Hamming(p, p_1)$$
$$= Hamming(p, p_2),$$

then

$$2d \geq Hamming(p_1, p_2).$$

The proof is straightforward and is left as an exercise for the reader. How is this fact used to solve the problem?

In this approach, we take an l-mer at some position j on s_i and associate this with an l-mer on some position j' on $s_{i'}$, $i' \neq i$ for each i' when it is plausible that the two l-mers are at a distance d from some hypothetical signal p. Since we do not know what p is, we use the above observation to associate the two l-mers only when the distance between them is no more than $2d$. Next, we seek a collection of t l-mers where every pair is associated with each other and each l-mer is from a distinct input sequence. In other words, if each l-mer is a vertex in a graph and the pair-wise association is an edge, we seek a clique of size t.

Formally put, a t-partite graph $G(V, E)$ is generated in the following two steps:

1. Each distinct l-mer in s_i, given by

$$s_i[j \ldots (j + l - 1)],$$

is mapped to a vertex v_{ij}. Each partition V_i is defined as

$$V_i = \{v_{ij} \mid \text{for some } j\},$$

and

$$V = V_1 \cup V_2 \cup \ldots \cup V_t.$$

Also note that

$$V_{i_1} \cap V_{i_2} = \emptyset,$$

for $i_1 \neq i_2$.

2. For $v_{i_1 j_1} \in V_{i_1}$ and $v_{i_2 j_2} \in V_{i_2}$,

$$v_{i_1 j_1} v_{i_2 j_2} \in E$$

if and only if

$$Hamming(s_{i_1}[j_1 \ldots (j_1 + l - 1)], s_{i_2}[j_2 \ldots (j_2 + l - 1)]) \leq 2d.$$

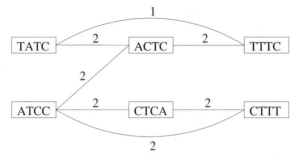

FIGURE 9.4:   The tripartite graph with each partition (of vertices) arranged along a column. The two cliques are the top row and bottom row respectively of vertices.

It is easy to see that $G(V, E)$ is a t-partite graph. Next the task is to find all cliques of size t in the graph.

Each such clique gives an alignment of the input sequences and a consensus motif p. This p is checked to see if the problem constraints are satisfied.

Example (2).   The 3-partite graph is constructed as follows (see Figure 9.4). The vertex set is

$$V = V_1 \cup V_2 \cup V_3,$$

where each V_i is defined as follows.

1. $V_1 = \{v_{11}, v_{12}\}$,
    where $TATC$ is mapped to v_{11} and $ATCC$ is mapped to v_{12}.

2. $V_2 = \{v_{21}, v_{22}\}$,
    where $ACTC$ is mapped to v_{21} and $CTCA$ is mapped to v_{22}.

3. $V_3 = \{v_{31}, v_{32}\}$,
    where $CTTT$ is mapped to v_{31} and $TTTC$ is mapped to v_{32}.

The following upper diagonal matrix shows the Hamming distance between two l-mers mapped to the two vertices. Each nonzero distance gives an edge in the graph.

		v_{11} ($TATC$)	v_{12} ($ATCC$)	v_{21} ($ACTC$)	v_{22} ($CTCA$)	v_{31} ($CTTT$)	v_{32} ($TTTC$)
v_{11}	($TATC$)	X		2	0	0	1
v_{12}	($ATCC$)			2	2	0	2
v_{21}	($ACTC$)			X		0	2
v_{22}	($CTCA$)					2	0
v_{31}	($CTTT$)					X	
v_{32}	($TTTC$)						

The two cliques in this graph are:

1. $Clq_1 = \{v_{11}, v_{21}, v_{32}\}$ and

2. $Clq_2 = \{v_{12}, v_{21}, v_{32}\}$.

The two alignments corresponding to the two cliques are:

	Using Clq_1 :		Using Clq_2 :
$s_1(v_{11}) =$	$T\ A\ T\ C\|C$	$s_1(v_{12}) = T$	$A\ T\ C\ C$
$s_2(v_{21}) =$	$A\ C\ T\ C\|A$	$s_2(v_{21}) =$	$A\ C\ T\ C\|A$
$s_3(v_{32}) = C$	$T\ T\ T\ C$	$s_3(v_{32}) = C$	$T\ T\ T\ C$
consensus	$T\ C\ T\ C$	consensus	$A\ T\ T\ C$

Thus there are two embedded signals,

$$TCTC \text{ and } ATTC,$$

that satisfy the $l = 4$, $d = 1$ constraints.

This concludes the discussion on the combinatorial approach of this section.

Other possible approaches can be based on enumerating possible patterns and checking their candidacy for being the subtle pattern using clever heuristics and an exhaustive search in a reduced space. Similar approaches, with different heuristics are presented in [PRP03, KP02a, EP02].

9.8   A Probabilistic Algorithm

What is the issue with the combinatorial approach? Recall that the clique computation is not easy. Most of the times this approach fails on nontrivial data since the heuristics are unable to extract the clique. The trouble is with enumerating (scanning) the solution space.

So we pursue a method that randomly samples this solution space. However to be effective, we carefully orchestrate the sampling as described below. The method is named *Projections* by its creators/designers Buhler and Tompa.

We first define a *condensed submotif*. Given a motif p, a submotif of p is a sequence with one or more matching characters with p. If k characters match with p then a k-mer is obtained by simply removing the nonmatching spaces. For example, if

$$p = ACGCCT,$$

Then some condensed submotifs of p are:

submotif		k	condensed submotif k-mer
$(p = p_0)$	$= A\ C\ G\ C\ C\ T$	6	$p_0' = ACGCCT$
p_1	$= A\ C\ .\ C\ C\ T$	5	$p_1' = ACCCT$
p_2	$= A\ C\ G\ .\ C\ T$	5	$p_2' = ACGCT$
p_3	$= A\ .\ G\ C\ .\ T$	4	$p_3' = AGCT$
p_4	$= .\ C\ G\ C\ .\ .$	3	$p_4' = CGC$
p_5	$= .\ C\ G\ .\ C\ .$	3	$p_5' = CGC$
p_6	$= .\ C\ .\ .\ .\ .$	1	$p_6' = C$
p_7	$= .\ .\ .\ C\ .\ .$	1	$p_7' = C$

Note that distinct condensed submotifs could give rise to the same k-mers. For example,

$$p_4 \neq p_5, \text{ but } p_4' = p_5' \text{ and}$$
$$p_6 \neq p_7, \text{ but } p_6' = p_7'.$$

This may affect the method, but we ignore this.

Most probabilistic algorithms perform a number of independent trials of a basic iterant. In this case each iterant is exactly along the lines the *Submotif enumeration scheme* for some k (Section 9.6.2) picked randomly. Also, condensed submotifs are used instead of submotifs.

Recall that Step 3 of the enumeration scheme tracks the occurrence of the l-mer (k-mer for the condensed submotif). In practice, a *hash table* is used to store the details of the occurrences. Roughly speaking, a hash table is indexed by an *attribute* (or a *key*) and usually takes only

$$\mathcal{O}(1)$$

time to access the entry in the table. Thus the tables shown in Step 3 of the enumeration scheme of Section 9.6.2 can be efficiently constructed, or filled in. Note that in this case a condensed submotif is a key to the hash table, which results in considerable reduction in the size of the hash table. Also, at each iterant a distinct value of k is used and the same hash table is fortified with more entries. Thus each iteration strengthens the table.

At the end of this process, each entry in the table that shows significant support is picked up for further scrutiny and the hidden signal is extracted from the local alignment suggested by the support. In fact this step uses the learning algorithms discussed in Chapter 8. The reader is directed to the paper by Buhler and Tompa [BT02] for further details of this algorithm.

9.9 A Modular Solution

We conclude the chapter by discussing a method that combines the solution from two well-studied problems

1. unsupervised (combinatorial) pattern discovery and

2. sequence alignment

By delegating the task to these subproblems, the method can also handle deletions and insertions (called *indel*) in the embedded signal.

Problem 11 (The indel consensus motif problem): *Given t sequences s_i on an alphabet Σ, a length $l > 0$ and a distance $d \geq 0$, the task is to find all patterns p, of length l that occur in each s_i such that each occurrence p'_i on s_i is at an edit distance (mutation, insertion, deletion) at most d from p.*

This approach uses unsupervised motif discovery to solve Problem 10 and also works well for the more general Problem 9.[3]

Recall that the signal ('subtle motifs') is embedded in t random sequences. The problem is compounded by the fact that although the consensus motif is solid (i.e., an l-mer without wild cards or dont-care characters), it is not necessarily contained in any of the t sequences. However, given an alignment, the consensus motif satisfying the (l, d) constraint may be extracted with ease.

In other words, one of the difficulties of the problem is that the sequences are unaligned. The extent of similarity across the sequences is so little that any global alignment scheme cannot be employed.

This method employs two step:

1. First, *potential signal (PS)* segments of interest are identified in the input sequences. This is done by using the imprints of the discovered motifs on the input.

2. Second, amongst these segments, exhaustive comparison and alignment is undertaken to extract the consensus motif.

This delineation into two steps also helps address the more realistic version of the problem that includes insertion and deletion in the consensus motif (Problem 11). The main focus of this method is in obtaining good quality PS segments and restricting the number of such segments to keep the problem tractable.

---

[3] A similar approach of using pattern discovery on a simpler problem of finding similarities in protein sequences has been attempted in [FRP+99].

The Type I error or false negative errors, in detecting PS segments, are reduced by using appropriate parameters for the discovery process based on the statistical analysis of consensus motifs discussed in Section 9.4.

The Type II error or false positive errors are reduced by using irredundant motifs [AP04] and their statistical significance measures [ACP05] discussed in Chapter 7. Loosely speaking, irredundancy helps to control the extent of over-counting of patterns and the pattern-statistics helps filter the true signal from the signal-like-background. In the scenario where indels (insertions and/or deletes) are permitted along with mutations, the unsupervised discovery process detects *extensible* motifs (instead of *rigid* motifs that have a fixed imprint length in all the occurrences). Also, the second step uses *gapped* alignments.

All nonexact methods are based on profiles or on k-mers, both of which are rigid. It is reasonable to say that, if the number of indels is much smaller than the size of the consensus, the chance of recovering the signal by such methods may be high. However, when the number of indels grow, it is unclear how these methods would work. Also, it is not immediately apparent how these methods can accommodate indels, since the rigidity in the profiles or k-mers is intrinsic to the method. On the other hand, the modular approach of using extensible motifs is one possible solution to overcome this bottleneck.

Rationale for using combinatorial unsupervised motif discovery. A motif of length l that occurs across $t' \leq t$ sequences provides a local alignment of length l for the t' sequences. The best case scenario, for the problem, is when the embedded motif m is identical in all t sequences and the discovery process detects this *single* maximal (combinatorial) motif with quorum t. So the scenarios closer to the best case should have fewer (but important) maximal motifs. Figure 9.2(a) shows the expected number of motifs with different values of q and quorum K. Notice that the expected number of motifs saturates for small values of K and falls dramatically as K increases. The saturation at lower values occurs since *maximal* motifs are being sought. Thus as q increases the saturation occurs at a higher value of K. Figure 9.2(b) shows the variation of the expected number of maximal motifs with q which is unimodal, for different values of K. The value of q is determined by the given problem scenario and thus a large value of K is a good handle on controlling the number and 'quality' of maximal motifs.

The *signal* is embedded in the *background* and it is important to exploit the characteristics that distinguishes one from the other. The background is assumed to be random. Under this condition, it is easy to see that

$$q = \frac{1}{4}$$

(see Section 9.4). Thus the need is to compare

$$E[Z_{K,q}]$$

with

$$E\left[Z_{K,\frac{1}{4}}\right],$$

the expectation for the random case. See Figure 9.3 for the plots of

$$\log(E[Z_{K,q}])$$

against quorum K to compare these expectation curves, particulary around small values (close to 1 in the Y-axis).

For example, consider the case when

$$q = 0.75,$$

this is the approximate value of q for the *challenge problem* of Section 9.1. In Figure 9.3, this is shown by the red curve and for large K, say

$$K \geq 16,$$

the expected number of motifs is small. Also, the corresponding expected numbers for the random case is extremely low, thus providing a strong contrast in the number of expected motifs. Hence the reasonable choice for the quorum parameter K is 16 or more, in the unsupervised discovery process.

It must be pointed out that in the case where the embedded motif is changed with insertions and/or deletions (indels), the q value is computed appropriately using Equation (9.1) and the corresponding expectation curve in Figure 9.3 must be studied. However, the burden is heavier on the unsupervised discovery process and an extensible (or, variable-sized gaps) motif discovery capability can be used.[4]

9.10 Conclusion

We have discussed several strategies to tackling the problem of finding subtle signals across sequences. This continues to be an active area of research with very close interaction between biologists, computer scientists and mathematicians.

---

[4]Varun [ACP05] is available at:
www.research.ibm.com/computationalgenomics.

9.11   Exercises

Exercise 93 (Distance) *Let* $\Sigma = \{0, 1\}$. *A pattern* p *of size* l *is defined on* Σ.

1. *Enumerate all* p', *defined on* Σ, *at a Hamming distance*

   (i) *exactly* d *from* p *and*

   (ii) *at most* d *from* p.

2. *Enumerate all* p', *defined on* Σ, *at an edit distance*

   (i) *exactly* d *from* p *and*

   (ii) *at most* d *from* p,

   where the edit operations allowed are (a) mutation, (b) insertion and (c) deletion.

Hint: Design an 'enumeration tree' particularly for 1(ii) and 2(ii) to avoid multiple enumerations of the same patterns.

Exercise 94 (Exact neighbor enumeration)

1. *Devise an efficient algorithm to generate the neighbors of a string* C_i' *of Equation (9.7) from* C_i.

2. *What is the running time complexity of the enumeration scheme of Section 9.6?*

Hint: Note that duplicates must be removed to compute the sets $C_1, C_2, \ldots, C_t$ and $C_1', C_2', \ldots, C_t'$.

Exercise 95 *Show that if*

$$d = Hamming(p, p_1)$$
$$= Hamming(p, p_2),$$

then

$$2d \geq Hamming(p_1, p_2).$$

Exercise 96 (Enumerations) *Let* $p = ACGCCT$.

1. *How many submotifs does* p *have?*

2. *How many distinct k-mers (k = 1, 2, ..., 6) does p have?*

Hint: Note that C is repeated in p that can give the same k-mer from distinct submotifs.

Exercise 97 *What is inexact about the enumeration scheme of Section 9.6.2?*

Hint: If d is the edit distance, then how many dot characters, d', does the mask have? Is there a formula for d'?

Exercise 98 *Modify the combinatorial algorithm of Section 9.7 to utilize the Hamming distance between vertices. The current algorithm simply uses nonzero distance.*

Hint: See also SP-Star [PS00] and patternbranching [PRP03, KP02a] for effective utilization of these weights and more.

Comments

The topic of this chapter exemplifies the difficulties with biological reality. Elegant combinatorics and practical statistical principles, along with biological wisdom may be sometimes required to answer innocent-looking questions.

Part III

Patterns on Meta-Data

Chapter 10

Permutation Patterns

Out of clutter, find simplicity;
from discord, find harmony.
- attributed to A. Einstein

10.1 Introduction

In this chapter we deal with a different kind of motif or pattern: one that is defined by merely its composition and not the order in which they appear in the data. For example, consider two chromosomes in different organisms. We study gene orders in a section of the chromosomes of two organisms as shown below:

$$s_1 = \ldots g_1 \boxed{g_2\ g_3\ g_4\ g_5}\ g_6\ g_7 \cdots$$

$$s_2 = \ldots g_8 \boxed{g_5'\ g_2'\ g_4'\ g_3'}\ g_9\ g_0 \cdots$$

Genes g_i (in s_1) and g_i' (in s_2) are assumed to be *orthologous* genes. Clearly, it is of interest to note that the block of genes g_2, g_3, g_4, g_5 appear together, albeit in a different order in each of the chromosomes. This collection of genes is often called a *gene cluster*. The size of the cluster is the number of elements in it and in this example the size is 4. Such clusters or sets of objects are termed *permutation patterns*.[1] They are called so because any one of the patterns can be numbered 1 to L where L is the size of the pattern and every other occurrence is a permutation of the L integers. For example in s_1, the pattern can be numbered as

$$1\ 2\ 3\ 4,$$

and in s_2 the occurrence is a permutation given as

$$4\ 1\ 3\ 2.$$

---

[1] This cluster or set is also called a *Parikh vector* (Section 10.4) or a *compomer* (Exercise 118)

10.1.1   Notation

Recall from Chapter 2 (Section 2.8.2) that $\Pi(s)$ denotes the set of all characters occurring in a sequence s. For example, if

$$s = a\,b\,c\,d\,a,$$

then

$$\Pi(s) = \{a, b, c, d\}.$$

However s may have characters that appear multiple times (also referred to as the *copy number*). Then we use a new notation $\Pi'(s)$. In this notation, each character is annotated with the number of times it appears. For example,

$$s = a\,b\,b\,c\,c\,b\,d\,a\,c\,b,$$
$$\Pi(s) = \{a, b, c, d\},$$
$$\Pi'(s) = \{a(2), b(4), c(3), d\}.$$

Thus element a has copy number 2, b has copy number 4 and so on. Note that d appears only once and the copy number annotation is omitted altogether.

Given an input string s on a finite alphabet Σ, a *permutation pattern* (or *π pattern*) is a set $p \subseteq \Sigma$. p occurs at location i on s if

$$p = \Pi\left(s\left[i, i+1, \ldots, i+L\text{-}1\right]\right),$$

where $L = |p|$. p has *multiplicity*, if we are interested in the multiple occurrences of a $\sigma \in p$. Although strictly speaking, p is not a set anymore, to avoid clutter we do not distinguish the two forms of p. For example, consider

$$s = \boxed{\text{a a c b b b}}\,x\,x\,\boxed{\text{a b c b a b}}.$$

The two occurrences of a pattern are shown in boxes. We represent these occurrences as o_1 and o_2 and for convenience write them as

$$o_1 = a\,a\,c\,b\,b\,b,$$
$$o_2 = a\,b\,c\,b\,a\,b.$$

If we are interested in multiplicity, or in counting copy numbers, then

$$p = \{a(2), b(3), c\}.$$

Otherwise,

$$p = \{a, b, c\}.$$

The size of the pattern p is written as $|p|$. In the first case $|p| = 6$ and in the second case $|p| = 3$. Note that in both cases, the length at each occurrence of the pattern p must be the same.

Further, p satisfies a *quorum* K, if it occurs at some $K' \geq K$ distinct locations on s given as

$$\mathcal{L}_p = \{i_1, i_2, \ldots, i_{K'}\}.$$

$\mathcal{L}_p$ is the *location list* of p. In the running example, assuming a quorum $K = 2$, permutation pattern p occurs at locations 1 and 9 on s, written as

$$\mathcal{L}_p = \{1, 9\}.$$

10.2 How Many Permutation Patterns?

It is important to know the total number of permutation patterns that may occur on a string s for at least two reasons. Firstly, if it is unduly large, one needs to explore the possibility of reducing this number without compromising the pattern definition. Secondly, the number is useful in the statistical analysis of permutation patterns.

$$
\begin{aligned}
p_1 &= \{a, b\} & \mathcal{L}_{p_1} &= \{1, 6, 11\} \\
p_2 &= \{b, c\} & \mathcal{L}_{p_2} &= \{2, 7, 12\} \\
p_3 &= \{c, d\} & \mathcal{L}_{p_3} &= \{3, 8, 13\} \\
p_4 &= \{d, e\} & \mathcal{L}_{p_4} &= \{4, 9, 14\} \\
p_5 &= \{a, b, c\} & \mathcal{L}_{p_5} &= \{1, 6, 11\} \\
p_6 &= \{b, c, d\} & \mathcal{L}_{p_6} &= \{2, 7, 12\} \\
p_7 &= \{c, d, e\} & \mathcal{L}_{p_7} &= \{3, 8, 13\} \\
p_8 &= \{a, b, c, d\} & \mathcal{L}_{p_8} &= \{1, 6, 11\} \\
p_9 &= \{b, c, d, e\} & \mathcal{L}_{p_9} &= \{2, 7, 12\} \\
p_{10} &= \{a, b, c, d, e\} & \mathcal{L}_{p_{10}} &= \{1, 6, 11\}
\end{aligned}
$$

FIGURE 10.1: The exhaustive list of permutation patterns p, with $|p| > 1$, occurring on $s = a\,b\,c\,d\,e\,a\,b\,c\,d\,e\,a\,b\,c\,d\,e$ satisfying quorum $K = 3$.

Let P be the collection of all permutation patterns on a given input string s of length. What is the size of P? In other words, what is the maximum number of elements in P?

Assuming that a permutation pattern can start at an arbitrary position i on s and end at another position $j > i$ on s. Thus the number of permutation patterns is

$$\mathcal{O}(n^2).$$

In other words, we estimate an upper bound of n^2 on the total number of patterns, in the worst case. But is this number actually attained? Consider the following example:

$$s = a\,b\,c\,d\,e\,a\,b\,c\,d\,e\,a\,b\,c\,d\,e,$$

and quorum $K = 3$. The collection of permutation patterns, p with $|p| > 1$, on s is listed in Figure 10.1. This construction shows that the number of such patterns is

$$\frac{m(m+1)}{2},$$

where

$$m = \frac{n}{3} - 1.$$

This construction shows that such a number can indeed be attained.

We next explore the possibility of refining the definition of the pattern to reduce their number.

10.3   Maximality

In an attempt to reduce the number of permutation patterns in an input string s, without any loss of information, we use the following definition of a maximal pattern [LPW05].

Let P be the set of all permutation patterns on a given input string s. $(p_1 \in P)$ is *nonmaximal* with respect to $(p_2 \in P)$ if both of the following hold.

(1) Each occurrence of p_1 on s is covered by an occurrence of p_2 on s. In other words, each occurrence of p_1 is a substring in an occurrence of p_2.

(2) Each occurrence of p_2 on s covers $l \geq 1$, occurrence(s) of p_1 on s.

A pattern $(p_2 \in P)$ is *maximal*, if there exists no $(p_1 \in P)$ such that p_2 is nonmaximal w.r.t. p_1.

It is straightforward to verify the following and we leave the proof as an exercise for the reader (Exercise 99). Note that this directly follows from the framework presented in Chapter 4.

LEMMA 10.1
(Maximal lemma) *If p_2 is nonmaximal with respect to p_1, then $p_1 \subset p_2$.*

When p_2 is nonmaximal with respect to p_1 does the following hold

$$|\mathcal{L}_{p_1}| = |\mathcal{L}_{p_2}| \ ?$$

The sizes of the location lists must be the same when each element of the pattern p_1 and p_2 has a copy number 1. See Exercise 100 for the possible relationship between $|\mathcal{L}_{p_1}|$ and $|\mathcal{L}_{p_2}|$ when copy number of some elements > 1.

However, to show that maximality as defined here is valid, it is important to show the uniqueness of the set of maximal permutation patterns. Again, this also follows from the framework presented in Chapter 4.

THEOREM 10.1

(Unique maximal theorem) *Let M be the set of all maximal permutation patterns, i.e.,*

$$M = \{p \in P|\ \text{there is no } (p' \in P)\ \text{maximal w.r.t } p\}$$

Then M is unique.

PROOF   We prove this by contradiction. Assume that M is not unique, i.e. there exist at least two distinct maximal collections M_1 and M_2 ($M_1 \neq M_2$) satisfying the definition. Without loss of generality, let

$$p \in M_1 \text{ and } p \notin M_2.$$

Since $p \notin M_2$, there must exist $p' \in M_2$ such that p is nonmaximal with respect to p'. In other words,

$$p' \notin M_1 \text{ and } p' \in M_2,$$

and p is nonmaximal with respect to p'. This contradicts the assumption that M_1 is a maximal collection. Hence the assumption must be wrong and $M_1 = M_2$. □

It is easy to see that a nonmaximal pattern p_1 can be 'deduced' from p_2 and the occurrences of p_1 on s can be estimated to be within the occurrences of p_2.

10.3.1 $P_{=1}$: Linear notation & PQ trees

Given an input s and a quorum K, let

$$P_{=1}$$

be the set of all permutation patterns on s where each element in any $p \in P_{=1}$ has a copy number 1. In other words each occurrence of the pattern contains exactly one instance of each character in p.

Recall that in case of substring patterns, the maximal pattern very obviously indicates the nonmaximal patterns as well. For example a maximal pattern of the form $a\,b\,c\,d$ implicates

$$a\,b,\ b\,c,\ c\,d,\ a\,b\,c,\ \text{and } b\,c\,d$$

as possible nonmaximal patterns, unless they have occurrences not covered by $a\,b\,c\,d$.

Do maximal permutation patterns (in $P_{=1}$) have such a simple form? For $p \in P_{=1}$, let

$$M(p) = \{p' \in P_{=1} |\; p' \text{ is nonmaximal w.r.t } p \; \}.$$

Can p have a representation that captures $M(p)$ (without explicitly enumeration)?

Note that the K occurrences on the input

$$o_1, o_2, \ldots, o_K$$

of p are simply different permutations of the elements of p. So the question is:

> *Is there is a representation that captures the 'commonality' (or $M(p)$) in all of these K occurrences?*

To answer this question, we study the solution to a classic problem in combinatorics, the *general consecutive arrangement problem*.

Problem 12 *(The general consecutive arrangement (GCA) problem)*
Given a finite set Σ and a collection $\mathcal{I}$ of subsets of Σ, does there exist a permutation s of Σ in which the members of each subset $I \in \mathcal{I}$ appear as a consecutive substring of s?

Mapping the GCA problem to our setting:

> Σ is the elements of p,
> $\mathcal{I}$ is $M(p)$, and
> we know that $o_1, o_2, \ldots, o_K$ are
> consecutive (linear) arrangements of the elements of p.

Thus the data structure (called a *PQ Tree*) used in the solution to the GCA problem can be used as the representation to capture $M(p)$. See Chapter 13 for an exposition on this.

Consider a pattern p and its collection of nonmaximal patterns $M(p)$ given in Figure 10.2. The PQ tree representation of the maximal pattern in shown in Figure 10.2. The root node represents the maximal permutation pattern given as set (10.1), the Q node represents the nonmaximal patterns given as set (10.2) and the internal P node represents the nonmaximal pattern given as set (10.3).

Using the symbol '-' to denote *immediate neighbors*, since the PQ tree is a hierarchy, it can also be written linearly as:

$$((a, b, c, d)\text{-}(e\text{-}f\text{-}g)).$$

This is the maximal notation of the pattern $\{a, b, c, d, e, f, g\}$ (set (10.1)).

$$p = \{a, b, c, d, e, f, g\}, \tag{10.1}$$
$$M(p) = \{\{e, f\}, \{f, g\}, \{e, f, g\}, \tag{10.2}$$
$$\{a, b, c, d, \}\}. \tag{10.3}$$

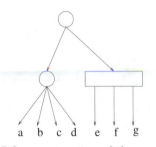

FIGURE 10.2: The PQ tree notation of the maximal pattern p.

10.3.2 $P_{>1}$: Linear notation?

A PQ tree captures the internal structure (in terms of its nonmaximal permutation components) of a pattern and is an excellent visual representation of a maximal pattern. However, even an elegant structure such as this has its limitations: we describe such a scenario where we must use multiple PQ trees to denote a single maximal pattern.

Given an input s and a quorum K, let $P_{>1}$ be the set of all permutation patterns on s where there is some $p \in P_{>1}$, which has at least one element that has a copy number > 1.

Let $p \in P_{>1}$ be as follows:

$$p = \{a, b, c(2), d, e, x\}. \tag{10.4}$$

Also, let p have exactly three occurrences on s given as

$$o_1 = d\,e\,a\,b\,c\,x\,c,$$
$$o_2 = c\,d\,e\,a\,b\,x\,c,$$
$$o_3 = c\,x\,c\,b\,a\,e\,d.$$

Assume that none of the elements of p appear elsewhere in the input. What are the nonmaximal patterns?

Recall that the leaves of a PQ tree are labeled bijectively by the elements of p. Since p has at least one element σ with copy number $c > 1$, then the tree must have c leaves labeled by σ. Assuming we can abuse a PQ structure thus, can a PQ tree represent all the nonmaximal patterns?

Can we simply rename the two c's as c_1 and c_2? We can fix this in o_1, but which c is c_1 and which one is c_2 in o_2 and in o_3? We must take all possible renaming into account.

We take a systematic approach and rename the elements of o_1 as integers $1, 2, \ldots, 7$ and using this same scheme, we rename o_2 and o_3. Since c has a copy number of 2, c is renamed as two integers 5 and 7. So a c in o_2 and o_3 is renamed either as 5 or as 7 and is written as [57]. Then

$$
\begin{aligned}
o_1 &= d\ e\ a\ b\ c\ x\ c &=&\quad 1\ 2\ 3\ 4\,5\,6\ 7, \\
o_2 &= c\ d\ e\ a\ b\ x\ c &=&\quad [57]\ 1\ 2\ 3\,4\,6\ [57], \\
o_3 &= c\ x\ c\ b\ a\ e\ d &=&\quad [57]\ 6\ [57]\ 4\,3\,2\ 1.
\end{aligned}
$$

The two renamed choices for o_2 are

1. $o_2 = 5123467$, hence $o_3 = 5674321$ or $o_3 = 7654321$, and

2. $o_2 = 7123465$, hence $o_3 = 5674321$ or $o_3 = 7654321$.

Thus the four possible scenarios are:

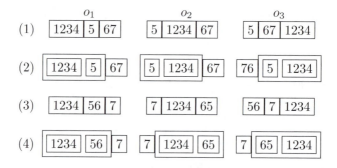

The nonmaximal patterns are shown as nested boxes. The following trees capture the nonmaximal patterns: $T_{1,3}$ represents the first and third cases, T_2 and T_4 represent the second and fourth cases respectively.

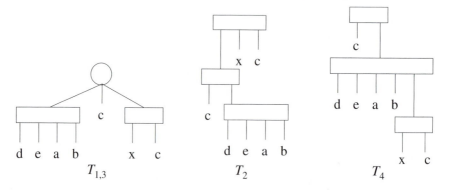

Note that a nonmaximal pattern may be represented in more than one PQ tree. For this example, is it possible to construct a single PQ tree that captures all the non-maximal patterns? Given any maximal $p \in P_{>1}$, in the worst case how many PQ trees can represent all the nonmaximal patterns? See Exercise 102.

10.4 Parikh Mapping-based Algorithm

We next explore the problem of automatically discovering all the permutation patterns in one or more strings. We formalize the problem as follows.

Problem 13 *(The permutation pattern discovery problem)* *Given a string s of length n, over a finite alphabet Σ, and a quorum $K < n$, find all permutation patterns p (and the location list $\mathcal{L}_p$) that occur at least K times.*

A permutation pattern of size L cannot necessarily be built from a pattern of size

$$L' < L,$$

as in the case of substring patterns. Hence this problem is considered harder than the substring pattern discovery problem and to date there is no clever way of exhaustively discovering these patterns other than finding patterns of size

$$2, 3, \ldots, L^*$$

where L^* is the size of the largest pattern in s.

Overview of the method. We present a rather straightforward discovery process: in this algorithm a window of size L is scanned over the input, observing the characters of the scanned window. This tracks the new as well as previously seen patterns.

If the size of the alphabet is small, i.e.,

$$|\Sigma| = \mathcal{O}(1),$$

then in $\mathcal{O}(1)$ time each new or old pattern can be accounted for (using an appropriate hash function), giving an overall $\mathcal{O}(n)$ time algorithm, for a fixed L. However, this assumption may not be realistic and in general

$$|\Sigma| = \mathcal{O}(n).$$

Then the approach needs some more care for efficiency and the discussion here using Parikh Mapping is adapted from an algorithm given by Amir *et al* [AALS03]. The reader is directed to [Did03, SS04, ELP03] for a discussion on other approaches to this problem.

However, the discovered patterns in the algorithm are not in the maximal notation. We postpone this discussion to Section 10.5 where the *Intervals Problem* is presented. We take all the occurrences,

$$o_1, o_2, \ldots o_k$$

of a pattern p, obtained by the algorithm of this section and solve an instance of the Intervals Problem using the k occurrences as input. In Section 10.5 these occurrences are denoted as

$$s_1, s_2, \ldots s_k,$$

and the output is the maximal notation of p in terms of a PQ tree. Note that if p has multiplicities, then the Intervals Problem is invoked multiple times: see Section 10.3.2 for a detailed discussion on this.

Computing Ψ. Parikh mapping is an important concept in the theory of formal languages [Par66]. [2] Given an alphabet of size m,

$$\Sigma = \{\sigma_1 < \sigma_2 < \ldots < \sigma_m\},$$

let w_{σ_i} be the number of occurrences of σ_i in $(w \in \Sigma^*)$. Parikh mapping is a morphism

$$\Psi : \Sigma^* \mapsto N^k,$$

where N denotes nonnegative integers and

$$\Psi(w) = (w_{\sigma_1}, w_{\sigma_2}, \ldots, w_{\sigma_m}).$$

In this section we discuss an efficient algorithm based on this mapping. The algorithm maintains an *array*

$$\Psi[1 \ldots |\Sigma|],$$

where $\Psi[q]$ keeps count of the number of appearances of letter q in the current window. Hence, the sum of the values of the elements of Ψ is L. In each iteration the window shifts one letter to the right, and at most 2 variables of Ψ are changed:

(1) one variable is increased by one (adding the rightmost letter) and

(2) one variable is decreased by one (deleting the leftmost letter of the previous window).

Note that for a given window of size L on s, $s_a s_{a+1} \ldots s_{a+L-1}$, Ψ represents

$$\Pi'(s_a s_{a+1} \ldots s_{a+L-1}).$$

---

[2]More than 40 years after the appearance of this seminal paper, during an informal conversation, Rohit Parikh, a logician at heart, told me that he had done this work on formal languages *for money!* He explained that as a graduate student he was compelled to take a summer job that produced this work.

10.4.1 Tagging technique

It is easy to see that substrings of s, of length L, that are permutations of the same string are represented by the same Ψ. Each distinct Ψ is assigned a unique tag - an integer in the range 0 to $2n$. The tags are given by using the *naming* technique [AIL$^+$88, KLP96], which is a modified version of the algorithm of Karp, Miller and Rosenberg [KMR72].

Assume, for the sake of simplicity, that $|\Sigma|$ is a power of 2. If $|\Sigma|$ is not a power of 2, Ψ can be extended to an appropriate size. The size of the resulting array is no more than twice the size of the original array.

A tag is completely determined by a pair of previous tags. At level j, the tag of subarray $\Psi_1\Psi_2$ of size 2^j is assigned, where Ψ_1 and Ψ_2 are consecutive subarrays of size 2^{j-1} each. The tags are natural numbers in increasing order. The process may be viewed as constructing a complete binary tree, which is the *binary tagging tree*. Notice that every level only uses the tags of the level below it, thus the tags are nonnegative integers that can be bounded as discussed below.

We next illustrate this elegant algorithm using a simple example. Consider the following example with $K = 2$ and $L = 4$:

$$\Sigma = \{a < b < c < d < e < f\},$$
$$s = bbacfbcbaa.$$

Note that

$$|\Sigma| = 6$$

and the Parikh Mapping array Ψ is padded with the $\bullet$ character so as to make it a power of 2. This complete example is described in Figure 10.3.

10.4.2 Time complexity analysis

We begin by bounding t, the number of distinct tags generated for a given input s. This bound not only estimates the space required by the algorithm but also helps in estimating the time required to search for the existence of each pair of tags, thus giving a more accurate bound on the running time.

Although it is tempting to compute this number as a function of the window size L, using the alphabet size $|\Sigma|$ gives a better bound as shown below.

LEMMA 10.2

The maximum number of distinct tags generated by the algorithm's tagging scheme, using a window of size L on a text of length n is

$$\mathcal{O}(|\Sigma| + n\log|\Sigma|),$$

where $|\Sigma|$ is the size of the alphabet.

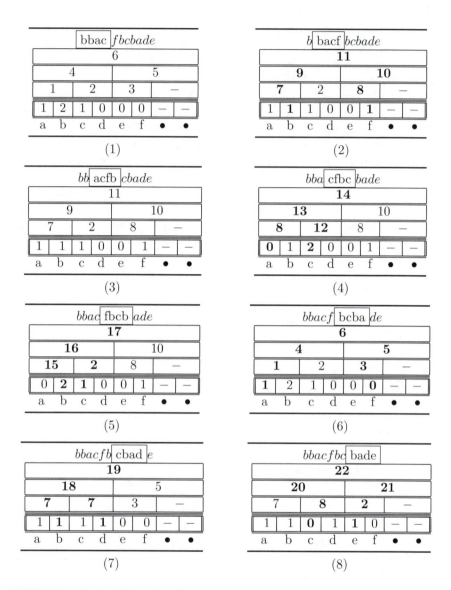

FIGURE 10.3: The algorithm run on $s = bbacfbcbade$ with $L = 4$, $K = 2$. (1)-(8) shows the sliding of the window on the input and the change in the binary name tree at each stage. The tags shown in bold are the ones that change at each iteration from the previous ones. The run shows that there are two permutation patterns of size 4 on s: (1) $p_1 = \{a, b, c, f\}$, tagged 11, with $\mathcal{L}_{p_1} = \{2, 3\}$, (2) $p_2 = \{a, b(2), c\}$, tagged 6, with $\mathcal{L}_{p_2} = \{1, 6\}$.

PROOF Consider the very first window of size L at position $j = 1$ on the string with the corresponding Parikh Mapping array Ψ^1. The number of distinct tags in the *binary tagging tree* of Ψ^1 is

$$\mathcal{O}(|\Sigma|).$$

The height of this tree is
$$\mathcal{O}(\log|\Sigma|).$$

The total number of iterations is

$$n - L + 1.$$

At each iteration, at most $\log|\Sigma|$ changes are made due to addition of a new character to the right and at most $\log|\Sigma|$ changes are made due to the removal of an old character to the left. Thus at each iteration j no more than

$$2\log|\Sigma|$$

new tags are generated in the binary tagging tree of Ψ^j. Thus the number of distinct tags is
$$t = \mathcal{O}(|\Sigma| + n\log|\Sigma|).$$

$\square$

To give a subarray at level > 1 a tag, we need only to know if the pair of tags of the composing subarrays has appeared previously. If it did, then the array gets the tag of this pair. Otherwise, it gets a new tag. Assume that the first elements of the tag pairs is stored in a balanced tree $\mathbf{T}_1$. Further the pairs are gathered and yet another tree $\mathbf{T}_2^v$ is stored at each node v of $\mathbf{T}_1$ which is also balanced. Thus it takes

$$\mathcal{O}((\log t)^2)$$

time to access a tag pair where both $\mathbf{T}_1$ and $\mathbf{T}_2^v$ are binary searched and t is the number of distinct tags.
To summarize, it takes
$$\mathcal{O}(|\Sigma|)$$

time to initialize the binary tagging tree of Parikh Mapping array Ψ. The number of iterations is
$$\mathcal{O}(n)$$

and at each iteration
$$\mathcal{O}(\log|\Sigma|)$$

changes are made, each of which takes

$$\mathcal{O}((\log t)^2)$$

time. Thus the algorithm takes

$$\mathcal{O}(|\Sigma| + n(\log t)^2 \log |\Sigma|)$$

time, for a fixed L. If L^* the size of the largest pattern on s is not known, then this algorithm is iterated $\mathcal{O}(n)$ times.

10.5   Intervals

The last section gives an algorithm to discover all permutation patterns in a given string s. We now take a look at a relatively simple scenario: Given K sequences where n characters appear exactly once in each sequence and each sequence is of length n, the task is to discover common permutation patterns that occur in *all* the sequences.

In other words, s_1 can be viewed as the sequence of integers $1, 2, 3, \ldots, n$ and each of

$$s_2, s_3, \ldots, s_K$$

is a permutation of n integers. See Figure 10.4 for an illustrative example. Why is this problem scenario any simpler? For input sequences $s_2, s_3, \ldots s_K$, the encoding to integers allows us to simply study the integers and deduce if they are potential permutation patterns or not. For example a sequence of the form

$$4\,6,$$

can never contribute to a common permutation pattern of size 2 since, in s_1 the two are not immediate neighbors. By the same argument, the subsequence

$$6\,4\,5$$

can potentially be a permutation pattern. Thus for this problem scenario, it suffices to store integer intervals, rather than Parikh-maps as was discussed in the previous section. In Figure 10.4, the interval 1-2 appears in all the last three sequences. Thus a common permutation pattern is

$$\{1, 2\} \text{ or } \{a, d\}.$$

Thus for a pair of integers $1 \leq i < j \leq n$, let

$$I = [i, j] = \{s[i], s[i+1], \ldots, s[j\text{-}1], s[j]\}.$$

Then the following check[3] yields I's potential to be a *permutation pattern*.

---

[3]Heber and Stoye call the difference between the maximum and minimum values the *interval defect*.

$$a\;d\;c\;b\;e \;\Rightarrow\; 1\;2\;3\;4\;5$$
$$c\;e\;d\;a\;b \;\Rightarrow\; 3\;5\;2\;1\;4$$
$$d\;a\;b\;e\;c \;\Rightarrow\; 2\;1\;4\;5\;3$$
$$b\;c\;e\;a\;d \;\Rightarrow\; 4\;3\;5\;1\;2$$

FIGURE 10.4: A collection of four strings, each defined on the alphabet $\{a, b, c, d, e\}$. The first string is encoded by consecutive integers from 1 to 5, giving the mappings $a \Leftrightarrow 1$, $d \Leftrightarrow 2$, and so on. Thus the remaining three strings are encoded as shown.

For each $1 < k \le K$ and each pair of integers, $1 \le i < j \le n$, if

$$\max(\pi_{k,i,j}) - \min(\pi_{k,i,j}) \ne (j - i),$$

then $I = [i...j]$ cannot be a permutation pattern, where

$$\pi_{k,i,j} = \Pi(s_k[i...j]).$$

In this section we focus on the following problem.

Problem 14 *(The intervals problem)* Given a sequence s on integers 1, 2, ..., n, $I = [i, j]$ is an interval if for some l,

$$\Pi(s[i...j]) = \{l, l+1, l+2, \dots, l+(j\text{-}i)\}.$$

The task is to find all such intervals I on s.

Given $s = 2\,1\,4\,5\,3$, the four intervals I_0, I_1, I_2, I_3 are:

$I_0 = [1, 5]$, marked on s as $\boxed{2\,1\,4\,5\,3}$.

$I_1 = [1, 2]$, marked on s as $\boxed{2\,1}\,4\,5\,3$.

$I_2 = [3, 4]$, marked on s as $2\,1\,\boxed{4\,5}\,3$.

$I_3 = [3, 5]$, marked on s as $2\,1\,\boxed{4\,5\,3}$.

Clearly, intervals are an alternative representation for permutation patterns as shown in the following example.

Example 3 Let $s = 3\,5\,2\,4\,7\,6\,8\,1$. Then

permutation patterns	intervals
$p_0 = \{1, 2, 3, 4, 5, 6, 7, 8\}$	$I_0 = [1, 8]$
$p_1 = \{2, 3, 4, 5, 6, 7, 8\}$	$I_1 = [1, 7]$
$p_2 = \{2, 3, 4, 5\}$	$I_2 = [1, 4]$
$p_3 = \{6, 7, 8\}$	$I_3 = [5, 7]$
$p_4 = \{6, 7\}$	$I_4 = [5, 6]$

How many intervals?   Consider

$$s = 12\,34\ldots n.$$

Clearly, each of

$$\{1,2\}, \qquad \{2,3\} \qquad \cdots \qquad\qquad \cdots \qquad\qquad \cdots \qquad \{n\text{-}1,n\},$$
$$\{1,2,3\}, \quad \{2,3,4\} \quad \cdots \qquad\qquad \cdots \qquad \{n\text{-}2,n\text{-}1,n\},$$
$$\{1,2,3,4\}, \{2,3,4,5\} \ldots \{n\text{-}3,n\text{-}2,n\text{-}1,n\},$$
$$\cdots$$
$$\{1,2,\ldots,n\},$$

is an interval. Thus given s of length n, it is possible to have

$$\mathcal{O}(n^2)$$

intervals.

10.5.1   The naive algorithm

Given an instance of Problem (14), this algorithm discovers all the intervals by a single scan of s from right to left.

At each scan i, we move a pointer j from $i+1$ up to n and check if $I = [i, j]$ is an interval. We keep track of the highest and the lowest value of I, and that is sufficient to check if it is an interval. Thus the interval checking can be done in $\mathcal{O}(1)$ time.

For $1 \le i < j \le n$, recall

$$[i, j] = \Pi(s[i \ldots j]).$$

Then we define the following:

1. $l(i, j) = \min[i, j]$ and $u(i, j) = \max[i, j]$,

2. $R(i, j) = u(i, j) - l(i, j)$ and $r(i, j) = j - i$,

3. $f(i, j) = R(i, j) - r(i, j)$

The following is rather a straightforward observation but critical in designing an efficient algorithm.

LEMMA 10.3
Let s be a sequence of n integers where each number appears exactly once. Then for all $1 \le i < j \le n$, the following statements hold.

1. *$f(i, j) \ge 0$. In other words, it cannot take negative values.*

2. *If $f(i, j) = 0$, then $[i, j]$ is an interval.*

Algorithm 7 *The Intervals Extraction*

```
(1) FOR i = n − 1 DOWNTO 1 DO
(2)     u ← s[i], l ← s[i]
(3)     FOR j = i + 1 TO n DO
(4)         IF s[j] > u, u ← s[j]
(5)         IF s[j] < l, l ← s[j]
(6)         f ← (u − l) − (j − i)
(7)         IF (f = 0), OUTPUT [i, j]
(8)     ENDFOR
(9) ENDFOR
```

10.5.1.1   Analysis of algorithm (7)

The algorithm has two loops:

1. Lines (1)-(9) is the main loop (that is executed $n − 1$ times), and

2. Lines (3)-(8) is the inner loop ($n − i$ times).

Lines (2), (4), (5), (6), (7) take $\mathcal{O}(1)$ time each. Lines (4), (5), (6), (7) are executed

$$1 + 2 + \ldots + (n − 2) + (n − 1) = \frac{n(n − 1)}{2}$$

times. Thus the entire algorithm takes

$$\mathcal{O}(n^2)$$

time.

Notice that the number of intervals in s could be $\mathcal{O}(n^2)$. Thus an algorithm that outputs all the intervals must do at least $\mathcal{O}(n^2)$ work. But what if s has only $\mathcal{O}(n)$ intervals, can we do better?

An algorithm whose time complexity is a function of the output size is called an *output sensitive algorithm*. Let N_O be the number of intervals in a string s of length $n = N_I$. We next describe an output sensitive algorithm that takes time

$$\mathcal{O}(N_O + N_I).$$

10.5.2   The Uno-Yagiura RC algorithm

One man's data structure is another man's algorithm, goes an old computer science adage. Here we discuss an algorithm that crucially depends on the linked list data structure that was briefly discussed in Section 2.6: this enables the algorithm to add elements to a list and access them in a LIFO (Last In First Out) order that makes the overall algorithm efficient.

This is an output sensitive algorithm which we call the Uno-Yagiura RC algorithm [UY00]. This follows the basic structure of Algorithm (7), but cuts down the number of candidates for the checking at lines (6) and (7). Hence, the authors Uno and Yagiura call this the Reduce Candidate (RC) algorithm. We give the pseudocode of the algorithm as Algorithm (8).

Algorithm 8 *The RC Intervals Extraction*

$(0\text{-}1)\,CreateList(n-1,n,s[n],L,NIL)$
$(0\text{-}2)\,CreateList(n-1,n,s[n],U,NIL)$

$(1)\,FOR\ i=n-1\ DOWNTO\ 1$
$(2)\qquad InsertLList(i,i+1,s[i],L)$
$(3)\qquad InsertUList(i,i+1,s[i],U)$
$(4)\qquad ScanpList(i,U,L)$
$(5)\,ENDFOR$

Algorithm 9 *The LIFO List Operations*

	$InsertUList(i,\ j,\ v,\ Hd)$
	$t\leftarrow Hd$
	$WHILE\ (t\neq NIL)\ \&\ (v>t.val)$
$CreateList(i,\ j,\ val,\ ptr,\ nxt)$	$\qquad t\leftarrow t.next$
$\quad NEW(ptr)$	$Create(i,j,v,Hd,t)$
$\quad ptr.i\leftarrow i,\ ptr.j\leftarrow j$	
$\quad ptr.val\leftarrow val$	$InsertLList(i,\ j,\ v,\ Hd)$
$\quad ptr.next\leftarrow nxt$	$\quad t\leftarrow Hd$
	$\quad WHILE\ (t\neq NIL)\ \&\ (v<t.val)$
	$\qquad t\leftarrow t.next$
	$Create(i,j,v,Hd,t)$

We describe the algorithm and its various aspects in the following five parts. We conclude with a concrete example.

1. Identification of *potent* indices.

2. Construction of two sequence of functions:

$$u(n-1,j),\ \ u(n-2,j),\ \ \ldots,\ \ u(2,j),\ \ u(1,j),\ \ \text{and}$$
$$l(n-1,j),\ \ l(n-2,j),\ \ \ldots,\ \ l(2,j),\ \ l(1,j).$$

Each function, $u(i,j)$ and $l(i,j)$ is defined over $j=i+1$ up to n.

3. **p**-list of potent indices.

4. Correctness of the RC algorithm.

5. Time complexity of the RC algorithm. Let N_O be the number of intervals in s.

    (a) The $u(\cdot, \cdot)$ function list U is processed in $\mathcal{O}(n)$ time.
    (b) The $l(\cdot)$ function list L is processed in $\mathcal{O}(n)$ time.
    (c) The list of potent indices **p**-list is processed in $\mathcal{O}(N_O)$ time.

1. Potent indices. We first identify certain j's (index) called *potent*.[4] For a fixed i, for some $i < j_p \leq n$, let

$$u_{j_p} = u(i, j_p) \text{ and } l_{j_p} = l(i, j_p).$$

j_p is *potent* with respect to (w.r.t.) i if and only if j_p is the largest possible j satisfying

$$u(i, j) = u_{j_p} \text{ and } l(i, j) = l_{j_p}.$$

Consider the following example where the input s is shown in bold below. Let $i = 2$.

$$
\begin{array}{c|ccccccc}
j & 1 & 2 & 3 & 4 & 5 & 6 & 7 \\
s[j] & \mathbf{2} & \mathbf{4} & \mathbf{3} & \mathbf{7} & \mathbf{6} & \mathbf{1} & \mathbf{5} \\
 & \uparrow & & & & \uparrow & \uparrow & \\
 & i & & & & j_1 & j_2 &
\end{array}
$$

1. Let $j_1(> i) = 5$. Is j_1 potent w.r.t. $i = 2$?
   To answer this, we compute the following:

$$u(2, 5) = 7 \text{ and } l(2, 5) = 3.$$

   Also, j_1 is the largest possible value of j with

$$u(2, j) = 7 \text{ and } l(2, j) = 3.$$

   Hence $j_1 = 5$ is potent w.r.t. $i = 2$.

2. Let $j_1(> i) = 6$. Is j_2 potent w.r.t. $i = 2$?
   We again compute the following:

$$u(2, 6) = 7 \text{ and } l(2, 6) = 1.$$

   But,

$$u(2, 7) = 7 \text{ and } l(2, 7) = 1.$$

   But $j = 7 > j_2$, hence $j_1 = 6$ is not potent w.r.t. $i = 2$.

---

[4]Uno and Yagiura in their paper use *unnecessary* j's, which in a sense is complementary to the idea of *potent* j. I define potent j's for a possible simpler exposition.

In other words, a j is potent if $[i', j]$ is potentially an interval, i.e., $f(i', j) = 0$, for some $i' \leq i$, .

LEMMA 10.4

If $[i, j]$ is an interval, then j must be potent w.r.t. i.

PROOF   It is easier to prove the contrapositive:

If $j > i$ is not potent w.r.t. i, then $[i, j]$ is not an interval.

Assume the contrary, i.e., $[i, j]$ is an interval. If j is not potent with respect to i, then there exists a $j' > j$ such that

$$u(i, j) = u(i, j') \text{ and } l(i, j) = l(i, j').$$

Then clearly

$$l(i, j) \leq s[j'] \leq u(i, j)$$

which leads to a contradiction. Hence the assumption must be wrong.   ▯

Thus, in conclusion, only the potent j's are sufficient to extract all the intervals in s. In the algorithm, **p**-list is the list of potent j's (in increasing value of the index).

We begin by studying some key properties of $u(i, j)$, $l(i, j)$ and $f(i, j)$ functions.

LEMMA 10.5
(Monotone functions lemma) *Let $i \geq 1$ be fixed and for $i < j_1 < j_2 \leq n$, the following hold.*

- **(U.1)** $u(\cdot, \cdot)$ *is a nonincreasing function, i.e., $u(i, j_1) \leq u(i, j_2)$.*

- **(L.1)** $l(\cdot)$ *is a nondecreasing function, i.e., $l(i, j_1) \geq l(i, j_2)$.*

It is straightforward to verify the statements and we leave this as an exercise for the reader (Exercise 108).

LEMMA 10.6

- **(F.1)** *Let $1 \leq i < j_1 < j_2 \leq n$. If*

$$f(i, j_1) > 0, \text{ and}$$
$$f(i, j_1) > f(i, j_2),$$

then $[i', j_1]$ is not an interval for any $1 \leq i' \leq i$.

- **(F.2)** *Let $1 \leq i_1 < i_2 < j_1 < j_2 \leq n$. Further, let the following hold.*

$$u(i_1, j_1) = u(i_2, j_1) \quad and \quad l(i_1, j_1) = l(i_2, j_1) \quad and$$
$$u(i_1, j_2) = u(i_2, j_2) \quad and \quad l(i_1, j_2) = l(i_2, j_2).$$

Then

$$f(i_1, j_1) - f(i_1, j_2) = f(i_2, j_1) - f(i_2, j_2).$$

PROOF   *Proof of statement (F.1):* Let

$$S(i, j) = \{k \mid l(i, j) \leq k \leq u(i, j)\},$$

then

$$f(i, j) = |S(i, j) \setminus [i, j]|,$$

i.e., $f(i, j)$ is the number of elements of $S(i, j)$ missing in $[i, j]$. Further since,

$$[i, j_1] \subset [i, j_2] \text{ and } f(i, j_1) > f(i, j_2),$$

then there must be some $j_1 < j'' \leq j_2$ such that

$$l(i, j_1) < s[j''] < u(i, j_1).$$

In other words, $s[j'']$ lies within the minimum and maximum values of the interval. Thus

$$l(i', j_1) \leq l(i, j_1) < s[j''] < u(i, j_1) \leq u(i', j_1),$$

and since

$$s[j''] \notin [i', j_1],$$

$[i', j_1]$ can never be an interval. This ends the proof of statement (F.1). We leave the proof of statement (F.2) as Exercise 109 for the reader.   ⬜

Figure 10.5 illustrates the facts of the lemma for a simple example. Notice that, for a fixed i (=1), the function $u(i, j)$ is nondecreasing and $l(i, j)$ is nonincreasing as j increases. The **p**-list of potent j's is shown at the bottom. We explain a few facts here.

1. $j = 3$ is potent since it is the largest j with

$$u(i, j) = 6 \text{ and } l(i, j) = 4.$$

2. $j = 5$ is potent since it is the largest j with

$$u(i, j) = 8 \text{ and } l(i, j) = 4.$$

3. By (F.1) of the lemma,

   (a) there is no interval of the form $[\cdot, 2]$ since

   $$(f(1, 2) = 1) > (f(1, 3) = 0), \text{ and}$$

   (b) there is no interval of the form $[\cdot, 4]$ since

   $$(f(1, 4) = 1) > (f(1, 5) = 0).$$

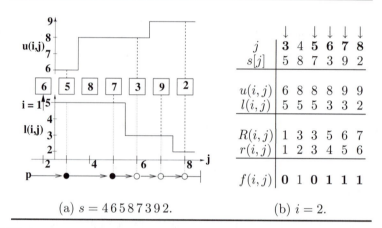

(a) $s = 4\,6\,5\,8\,7\,3\,9\,2.$	(b) $i = 2.$

FIGURE 10.5:   Illustration of Lemmas (10.5) and (10.6). (a) The input string s is shown in the center. The figure shows the snapshot of the $u(i, j)$ and $l(i, j)$ functions when index $i = 2$ pointing to 6 in s. As j goes from 3 to 8: (1) $u(i, j)$ is the nondecreasing function shown on top (U.1 of the lemma), (2) $l(i, j)$ is the nonincreasing function shown at the bottom (L.1 of the lemma), and (3) each of five potent indices ($j = 3, 5, 6, 7, 8$) of the **p**-list are shown as little hollow circles in the bottom row. Only two of the potent j's, $j = 3, 5$ evaluate $f(i, j)$ to 0. These are shown as dark circles. (b) The tabulated values of the functions. The potent j's are marked by arrows.

2. List of $u(\cdot, \cdot)$, $l(\cdot)$ functions. Consider the task of constructing the list of $u(\cdot, \cdot)$ and $l(\cdot)$ functions:

> For $i = (n-1)$, down to 1,
>   construct $u(i, j)$ and $l(i, j)$, for $i < j \le n$.

At iteration i, the function $u(i, j)$ and $l(i, j)$ is evaluated (or constructed for the algorithm). At i, a straightforward (say like that of the algorithm of Section 10.5.1) process scans the string from n down to i, taking $\mathcal{O}(i)$ time to compute $u(\cdot, \cdot)$ and $l(\cdot, \cdot)$. Since there are $n-1$ iterations and

$$1 + 2 + 3 + \ldots + (n-1) = \frac{(n-1)n}{2},$$

this task takes $\mathcal{O}(n^2)$ time for all the $n-1$ iterations.

The RC algorithm performs the above task in only $\mathcal{O}(n)$ time. This is done by a clever update at each iteration in the following manner.

1. The $u(\cdot, \cdot)$ and $l(\cdot, \cdot)$ function is stored as a list with the ability to add and remove from one of the list, called the *head* of the list. This is also called the Last In First Out (LIFO) order of accessing elements in a list.

   The algorithm maintains a U list to store values of $u(\cdot, \cdot)$ and an L list to store $l(\cdot, \cdot)$. However, only distinct elements are stored, along with the largest index j that has the value. Thus, if

   $$u(i, j-1) < u(i, j) = u(i, j+1) = \ldots = u(i, j+l) < u(i, j+l+1),$$

   for some l, then $s[j+l]$ is stored (along with the index $(j+l)$) in the list. By the same reasoning, $s[j-1]$ is stored (along with the index $(j-1)$) and is the head of the list if $i = j - 2$.

   For example consider the following segment of s and let $i = 2$.

j	2	3	4	5	6	7
$s[j]$	4	3	7	6	1	5
$u(2, j)$	$\rightarrow$ 7 $\rightarrow$	7 $\rightarrow$	7 $\rightarrow$	6 $\rightarrow$	5 $\rightarrow$	5
U		$\rightarrow$	$\boxed{7} \rightarrow$	$\boxed{6}$	$\longrightarrow$	$\boxed{5}$

   Note that U has only three elements. The head of the list points to element 7 (with index $j = 4$).

j	2	3	4	5	6	7
$s[j]$	4	3	7	6	1	5
$l(2, j)$	$\rightarrow$ 1 $\rightarrow$	1 $\rightarrow$	1 $\rightarrow$	1 $\rightarrow$	1 $\rightarrow$	5
L		$\longrightarrow$			$\boxed{1} \rightarrow$	$\boxed{5}$

   Note that L has only two elements. The head of the list points to element 1 (with index $j = 6$).

2. At each iteration, an element may be is added to the list (U or L or both), and zero, one or more consecutive elements may be removed in order from the head of the list.

This follows from Lemmas (10.7) and (10.8): the first deals with the U list and the second is an identical statement for the L list. The following can be verified and we leave the proof of these lemmas as an exercise for the reader.

LEMMA 10.7

For a fixed i, consider the two functions

(a) $u(i, j)$ *defined over $i < j \le n$, and*

(b) $u(i - 1, j)$ *defined over $(i - 1) < j \le n$.*

Then $u(i - 1, j)$ is defined in terms of $u(i, j)$ as follows:
If $s[i - 1] < u(i, i + 1)$, then

$$u(i - 1, j) = \begin{cases} u(i, i + 1) & \text{if } j = i \\ u(i, j) & \text{otherwise.} \end{cases}$$

If $s[i - 1] > u(i, i + 1)$, then

$$u(i - 1, j) = \begin{cases} s[i - 1] & \text{if } j = i \text{ or } s[j] > u(i, j) \\ u(i, j) & \text{otherwise.} \end{cases}$$

An identical result holds for the $l(\cdot, \cdot)$ function.

LEMMA 10.8

For a fixed i, consider the two functions

(a) $l(i, j)$ *defined over $i < j \le n$, and*

(b) $l(i - 1, j)$ *defined over $(i - 1) < j \le n$.*

Then $l(i - 1, j)$ is defined in terms of $l(i, j)$ as follows:
If $s[i - 1] > l(i, i + 1)$, then

$$l(i - 1, j) = \begin{cases} l(i, i + 1) & \text{if } j = i \\ l(i, j) & \text{otherwise.} \end{cases}$$

If $s[i - 1] < l(i, i + 1)$, then

$$l(i - 1, j) = \begin{cases} s[i - 1] & \text{if } j = i \text{ or } s[j] > l(i, j) \\ l(i, j) & \text{otherwise.} \end{cases}$$

3. p-list of potent indices. Note that the lists U and L are already sorted by j. Merging the two lists, gives the **p**-list or the list of potent j's. For example,

j	2	3	4		5		6	7
$s[j]$	4	3	7		6		1	5
U		$\longrightarrow$	$\boxed{7}$	$\to \boxed{6}$		$\longrightarrow$		$\boxed{5}$
L				$\longrightarrow$			$\boxed{1}$	$\to \boxed{5}$
p		$\longrightarrow$	$\bigcirc$	$\to \bigcirc$		$\to \bigcirc$	$\to$	$\bigcirc$

p has four elements with the head pointing to index $j - 4$.

By Lemma (10.6), there is no interval of the form $[i', j_1]$ if

$$f(i, j_1) > f(i, j_2).$$

Hence, **p**-list can be *pruned* by removing j_1 from the head of the list. We make the following claim:

> A **p**-list that is pruned only at the head of the list, possibly multiple times, is such that for any two consecutive indices, j_1 and j_2, in the pruned list,
>
> $$f(i, j_1) \le f(i, j_2).$$

This observation is crucial in asserting both the correctness and in the justification of the output-sensitive time complexity of the algorithm.

4. Correctness of the RC algorithm. The correctness of the algorithm follows from Lemma (10.4) and (F.1) of Lemma (10.6). The former ensures that if $[i, j]$ is an interval then j must be potent w.r.t. i. Note that the converse (i.e., if j is potent w.r.t. i, then $[i, j]$ must be an interval) is not true and the latter claim gives a way of trimming the potential indices. Simultaneously, it also gives an efficient way of doing so by ensuring that the pruned **p**-list is traversed only as long as the $f(\cdot)$ values on the consecutive indices are nondecreasing.

Since for every other potent j (w.r.t. i), $[i, j]$ is explicitly checked to see if it is an interval, the algorithm does not miss any interval and the output is correct.

In other words, every interval is of the form $[i, j]$ where j is potent w.r.t i (captured in the potent list) and the potent j's that are not intervals are pruned using (F.1). This also ensures that multiple intervals can be found only as consecutive elements at the head of the list.

5. Time complexity of the RC algorithm.

1. An element is added at most once to the list (U or L list) and removed exactly once from the list. Since the total number of elements is n and

only consecutive elements are removed (without having to traverse the list in search of elements to be removed), all the $n - 1$ iterations take

$$\mathcal{O}(n)$$

time, for each list.

2. The elements in the **p**-list may be accessed multiple times. However, as discussed in the previous paragraphs, **p**-list is traversed from the head of the list only to report intervals (without having to search the entire list for intervals), thus it takes

$$\mathcal{O}(N_O)$$

time where N_O is the number of intervals in the data.

Let $n = N_I$, the size of the input. Thus the algorithm takes time

$$\mathcal{O}(N_I + N_O).$$

6. Concrete example. Figure 10.6 gives a complete concrete example. When the input is a string s of length n, the index i scans the input from $n - 1$ down to 1. At each scan i, the algorithm updates two lists (the U and the L list) and processes for a potential j (shown as **p**-list) and emits all intervals of the form $[i, \cdot]$. Notice that in Algorithm (10) this was an inner loop (taking time $\mathcal{O}(n)$) and the new procedure is outlined as Algorithm (8).

It begins by initializing two lists, U to store $u(i, j)$ and L to store $l(i, j)$ in lines (0-1) and (0-2) respectively. The U list stores the triplet:

$$(i,\ j,\ u(i, j)),$$

and the L list stores the triplet:

$$(i,\ j,\ l(i, j)).$$

Both the lists are initialized to point to the triplet:

$$(n - 1,\ n,\ s[n]).$$

The initialization is shown in Figure 10.6(1). To avoid clutter, only the value $u(i, j)$ and $l(i, j)$ is shown in the U and L lists respectively, and i and j are shown separately to track the iterations.

The algorithm loops through lines (1) through (5). At each iteration or position i, the algorithm maintains the upper bound information $u(i, j)$ and the lower bound information $l(i, j)$ for each $i < j \leq n$ in the two lists U and L respectively. The two lists store only distinct elements as shown in the figure. Recall from Lemma (10.5) that for a fixed i, as j goes from $(i + 1)$ to n,

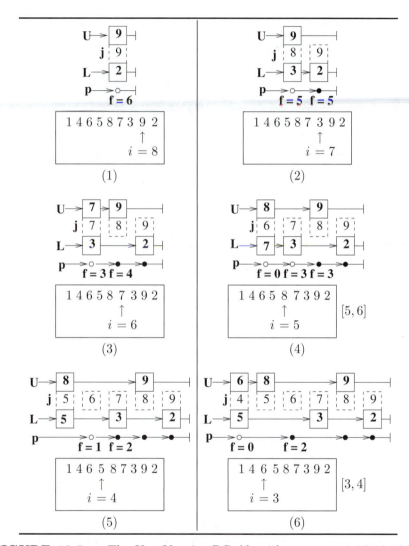

FIGURE 10.6: The Uno-Yaguira RC Algorithm on $s = 146587392$. The input is scanned right-to-left with the i index moving from 8 down to 1 as shown in each figure in (1)-(8). U is the u-list, L is the l-list and the of potent j's are shown in the **p**-list. Continued in Figure 10.7.

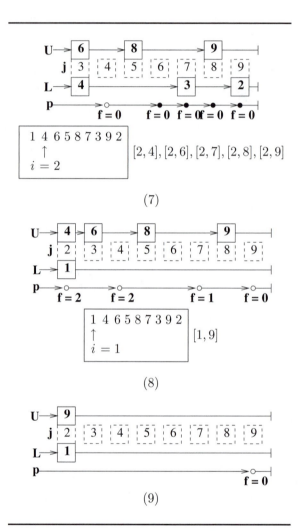

FIGURE 10.7:   Continued from Figure 10.6. Each potent j is shown by a circle: a hollow circles indicates a new potent j computed at that scan step and a solid j circle indicates an older potent j from earlier steps. The intervals are emitted at steps (4), (6), (7) and (8) are shown in the bottom row.

1. $u(i, j)$ is a nondecreasing function (U.1 of the lemma) and

2. $l(i, j)$ is a nonincreasing function (L.1 of the lemma).

Thus, by the monotonicity of functions $u(\cdot, \cdot)$ and $l(\cdot, \cdot))$, a new element in the list is only added at the *head* of the list. So this operation takes

$$\mathcal{O}(1)$$

time. The pseudocode for this operation is given in Algorithm (9) as *InsertLLList* and *InsertULList*.

$ScanpList(i, UHd, LHd)$ is an important component of the algorithm that 'scans the **p**-list': this checks for intervals and outputs them. We assume that routine also maintains the **p**-list of potent indices. Note that in practice **p**-list need not be explicitly maintained as a separate list but can be obtained by traversing the U and L lists. We describe the scanning process here through the concrete example and the algorithmic translation of this description is assigned as Exercise 111 for the reader.

In an attempt not to overwhelm the reader, we take up just a few scenarios from the concrete example to underline the essence of the approach. However, it is instructive to follow the example through in its entirety in Figures 10.6 and 10.7.

Scenario 1. When the scanning position is advanced from i to $(i - 1)$, the U and L lists are updated by inserting $s[i - 1]$ into the lists as shown in Figure 10.6, to maintain

1. U as a decreasing (to the left) list and

2. L as in increasing (to the left) list.

Thus if the new element $s[i - 1]$ is added to the list, it can only be the head of the list.

The list of potent j's, the **p**-list in Figure 10.6 can be computed from the U list and L list by traversing the two lists from the head and using the pair (i, j') where j' is the largest j such that

$$u(i, j) = u(i, j') \text{ or } l(i, j) = l(i, j').$$

For example, consider Figure 10.6(6). Here $i = 3$ and the four potent j's are:

1. $j = 4$ with $u(3, j) = 6$ and $l(3, j) = 5$,

2. $j = 6$ with $u(3, j) = 8$ and $l(3, j) = 3$,

3. $j = 8$ with $u(3, j) = 9$ and $l(3, j) = 3$ and

4. $j = 9$ with $u(3, j) = 9$ and $l(3, j) = 2$.

A potent j is *fresh* at iteration i, if it is computed during the iteration (scan) i and is shown as a hollow circle in the figure. The j's that are not fresh are shown as solid circles.

Only the first potent $j = 4$ is fresh and the other three had already been computed before and are shown as solid circles. The function $f(i,j)$ is computed for all the fresh potent j's only.

If $f(i,j) = 0$, the interval $[i,j]$ is emitted (as output). However,

$$\text{if } f(i,j_1) > f(i,j_2) \text{ for fresh potent } j\text{'s with } j_1 < j_2,$$

then by (F.1) of Lemma (10.6), the element j_1 is removed from the **p**-list, and also from U and L lists, if it belonged in these lists.

Scenario 2. When we encounter the first potent j', along the **p**-list satisfying

1. j' is not fresh, and

2. $f(i,j') \neq 0$,

then we stop the traversal of the list. This ensures that the list is being traversed only when an output is being emitted ($f(i,j) = 0$).

Consider Figure 10.7(7). At $i = 2$, potent $j = 4$ evaluates $f(i,j) = 0$, thus interval $[2,4]$ is emitted. Then the traversal of the **p**-list continues to $j = 6,7,8,9$ where in each case $f(i,j)$ evaluates to 0, thus further emitting the intervals

$$[2,6], \quad [2,7], \quad [2,8], \quad [2,9].$$

Thus the **p**-list is traversed only as long as an output is being emitted.

Scenario 3. Consider Figure 10.7(8). Here $i = 1$. The fresh potent j's are $2,4,7$ and 9. First

$$f(1,2), \quad f(1,4)$$

are each evaluated to be 2. Then $f(1,7)$ is computed to be 1, hence both $j = 4$ and then $j = 2$ are removed (by (F.1) of Lemma (10.6)): thus head of U list points to 8 and head of L list continues to point to 1.

Next $f(1,9)$ is evaluated to be 0, so the potent $j = 7$ is removed from the **p**-list and the head of U list is now 9 (head of L list continues to be 1). The interval $[1,9]$ is emitted and the U, L and p lists are as shown in Figure 10.7(9).

This concludes the description of the concrete example.

10.6   Intervals to PQ Trees

Here we discuss how to encode the intervals as PQ trees in time linear in the size of the interval. First we identify a special set of intervals called *irreducible* and then present the algorithm which uses this to give a very efficient algorithm.

10.6.1   Irreducible intervals

$(p \in \mathcal{P})$ is *reducible* if there exists $(p_1 \neq p), (p_2 \neq p) \in \mathcal{P}$ such that

$$p_1 \cap p_2 \neq \phi, \text{ and}$$
$$p_1 \cup p_2 = p.$$

A pattern $(p \in \mathcal{P})$ that is not reducible is *irreducible*. An interval $[i, j]$ is *reducible* if there exists

$$i < j_1 \leq j_2 < j \text{ such that } [i, j_2] \text{ and } [j_1, j]$$

are intervals. An interval that is not reducible is called *irreducible*.

Recall that patterns and intervals are two different representations of the same entity.

Example 4 *Let* $s = 3\,5\,2\,4\,7\,6\,8\,1.$ *Then*

irreducible permutation patterns	irreducible intervals
$p_0 = \{1, 2, 3, 4, 5, 6, 7, 8\}$	$I_0 = [1, 8]$
$p_1 = \{2, 3, 4, 5, 6, 7, 8\}$	$I_1 = [1, 7]$
$p_2 = \{2, 3, 4, 5\}$	$I_2 = [1, 4]$
$p_3 = \{6, 7, 8\}$	$I_3 = [5, 7]$
$p_4 = \{6, 7\}$	$I_4 = [5, 6]$

Note that in the example, $p_1 = p_2 \cup p_3$, but $p_2 \cap p_3 = \phi$, hence p_1 is irreducible.

Example 5 *Let* $s = 3\,5\,2\,4\,6\,7\,8\,1.$ *Then*

irreducible permutation patterns	irreducible intervals
$p_0 = \{1, 2, 3, 4, 5, 6, 7, 8\}$	$I_0 = [1, 8]$
$p_1 = \{2, 3, 4, 5, 6, 7, 8\}$	$I_1 = [1, 7]$
$p_2 = \{2, 3, 4, 5\}$	$I_2 = [1, 4]$
$p_3 = \{6, 7\}$	$I_3 = [5, 6]$
$p_4 = \{7, 8\}$	$I_4 = [6, 7]$

Note that

$$p = \{6, 7, 8\} \text{ (interval } [5, 7])$$

is not irreducible since

$$p = p_3 \cup p_4 \text{ with } p_3 \cap p_4 = \{7\}.$$

In other words, interval $[5, 7]$ is reducible since $[5, 6]$ and $[6, 7]$ are intervals.

Our next step is to design an algorithm that extracts the irreducible intervals. This algorithm is based on the Uno-Yaguira RC algorithm. We begin by identifying some special j's in the **p**-list:

$$j_{\min}^i \text{ is the minimum } j \text{ such that } f(i, j) = 0 \text{ and}$$
$$j_{\max}^i \text{ is the maximum } j \text{ such that } f(i, j) = 0.$$

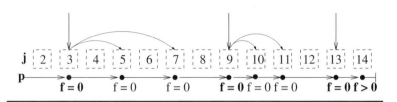

FIGURE 10.8:   Here $i = 1$ and the three irreducible intervals are shown by arrows at $j = 3$, $j = 9$ and $j = 13$ representing intervals $[1, 3]$, $[1, 9]$ and $[1, 13]$ respectively. Intervals $[1, 5]$ $[1, 7]$, $[1, 10]$ and $[1, 11]$ are not irreducible, however $f(1, j) = 0$, for $j = 5, 7, 10, 11$. While scanning **p** list for irreducible intervals of the form $[1, \cdot]$, the j's for which $f(i, j)$ is actually evaluated are $j = 3, 9, 13, 14$. The scanning terminates when $f(i, j) > 0$ (here at $j = 14$).

LEMMA 10.9
Let
$$1 \le i < j_1 < j < j_2 \le n.$$
Then if
$$[i, j_1] \text{ and } [j_1, j_2] \text{ are intervals,}$$
then
$$[i, j]$$
is not a irreducible interval.

PROOF   We first show that $[j_1, j]$ is an interval: the proof of this statement is not very difficult and left as an exercise for the reader (Exercise 115). Next, the interval $[i, j]$ cannot be irreducible, since there are two other intervals $[i, j_1]$ and $[j_1, j_2]$ that overlap and their union is $[i, j]$.   ☐

This simple observation helps design a very efficient algorithm to detect only the irreducible intervals. We explain this through an example. This is also termed the *ScanpListirreducible(·)* operation in Algorithm (10).

Consider Figure 10.8. The list of potent j's is

$$(3, 5, 7, 9, 10, 11, 13, 14)$$

marked by solid circles for $i = 1$. Scanning the **p**-list:

1. The head of the list, $j = 3$ evaluates $f(1, 3)$ to 0.
    $[1, j=3]$ is a irreducible interval since it is the smallest interval with $i = 1$.
    $j_{\min}^3 = 5$ and $j_{\max}^3 = 7$, shown by the curved segments in the figure, which had been computed in the previous steps.

2. The scanning of **p**-list now jumps to the element following $j^3_{\max} = 7$, which in this example is 9.

$f(1, 9)$ evaluates to 0.

Again $j^9_{\min} = 10$ and $j^9_{\max} = 11$, which had been computed before.

3. The scanning of **p**-list now jumps to the element following $j^9_{\max} = 11$, which here is 13.

$f(1, 13)$ evaluates to 0, but there are no intervals of the form $[13, \cdot]$.

4. So, the scanning continues to the next element on the list, 14.

$f(1, 14)$ evaluates to a nonzero value and the scanning stops.

Next, $j^1_{\min}$ is updated to 3 and $j^1_{\max}$ is updated to 13, for subsequent iterations.

Algorithm 10 *The Irreducible Intervals Extraction*

> *CreateList(n-1,n,s[n],LHd,NIL)*
> *CreateList(n-1,n,s[n],UHd,NIL)*
> $\Longrightarrow j^n_{\min} \leftarrow n, j^n_{\max} \leftarrow n$
>
> *FOR i = n − 1 DOWNTO 1 DO*
> *InsertLList(i,i+1,s[i],LHd)*
> *InsertUList(i,i+1,s[i],UHd)*
> $\Longrightarrow$ *ScanpListirreducible(i,UHd,LHd)*
> $\Longrightarrow$ *Update* $j^i_{\min}, j^i_{\max}$
> *ENDFOR*

To summarize, the RC intervals algorithm can be modified to compute the irreducible intervals and this is shown as Algorithm (10). The lines marked with right arrows on the left are the new statements introduced here.

The last paragraph summarized the *ScanpListirreducible(·)* routine, and the workings of the other routines are straightforward and are left as an exercise for the reader.

A complete example of computing irreducible intervals on $s = 3\,2\,4\,6\,5\,7\,8\,1\,9$ is shown in Figure 10.9.

Correctness of algorithm (10). The algorithm emits the same intervals as the RC algorithm except the ones suppressed by the scan *jumps*. This is straightforward to see from Lemma (10.9).

We first establish a connection between irreducible intervals and PQ trees. We postpone the analysis of the complexity of the algorithm to after this discussion.

10.6.2 Encoding intervals as a PQ tree

Consider a PQ tree T whose leaf nodes are labeled bijectively by the integers $1, 2, \ldots n$. For a node v let

$$I(v) = \{i \mid \text{the leaf node labeled by } i \text{ is reachable from node } v\}.$$

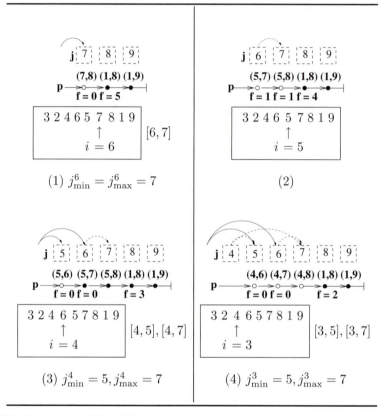

FIGURE 10.9:   The Heber-Stoye Algorithm on $s = 3\,2\,4\,6\,5\,7\,8\,1\,9$. The input is scanned right-to-left with index i and the upper and lower bound corresponding to each j is shown in round brackets. The irreducible intervals and the $j_{\min}^i, j_{\max}^i$ values at steps (3), (4) and (6) are also shown. See Figure 10.10 for continuation (for (5) and (6)) of the example and text for further details.

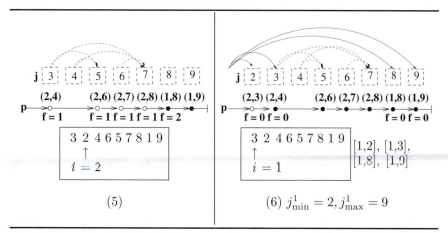

FIGURE 10.10: Continuation of the example of Figure 10.9.

1. For each P node v, the interval is given by $I(v)$.

2. Let v be a Q node with l children written as the sequence

$$v_1 v_2 \ldots v_l.$$

Then for $1 \leq j_1 \leq j_2 \leq l$,

$$I(v_{j_1, j_2}) = I(v_{j_1}) \cup I(v_{j_1+1}) \cup \ldots \cup I(v_{j_2}).$$

The set of intervals denoted by the Q node v are

$$\mathcal{I}(v) = \{I(v_{j_1, j_2}) \mid 1 \leq j_1 \leq j_2 \leq l\}.$$

Let V_1 be the set of P nodes and V_2 the set of Q nodes in T. Then the set of intervals encoded by this PQ tree T is:

$$\mathcal{I}(T) = \left(\bigcup_{v \in V_1} \{I(v)\} \right) \bigcup \left(\bigcup_{v \in V_2} \mathcal{I}(v) \right) \tag{10.5}$$

In this section we show that the collection of irreducible intervals can be organized as a PQ tree. Recall that any two irreducible intervals,

$$I_1 = [k_{11}, k_{12}], \quad I_2 = [k_{21}, k_{22}],$$

satisfy one of the following:

1. (contained or nested) $k_{21} \leq k_{11} < k_{12} \leq k_{22}$, without loss of generality, or,

2. (overlap) $k_{12} = k_{21}$, or,

3. (disjoint) $k_{11} < k_{12} < k_{21} < k_{22}$, without loss of generality.

Each leaf node of the PQ tree is labeled with an integer $1 \le i \le n$. The *overlapping* irreducible intervals can be arranged in the left-to-right order, since they overlap by a single element: this can be represented by a Q node whose children are ordered. If a Q node have l children, denoted as

$$j_1, j_2, \ldots, j_{l-1}, j_l,$$

then the irreducible intervals are

$$[j_1, j_2], [j_2, j_3], \ldots, [j_{l-1}, j_l].$$

The *contained* intervals can be represented by P nodes. Thus each P node corresponds to a irreducible interval I and $\Pi(I)$ is the set of leaf nodes reachable by this P node. Since $[1, n]$ is always an interval, we get a single (connected) PQ tree.

We next explore the construction of the PQ tree from the irreducible intervals. In fact, the irreducible interval algorithm (Algorithm (10)) can be modified to also produce the PQ tree representation of the interval, $[1, n]$.

The process involves constructing the PQ tree bottom-up. In the process more than one PQ tree may be under construction. At each stage, the root of each of the PQ tree under construction maintains information about the interval it represents, through pointers

u-ptr and *l-ptr*,

to the sequence being scanned. Note that it is adequate to maintain the pointers only of the roots. So when a node becomes the child of another node, the pointers become redundant and are removed.

This is best explained through an example illustrated in Figure 10.11 on

$$s = 8\,9\,1\,4\,6\,3\,5\,2\,7.$$

As i moves down from 9, the first irreducible intervals,

$$[4, 7], [4, 8] \text{ and } [4, 9]$$

are emitted at $i = 4$, which are shown as solid rectangles in Figure 10.11(1-3). Consider the irreducible intervals in increasing order of their sizes:

$$[4, 7], [4, 8] \text{ and } [4, 9].$$

1. First $[4, 7]$ is processed (Figure 10.11(1)). $j = 7$ has no pointers, $s[7]$ is collected as a child node. Further,

$$j = 6 \text{ down to } j = 4$$

have no pointers, so

$$s[6], \ s[5], \ s[4]$$

are collected as children. These 4 children are assembled together as a
P node as shown in (1).

$j = 7$ maintains a unidirectional pointer, *u-ptr*, to this P node and
$j = 4$ maintains a bidirectional pointer, *l-ptr*. Both are shown as dashed
curves in the figure.

In other words the interval spanned by the P node is captured through
the *u-ptr* and the *l-ptr*. The *u-ptr* of $j = 7$ and the *l=ptr* of $j = 4$ are
updated to point to the constructed P node as shown in (1).

2. Next consider irreducible interval $[4, 8]$ (Figure 10.11(2)). $j = 8$ has no
   pointers, so $s[8]$ is collected as a child node. But $j = 7$ has a *u-ptr*
   pointing to the P node which points to $j = 4$ via the bidirectional *l-ptr*.
   Hence the P node and $s[8]$ are collected as children.

   As there are only two elements a Q node is constructed with these two
   as children. This is shown in (2). The *u-ptr* of $j = 8$ and the *l=ptr* of
   $j = 4$ are updated to point to the constructed Q node as shown in (2).

3. Similarly irreducible interval

$$[4, 9]$$

is processed and is shown in Figure 10.11(3).

Next at $j = 1$, two irreducible intervals $[1, 2]$ and $[1, 9]$ are emitted. They
are also considered in the increasing order of their sizes.

1. First $[1, 2]$ is processed.

   $j = 2$ has no pointers and clearly $j = 1$ has no pointers either, so $s[2]$
   and $s[1]$ are collected as children. Since there are only two children, a
   Q node is constructed with these two as children as shown in (4).

   The *u-ptr* of $j = 2$ and the *l=ptr* of $j = 1$ are updated to point to the
   freshly constructed Q node.

2. Next $[1, 9]$ is processed.

   $j = 9$ has a *u-ptr* to a Q node that points to $j = 4$ via the *l-ptr*. So the
   Q node is assembled as a child.

   The next considered is $j = 3$ (to the immediate left of the *l-ptr* of the
   Q node). This has no pointers, $s[3]$ is assembled as a child.

   Next $j = 2$ (immediate left of $j = 3$) is considered. This has a *u-ptr*
   pointing to a Q node, whose *l-ptr* points to 1. Thus this Q node is
   assembled as a child and the scanning stops.

   Since there are three children a P node is constructed with these three
   children as shown in (4).

This completes the example. Figure 10.13 describes another example. Here
we illustrate a case when $j = 2$ at Figure 10.13(4) has a *l-ptr* (but no *u-ptr*). In this case $s[2]$ will be collected as a sibling, not child, as shown in
Figure 10.13(5).

Algorithm 11 *The PQ Tree Construction*

$constructPQTree(i, k, J[])$

$\qquad\qquad\qquad\qquad\qquad$ //$J[]$ *is a k-dim array; the irreducible*
$\qquad\qquad\qquad\qquad$ //*intervals are* $[i, J[1]], [i, J[2]], \ldots, [i, J[k]]$

$\quad FOR\ l = 1\ TO\ k\ DO$
$\qquad Tmp \leftarrow \phi,\ j \leftarrow J[l],\ sblng \leftarrow FALSE$
$\qquad WHILE\ j \neq i\ DO$
$\qquad\quad IF\ s[J[l]]$'s *u-ptr* $\neq NIL$
$\qquad\qquad$ *Place the node N, the ptr points to, in Tmp*
$\qquad\qquad$ *Let* j_t *be cell pointed to by l-ptr of N*
$\qquad\qquad j \leftarrow j_t - 1$

$\qquad\qquad$ $\boxed{Remove\ pointers\ of\ N}$ $\qquad\quad$ //*now they are redundant*

$\qquad\quad ELSEIF\ s[J[l]]$'s *l-ptr* $\neq NIL$
$\qquad\qquad j \leftarrow J[l] - 1$ $\qquad\qquad$ //$s[J[l]$ *must point to a Q node*
$\qquad\qquad$ *Let the Q node be* N_Q; $sblng \leftarrow TRUE$
$\qquad\quad ELSEIF$ $\qquad\qquad\qquad\qquad$ //*all pointers are NIL*
$\qquad\qquad$ *Create a leaf node* $s[J[l]]$
$\qquad\qquad$ *Add this node to Tmp*
$\qquad ENDWHILE$
$\qquad IF\ |Tmp| > 2\ create\ a\ P\ node\ T$
$\qquad\quad$ *The pointers in Tmp are made the children of T*
$\qquad ELSE\ create\ a\ Q\ node\ T$
$\qquad\quad$ *The pointers in Tmp are made the ordered children of T*
$\qquad IF\ sblng{=}TRUE\ make\ T\ the\ leftmost\ child\ of\ N_Q$
$\quad ENDFOR$

The correctness of the algorithm follows from the following lemma.

LEMMA 10.10
At every iteration, j has no more than 1 pointer. The pointer is either a u-ptr or a l-ptr.

PROOF Assume cell j has two *u-ptr*s, then there are two irreducible intervals of the form $[\cdot, j]$. Clearly one is contained in the other, hence must be a child (or descendent) of the other. By the step shown as a boxed statement in the pseudocode of Algorithm (11), the pointers of the child (or descendent) have been removed, leading to a contradiction. Similarly j can not have multiple *l-ptr*s.

Next assume that cell j has a *u-ptr* and an *l-ptr*. Then they must have a parent Q node and by the boxed statement of the algorithm, the children's pointers are removed, leading to a contradiction. □

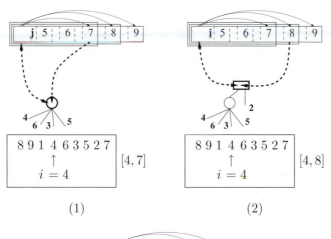

(1)   (2)

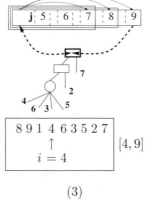

(3)

FIGURE 10.11:   The PQ Tree Algorithm on $s = 891463527$. See Figure 10.12 for continuation of this example and text for further details.

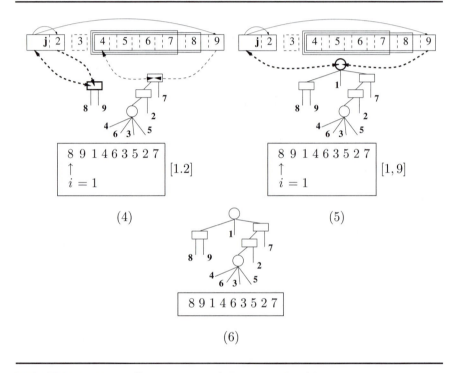

(4)   (5)

(6)

FIGURE 10.12:   Continuation of the example of Figure 10.11.

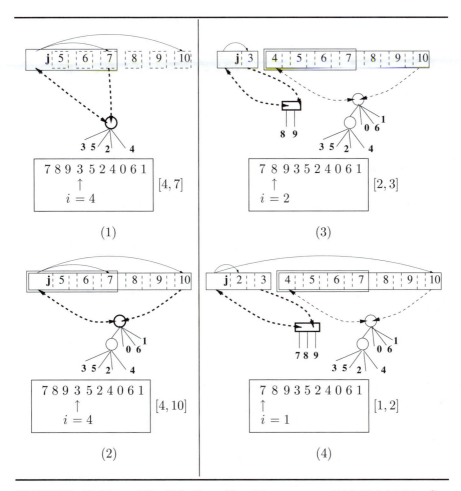

FIGURE 10.13: The PQ Tree Algorithm on $s = 7\,8\,9\,3\,5\,2\,4\,0\,6\,1$. See Figure 10.14 for the continuation of the example and see text for further details.

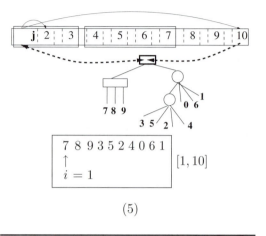

$$(5)$$

FIGURE 10.14:   Continuation of the example of Figure 10.13.

10.6.2.1   Time complexity of algorithms (10) and (11)

What is a irreducible interval good for? Firstly, the other (reducible) permutation patterns can be constructed from irreducible patterns and secondly there is only a small number of irreducible patterns. See Section 13.4 for an exposition on boolean closure. We leave the proof of the following as Exercise 116 for the reader.

THEOREM 10.2

(Irreducible intervals theorem)   *Consider s, a permutation of integers* $1, 2, \ldots, n$. *Let* $\mathcal{I}$ *be the set of all intervals on s and let M be the set of all irreducible intervals on s. Then the following statements hold.*

1. *M is the smallest set such that*

$$\mathcal{I} = B(M).$$

2. *M is unique. In other words there does not exist* M' *with* $|M'| \leq |M|$, *such that* $\mathcal{I} = B(M')$.

3. *The size of M is bounded by n, i.e.,* [5]

$$|M| < n.$$

---
[5] This was first proved by Heber and Stoye.

Is the bound of n on the size of M tight? We go back to the example at the beginning of this section (where we showed that $\mathcal{I} = \mathcal{O}(n^2)$):

$$s = 1\,2\,3\,4 \ldots n.$$

Then

$$M = \{\{1, 2\}, \{2, 3\}, \ldots, \{n-1, n\}\}.$$

Thus $|M| = (n-1)$ and the bound is tight.

A maximal permutation pattern is relevant in the context of multiply appearing characters or patterns that appear only in a subset (not necessarily all) of the collection of sequences. Now it is easy to see that both the algorithms take $\mathcal{O}(n)$ time. It is clear that Algorithm (10) is linear in the size of the output. Since the number of irreducible intervals is no more than n, the algorithm takes $\mathcal{O}(n)$ time.

It is easy to see in Algorithm (11) that each cell is scanned once. The number of internal nodes is bounded by n. Thus the algorithm takes $\mathcal{O}(n)$ time.

10.7 Applications

Genes that appear together consistently across genomes are believed to be functionally related: these genes in each others' neighborhood often code for proteins that interact with one another suggesting a common functional association. However, the order of the genes in the chromosomes may not be the same. In other words, a group of genes appear in different permutations in the genomes [MPN+99, OFD+99, SLBH00]. For example in plants, the majority of snoRNA genes are organized in polycistrons and transcribed as polycistronic precursor snoRNAs [BCL+01]. Also, the olfactory receptor(OR)-gene superfamily is the largest in the mammalian genome. Several of the human OR genes appear in cluster with ten or more members located on almost all human chromosomes and some chromosomes contain more than one cluster [GBM+01].

As the available number of complete genome sequences of organisms grows, it becomes a fertile ground for investigation along the direction of detecting gene clusters by comparative analysis of the genomes. A gene g is compared with its orthologs g' in the different organism genomes. Even phylogenetically close species are not immune from gene shuffling, such as in *Haemophilus influenzae* and *Escherichia Coli* [WMIG97, SMA+97]. Also, a multicistronic gene cluster sometimes results from horizontal transfer between species [LR96] and multiple genes in a bacterial operon fuse into a single gene encoding multi-domain protein in eukaryotic genomes [MPN+99].

If the function of genes say $g_1 g_2$ is known, the function of its corresponding ortholog clusters $g_2' g_1'$ may be predicted. Such positional correlation of genes as clusters and their corresponding orthologs have been used to predict functions of ABC transporters [TK98] and other membrane proteins [KK00].

Domains are portions of the coding gene (or the translated amino acid sequences) that correspond to a functional subunit of the protein. Often, these are detectable by conserved nucleic acid sequences or amino acid sequences. The conservation helps in a relative easy detection by automatic motif discovery tools. However, the domains may appear in a different order in the distinct genes giving rise to distinct proteins. But, they are functionally related due to the common domains. Thus these represent functionally coupled genes such as forming operon structures for co-expression [TCOV97, DSHB98].

Next we present two case studies: these were carried out mainly by Oren Weimann and discussed in [LPW05].

10.7.1   Case study I: Human and rat

In order to build a PQ tree for human and rat whole genome comparisons the output of a program called SLAM [ACP03] was used: SLAM is a comparative-based annotation and alignment tool for syntenic genomic sequences that performs gene finding and alignment simultaneously and predicts in both sequences symmetrically. When comparing two sequences, SLAM works as follows: Orthologous regions from the two genomes as specified by a homology map are used as input, and for each gene prediction made in the human genome there is a corresponding gene prediction in the rat genome with identical exon structure. The results from SLAM of comparing human (NCBI Build 31, November 2002) and rat (RGSC v2, November 2002) genomes, sorted by human chromosomes has been used in the following analysis. The data in every chromosome is presented as a table containing columns: *Gene name, rat coords, human coords, rat coding length, human coding length* and *# Exons.*

There were 25,422 genes predicted by SLAM, each gene appears exactly once in each of the genomes. Each one of the 25,422 genes is mapped to an integer, thus, the human genome becomes the identity permutation

$$1, 2, 3, \ldots, 25422,$$

and the rat genome becomes a permutation of

$$1, 2, 3, \ldots, 25422$$

obtained from the SLAM output table. The full mapping can be found in: http://crilx2.hevra.haifa.ac.il/~orenw/MappingTable.ps.

Ignoring the trivial permutation pattern involving all the genes, there are only 504 interesting maximal ones out of 1,574,312 permutation patterns in this data set. In Figure 10.15 a subtree of the Human-Rat whole genome PQ

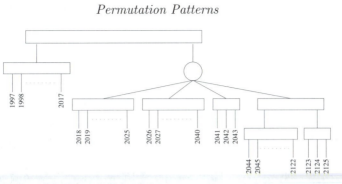

FIGURE 10.15: A subtree of the common maximal permutation pattern PQ tree of human and rat orthologous genes.

tree is presented. This tree corresponds to a section of 129 genes in human chromosome 1 and in rat chromosome 13. By the mapping, these genes appear in the human genome as the permutation:

$$(1997 - 2125)$$

and in the rat genome as the permutation:

$$(2043 - 2041, 2025 - 2018, 2123 - 2125, 2122 - 2044, 2040 - 2026, 2017 - 1997).$$

Another subtree of the Human-Rat whole genome PQ tree, corresponding to a section of 156 genes in human chromosome 17 and in rat chromosome 10 is

$$((21028 - 21061) - (21019 - 21027) - (21018 - 20906)).$$

The neighboring genes PMP22 and TEKTIN3 (corresponding to 21014 and 12015) are functionally related genes as explained in [BMRY04].

Figure 10.16 shows a few more common gene clusters of human and rat.

10.7.2 Case study II: *E. Coli K-12* and *B. Subtilis*

Here a PQ tree obtained from a pairwise comparison between the genomes of *E. Coli K-12* and *B. Subtilis* is discussed. The input data is from NCBI GenBank, in the form of the order of COGs (Clusters Of Orthologous Groups) and their location in each genome.

The data can be found in http://euler.slu.edu/~goldwasser/cogteams/data as part of an experiment discussed by He and Goldwasser in [HG04], whose goal was to find COG teams. They extracted all clusters of genes appearing in both sequences, such that two genes are considered neighboring if the distance between their starting position on the chromosome (in bps) is smaller than a chosen parameter $\delta > 0$. One of their experimental results, for $\delta = 1900$ was the detection of a cluster of only two genes: COG0718, whose product is an uncharacterized protein conserved in bacteria, and COG0353, whose product

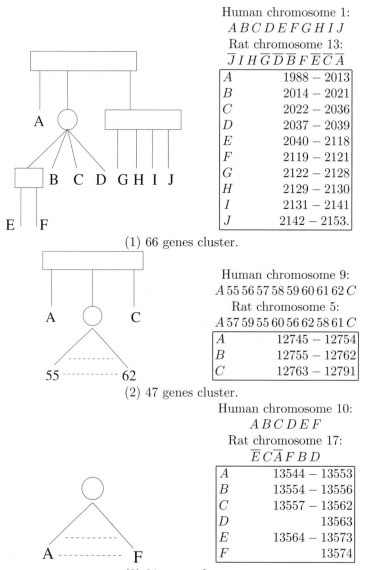

Human chromosome 1:
$$ABCDEFGHIJ$$
Rat chromosome 13:
$$\overline{JIHGD}\,\overline{B}\,F\,\overline{EC}\,\overline{A}$$

A	$1988 - 2013$
B	$2014 - 2021$
C	$2022 - 2036$
D	$2037 - 2039$
E	$2040 - 2118$
F	$2119 - 2121$
G	$2122 - 2128$
H	$2129 - 2130$
I	$2131 - 2141$
J	$2142 - 2153.$

(1) 66 genes cluster.

Human chromosome 9:
$$A\,55\,56\,57\,58\,59\,60\,61\,62\,C$$
Rat chromosome 5:
$$A\,57\,59\,55\,60\,56\,62\,58\,61\,C$$

A	$12745 - 12754$
B	$12755 - 12762$
C	$12763 - 12791$

(2) 47 genes cluster.

Human chromosome 10:
$$ABCDEF$$
Rat chromosome 17:
$$\overline{E}\,C\overline{A}\,F\,B\,D$$

A	$13544 - 13553$
B	$13554 - 13556$
C	$13557 - 13562$
D	13563
E	$13564 - 13573$
F	13574

(3) 31 genes cluster.

FIGURE 10.16: Examples of common gene clusters of human and rat. See text for details.

is a recombinational DNA repair protein. They conjecture that the function of COG0353 might give some clues as to the function of COG0718 (which is undetermined).

Here PQ trees of clusters of genes appearing in both sequences are built. Two genes are considered neighboring if they are consecutive in the input data irrespective of the distance between them. There are 450 maximal permutation patterns out of 15,000 permutation patterns.

DNA repair genes. Here we mention a particularly interesting cluster:

$$(COG2812 - COG0718 - COG0353).$$

The product of COG2812 is DNA polymerase III, which according to [BHM87] is also related to DNA repair. The PQ tree clearly shows that COG0718, whose function is undetermined is located between two genes whose function is related to DNA repair. This observation further contributes to the conjecture that the function of COG0718 might be also related to DNA repair. Note that the reason that COG2812 was not clustered with COG0718 and COG0353 in [HG04] is because the distance between COG2812 and COG0718 is 1984 ($> \delta = 1900$).

10.8 Conclusion

Although permutation patterns have been studied more recently than substring patterns, their usefulness can not be underestimated. The notion of maximality in this new context is particularly interesting since it provides a purely combinatorial way of cutting down on the output size without compromising on any information content. We end the chapter by reiterating the dramatic reduction in the output size simply by the use of maximality on two biological data sets.

	Number of all patterns	Number of maximal patterns
human & rat	$1,574,312$	504
E. Coli K-12 & B. Subtilis	$15,000$	450

More sophisticated models such as permutation patterns with fixed gaps [Par07b], patterns with gaps of bounded size are being studied with applications to phylogenetic studies [KPL06, Par06] and other interesting problems [LKWP05]. Again, the burning question continues to be:

> *How significant is the discovered permutation pattern?*

We address this in the next chapter.

10.9 Exercises

Exercise 99 (Maximality) *Prove that if p_2 is nonmaximal with respect to p_1, then $p_1 \subset p_2$. Is the converse true? Why?*

Exercise 100 (Multiplicity) *Is it possible that*

$$|\mathcal{L}_{p_1}| \neq |\mathcal{L}_{p_2}|,$$

when p_2 is nonmaximal with respect to p_1?

Hint: (1) Let quorum $K = 2$ and $s = abcdebca \ldots\ldots abcde$. Then consider

$$p_1 = \{d, e\} \text{ and } p_2 = \{a, b, c, d, e\}.$$

Is p_1 nonmaximal with respect to p_2? Observe that p_1 occurs only two times but p_2 occurs three times.

(2) Let $K = 2$ and $s = abcdbac \ldots\ldots abcabcd \ldots\ldots abcdabc$. Then consider

$$p_1 = \{a, b, c\} \text{ and } p_2 = \{a(2), b(2), c(2), d\}.$$

How many times does p_2 occur? p_1 occurs two times each in the first and third occurrence of p_2, and, four times in the second occurrence of p_2. Is p_1 nonmaximal with respect to p_2?

Exercise 101 *Consider the permutation patterns shown in Figure 10.1. Which of these are maximal? Give the PQ tree representation of the maximal patterns.*

Hint: $p = a\text{-}b\text{-}c\text{-}d\text{-}e$ occurs at locations 1, 6 and 11 on the input s. Can every other pattern be deduced from p?

Exercise 102 (Multiple PQ trees) *Let $p \in P_{>1}$ be maximal and be defined as*

$$p = \{\sigma_1(c_1), \sigma_2(c_2), \ldots, \sigma_l(c_l)\},$$

occurring in K' locations. Assuming the elements of p do not occur elsewhere in the input, how many PQ trees may be required to represent all the nonmaximal patterns? Discuss.

Hint: Does the following PQ tree capture all the nonmaximal patterns of p given by Equation (10.4) in Section 10.3.2?

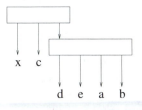

x   c   d e a b

Note that only one leaf node is labeled with c, although the multiplicity of c is 2 in p. In the following example, a single PQ tree cannot represent the two nonmaximal patterns $\{a, b, c, d\}$ and $\{c, d, e, f\}$ in the two occurrences:

$$o_1 = c\ b\ d\ \overline{a\ g\ e}\ \overline{f\ d\ c},$$

$$o_2 = d\ g\ c\ \overline{a\ b}\ \overline{c\ d\ e\ f}.$$

However, in general is the problem well-defined? Note that a universal PQ tree, i.e., a PQ tree with a single P node as the root node, represents all possible permutations of the elements of p.

Also, in general, a PQ tree represents more 'occurrences' than it actually encodes for. Can this 'excess' be minimized? This leads to the specification of the *Minimal Consensus PQ Tree Problem*.

Exercise 103 *Recall p of Equation (10.4):*

$$p = \{a, b, c(2), d, e, x\}.$$

p has exactly three occurrences on the input given as

$$o_1 = d\ e\ a\ b\ c\ x\ c,$$
$$o_2 = c\ d\ e\ a\ b\ x\ c,$$
$$o_3 = c\ x\ c\ b\ a\ e\ d.$$

Enumerate all the nonmaximal patterns with respect to p, assuming they do not occur elsewhere in the input.

Exercise 104 *Can the running time of the Parikh Mapping-based algorithm discussed in this chapter be improved?*

Hint: Is it possible to reduce factor $(\log t)^2$ to $\log t$ in the time complexity? Recall that the tags are assigned in increasing order. Let at stage j, t_j be the largest assigned integer to a tag. If the newly encountered tag (t'_1, t'_2) is such that, $t'_1, t'_2 \leq t_j$, then the first of the tag pair can be stored in an array and directly accessed in $\mathcal{O}(1)$ time reducing one of the $\mathcal{O}(\log t)$ factors to $\mathcal{O}(1)$.

This can be made possible if all the entries in the Ψ array are known in advance. This can be simply done by a linear scan of the input with the L-sized window and recording all the distinct numbers in Ψ that are generated by the L-sized window. Let the largest number encountered be t^*_0. Note that $t^*_0 \leq L$, by the choice of the window size. Then the tag values are assigned starting with t^*_0, thus every new number t_{new} encountered is such that $t_{new} \leq t^*_0$.

Exercise 105 (Algorithm generalization) *In some applications, the multiplicity of the permutation pattern is not used, thus a permutation pattern*

$$p = \{a, b, c\}$$

could have an occurrence o given

$$o = a\, b\, b\, c\, a.$$

In such cases, only

$$\Pi(o) = \{a, b, c\}$$

is of consequence. Under this condition Ψ is a binary *vector and the only possible pairs at level 0 of the naming tree are*

$$(0, 0), (0, 1), (1, 0), (1, 1).$$

How is the Parikh Mapping-based algorithm improved based on this simplifying assumption?

Exercise 106 *What is the maximum number of maximal permutation patterns in a sequence of length n with quorum $k = 3$? Give arguments to support your claim.*

Exercise 107   1. *Refer to the definitions in Section 10.5.1. Is it possible that*

$$l(i, j) = u(i, j)$$

for some $j > i$? Why?

2. *Given s and a fixed i, show that if $j(> i)$ is not potent with respect to i, then $[i', j]$ is not an interval for all $i' \leq i$.*

Hint: (1) Are all elements of s distinct? (2) See the proof of Lemma (10.4).

Exercise 108 (Monotone functions) *Prove statements (U.1),(L.1)of Lemma (10.5).*

- **(U.1)** *For $1 \leq i < n$, $u(i,j)$ is a nondecreasing function over $j = (i+1)$ up to n.*

- **(L.1)** *For $1 \leq i < n$, $l(i,j)$ is a nonincreasing function over $j = (i+1)$ up to n.*

Hint: Use proof by contradiction.

Exercise 109 *Give arguments to show that statement (F.2) of Lemma (10.6) is equivalent to the following statement: If*

$$u(i{-}1,j_1) = u(i,j_1) \quad and \quad l(i{-}1,j_1) = l(i,j_1) \quad and$$
$$u(i{-}1,j_2) = u(i,j_2) \quad and \quad l(i{-}1,j_2) = l(i,j_2),$$

for

$$1 < i < j_1 < j_2 \leq n,$$

then

$$f(i-1,j_1) - f(i-1,j_2) = f(i,j_1) - f(i,j_2).$$

Prove the above statement or (F.2).

Exercise 110 *Prove the following statements.*

- **(U.2)** *Let $1 < i < j_1 < j_2 \leq n$.. If*

$$u(i,j_1) < u(i,j_2), \quad and$$
$$u(i-1,j_1) = u(i-1,j_2),$$

  then, $f(i',j_1) > 0$ for $1 \leq i' < j_1$.

- **(L.2)** *Let $1 < i < j_1 < j_2 \leq n$.. If*

$$l(i,j_1) > l(i,j_2), \quad and$$
$$l(i-1,j_1) = l(i-1,j_2),$$

  then, $f(i',j_1) > 0$ for $1 \leq i' < j_1$.

Hint: Use proof by contradiction.

Exercise 111 *Refer to the algorithm discussion in Section 10.5.2.*

1. *Enumerate the steps involved in moving from $i = 2$, shown in Figure 10.5, to $i = 1$ shown here.*

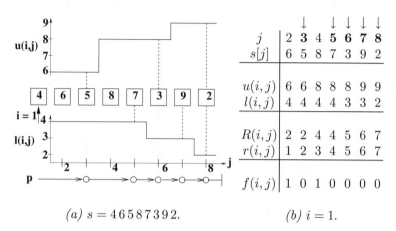

(a) $s = 46587392$.   (b) $i = 1$.

2. *Give a pseudocode for the algorithm of Section 10.5.2 assuming **p**-list is stored explicitly.*

3. *Give a pseudocode for the algorithm of Section 10.5.2 assuming **p**-list is computed on-the-fly from the U and L lists.*

Exercise 112 *Give a pseudocode description, along the lines of the subroutines in Algorithm (9), of the following three routines:*

1. *deleteU List(i, j, v, Hd),*

2. *deleteLList(i, j, v, Hd), and,*

3. *ScanpList(i, j, v, Hd).*

Exercise 113 *Let s of length n be such that each element is distinct and*

$$\Pi(s) \subset \{1, 2, \ldots, N\}$$

and $1, N \in \Pi(s)$, for some $N > n$. Does Algorithm (8) work for this input s?

Exercise 114 *(**Monge array**) An $m \times n$ matrix M, is said to be a Monge array if, for all i', i, j, j' such that*

$$1 \le i' < i \le m, \text{ and } 1 \le j < j' \le n,$$

the following holds (called the Monge property):

$$M[i', j] + M[i, j'] \ge M[i, j] + M[i', j'].$$

1. *For the upper-diagonal matrices below, does the Monge Property hold (when ever the matrix elements are defined)?*

		j'		j					
6	6	7	7	9	9	9	9	9	9
	4	7	7	9	9	9	9	9	9
		7	7	9	9	9	9	9	9
			2	9	9	9	9	9	9
				9	9	9	9	9	9
					1	8	8	8	
						8	8	8	
							3	5	
								5	

i' (at row 4), i (at row 6)

(a)M_u.

	j'		j					
6	4	4	2	2	1	1	1	1
	4	4	2	2	1	1	1	1
		7	2	2	1	1	1	1
			2	2	1	1	1	1
				9	1	1	1	1
					1	1	1	1
						8	3	3
							3	3
								5

i', i

(b)M_l.

	j'		j					
0	1	1	2	1	3	2	1	0
	0	2	3	4	4	3	2	1
		0	4	5	5	4	3	2
			0	6	6	5	4	3
				0	7	6	5	4
					0	6	5	4
						0	4	3
							0	1
								0

i', i

(c)M_f.

2. *For*

$$1 \leq i' < i < j < j' \leq n$$

show that

$$f(i', j) + f(i, j') \geq f(i, j) + f(i', j').$$

Hint: 1. A row i in M_u is $u(i, j)$, in M_l is $l(i, j)$, and in M_f is $f(i, j)$, for $j \geq i$ for some sequence s.
2. Show that

$$u(i', j) + u(i, j') \geq u(i, j) + u(i', j') \text{ and}$$
$$l(i', j) + l(i, j') \leq l(i, j) + l(i', j').$$

Exercise 115 *Let $[i, j_1]$ and $[j_1, j_2]$ be intervals. If*

$$[i, j], \quad j_1 < j < j_2,$$

is an interval, then show that $[j_1, j]$ is an interval.

Hint: Enumerate the cases possible and explore each case.

Exercise 116 (Irreducible intervals) *Let $\mathcal{I}$ be the set of all intervals on some s of length n. Let M be the the smallest set such that*

$$\mathcal{I} = B(M).$$

Then show the following statements hold.

1. M is the set of irreducible intervals, and

2. M is unique, and

3. $|M| < n$.

Hint: (1) Use the reduced partial order graph $G(\mathcal{I}, E_r)$.
(2) Use proof by contradiction.
(3) Use the PQ tree structure to prove the result. Consider a PQ tree T that uniquely captures the given permutations. Each Q node with k children can be replaced by a stack of $(k-1)$ nodes (shown as solid nodes below in (c)) to give a transformed PQ tree T'. An irreducible pattern corresponds to a node in this transformed tree. The number of internal nodes is bounded by the number of leaf nodes in a tree.

In the following figure (a) shows the PQ tree capturing Example (4). Each internal node corresponds to a irreducible pattern. (b) shows the PQ tree capturing Example (5) and (c) shows the transformed PQ tree of (b) where each dark node encodes a irreducible pattern. Each internal node corresponds to a irreducible pattern.

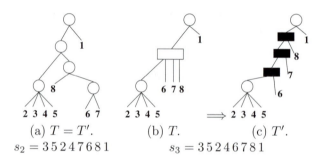

(a) $T = T'$.   (b) T.   (c) T'.
$s_2 = 3\,5\,2\,4\,7\,6\,8\,1$ $s_3 = 3\,5\,2\,4\,6\,7\,8\,1$

Exercise 117 *Consider Algorithm (10). If the scanning of the input is switched to left-to-right (instead of right-to-left as in the current description), does the algorithm emit the same irreducible intervals? Why?*

Exercise 118 *(Compomers) [Bö4] Consider a left-to-right ordered DNA sequence*

$$s = 0\,A\,C\,C\,G\,T\,T\,1$$

where symbols 0 and 1 denote the leftmost and rightmost end respectively of the sequence. If one or more C is removed from s, the following fragments arise:

$$f_1 = 0\,A, \ f_2 = C\,G\,T\,T\,1, \ f_3 = 0\,A\,C, \ f_4 = G\,T\,T\,1.$$

Here

$$\Pi'(f_1) = \{0, A\},$$
$$\Pi'(f_2) = \{C, G, T(2), 1\},$$
$$\Pi'(f_3) = \{0, A, C\},$$
$$\Pi'(f_4) = \{G, T(2), 1\}.$$

Similarly, one or more A, G and T can be cleaved giving rise to more fragments.

Assume an assay DNA technology (MALDI-TOF mass spectrometry [BÜ4]), that reads only $\Pi'(f)$ (also called a compomer*) for each fragment f. In this example, the complete collection of compomers is as follows:*

1. *(cleaved by C):* $\{0, A\}$, $\{C, G, T(2), 1\}$, $\{0, A, C\}$, $\{G, T(2), 1\}$,

2. *(cleaved by A):* $\{C(2), G(T)2, 1\}$,

3. *(cleaved by G):* $\{0, A, C(2)\}, \{T(2), 1\}$,

4. *(cleaved by T):* $\{0, A, C(2), G\}$, $\{T, 1\}$, $\{0, A, C(2), T, G\}$, $\{1\}$

Is it possible to reconstruct the original s from this collection of compomers?

Exercise 119 *(Local alignment of genomes)* *[OFG00] The local alignment of nucleic or amino acid sequences, called the* multiple sequence alignment *problem, is based on similar subsequences; however the local alignment of genomes is based on detecting locally conserved gene clusters. For example the chunk of genes*

$$g_1 g_2 g_3$$

may be aligned with

$$g_3' g_1' g_2'.$$

Such an alignment is never detected in subsequence alignments.

Give a formal definition of the local alignment of genomes problem and discuss a method to solve it.

Hint: Define a measure of gene similarity (to identify gene orthologs) to define the alignment problem.

Exercise 120 (Common connected components) *Consider the following application: In metabolic pathways different dependencies might exist with the enzymes, metabolites, proteins and so on whereas the players may still be the same. In other words, different organisms or tissue cultures may give evidence of different relations. Thus 'permutation patterns' on metabolic pathway networks yields a collection of enzymes and proteins that could possibly have different pathways.*

1. The task is to formulate the problem that can address the above.

2. Consider the following two examples. What are the common permutations (clusters) in both the graphs in (a) and (b)?

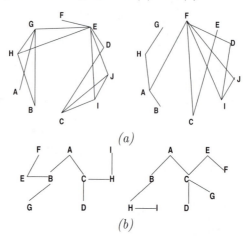

(a)

(b)

Hint: 1. This is a problem of finding permutations (or intervals) on graphs. The problem can be formalized as follows:

Common Connected Component Problem: Given n graphs

$$G(V, E_i), \quad 1 \leq i \leq n,$$

and a quorum $K > 1$ and a size m, the problem is to find all the maximal

$$V' \subseteq V \text{ with } |V'| \geq m,$$

such that for at least $k \geq K$ graphs,

$$G(V, E_{i_1}), G(V, E_{i_2}), \dots, G(V, E_{i_k}),$$

each induced graph, $1 \leq j \leq k$ is connected

$$G(V', (E'_{i_j} \subseteq E_{i_j})).$$

To date, the following best addresses this problem.

Partitive families and decomposition trees   [MCM81, CR04]: A family of subsets of a finite set V is *partitive* if and only if

a. it does not contain the empty set, but contains the trivial subsets of V (singletons and V itself), and

b. the union, intersection, and differences of any pair of *overlapping* subsets is in the family. (two subsets of a same set overlap if they intersect but neither is a subset of the other)

Although partitive families can be quite large (even exponentially large), they have a compact, recursive representation in the form of a tree, where the leaves are the singletons of V, namely, the *decomposition tree*:

THEOREM 10.3
(Decomposition theorem) *[MCM81] There are exactly three classes of internal nodes in a decomposition tree of a partitive family.*

  a. *A Prime node is such that none of its children belongs to the family, except for the node itself.*

  b. *A Degenerate node is such that every union of its children belongs to the family.*

  c. *A Linear node is given with an ordering on its children such that a union of children belongs to the family if and only if they are consecutive in the ordering.*

2.   The solutions to the two examples are shown below encoded as an annotated PQ tree. The P node has the same interpretation as in the regular PQ tree and the Q node is actually a graph as shown below. The set of leafnodes reachable from any subgraph of a graph of the Q node and a hollow P node is a solution to the problem.

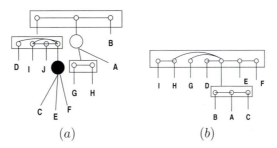

(a) (b)

Comments

I particularly like this chapter since it is a nice example of the marriage of elegant theory and useful practice. Usually, the idea of maximality of patterns is very important and in the case of permutation patterns, it is also nonobvious. Further, it beautifully fits in with the PQ tree data structure and Parikh mapping, both having been well studied independently in literature.

Chapter 11

Permutation Pattern Probabilities

> *Probability is like gravity,*
> *you can't fight it.*
> - Sonny in *Miami Vice*

11.1 Introduction

Just as it is reasonable to compute the odds of seeing a string pattern in a random sequence, so is the case with permutation patterns. We categorize permutation patterns as (1) unstructured and (2) structured.

The former usually refers to the case where these patterns (or clusters) are observed in sequences, usually defined on fairly large alphabet sets.

The structured permutations refer to PQ trees, that is the encapsulation of the common internal structure across all the occurrences of the permutation pattern. The question here is regarding the odds of seeing this structure (as a PQ tree) in a random sequence.

11.2 Unstructured Permutations

Consider the following problem: *What is the chance of seeing an n-mer consisting of*

i_1 *number of A's,*

i_2 *number of C's,*

i_3 *number of G's and*

i_4 *number of T's,*

with $i_1 + i_2 + i_3 + i_4 = n$,

in a random strand of DNA?

This is a pattern p where the order does not matter (also called a *permutation pattern* in Chapter 10) and is written as

$$p = \{A(i_1), C(i_2), G(i_3), T(i_4)\}.$$

We make the simplifying assumption that the multiple occurrences of p do not overlap.

We construct the discrete probability space (see Section 3.2.1 for the notation used here)

$$(\Omega, 2^\Omega, M_P)$$

as follows.[1] Let Ω be the set of all possible n-mers on $\{A, C, G, T\}$. Let

$$\omega_{i_1, i_2, i_3, i_4}$$

be an n-mer with

$\quad i_1$ number of A's,

$\quad i_2$ number of C's,

$\quad i_3$ number of G's and

$\quad i_4$ number of T's

and let p_X be the probability of occurrence of X where

$$X = A, C, G \text{ or } T$$

with

$$p_A + p_C + p_G + p_T = 1.$$

Then the probability measure function

$$M_P : \Omega \rightarrow \mathbb{R}_{\geq 0},$$

is defined as follows:

$$M_P(\omega_{i_1, i_2, i_3, i_4}) = \left(\frac{(i_1 + i_2 + i_3 + i_4)!}{i_1! \, i_2! \, i_3! \, i_4!} \right) (p_A)^{i_1} (p_C)^{i_2} (p_G)^{i_3} (p_T)^{i_4}$$

$$= \left(\frac{n!}{i_1! \, i_2! \, i_3! \, i_4!} \right) (p_A)^{i_1} (p_C)^{i_2} (p_G)^{i_3} (p_T)^{i_4}.$$

In particular, if the four nucleotides, A, C, G, T are equiprobable, then the formula simplifies to

$$M_P(\omega_{i_1, i_2, i_3, i_4}) = \left(\frac{n!}{i_1! \, i_2! \, i_3! \, i_4!} \right) \left(\frac{1}{4^n} \right).$$

---

[1]See Chapter 3 for the definitions of the terms.

How do we get this formula? And, does it satisfy the probability mass conditions (see Section 3.2.4)?

To address these curiosities, we pose the following general question where we use m instead of 4.

What is the number of distinct strings where each has exactly i_1 number of x_1's, i_2 number of x_2's, ..., i_m number of x_m's?

This is not a very difficult computation, but we also need to show its relation to a probability mass function. Hence we take a 'multinomial' view of the problem: It turns out that this number is precisely the *multinomial coefficient* in combinatorics. This is one of the easiest ways of computing this number and we study that in the next section. The summary of the discussion is as follows.

1. $M_P(\omega_{i_1,i_2,i_3,i_4})$ is computed from the multinomial coefficient (divided by m^n), and

2. $P(\Omega) = 1$ follows from Equation (11.3).

If

$$i_1 + i_2 + \ldots + i_m = n,$$

then each string is of length n. As an example, let

$$m = 2, n = 3 \text{ and } i_1 = 1.$$

Then $i_2 = 2$ and there are only three distinct strings given as

$$\omega_1 = x_1 x_2 x_2,$$
$$\omega_2 = x_2 x_1 x_2 \text{ and}$$
$$\omega_3 = x_2 x_2 x_1.$$

11.2.1 Multinomial coefficients

We briefly digress here to recall the *multinomial formula* which is the expansion of

$$(x_1 + x_2 + \ldots + x_m)^n.$$

Let Ψ be an m-dimensional array of nonnegative integers [2] such that

$$\sum_{i=1}^{m} \Psi[i] = n.$$

Let $Sig(m, n)$ be the set of all possible signatures for the given m and n. For instance,

$$Sig(2, 2) = \{[2,0], [1,1], [0,2]\}.$$

---

[2] This is also called the *Parikh vector* and is discussed in Chapter 10.

Further, let $\Psi[index]$ be denoted by i_{index}. Thus

$$\text{if } \Psi = [1,0], \text{ then } i_1 = 1 \text{ and } i_2 = 0.$$

Relating the terms to the discrete probability space, there is an injective mapping

$$I : \Omega \rightarrow Sig(m,n),$$

where

$$I(\omega_{i_1,i_2,\ldots,i_m}) = (\Psi = [i_1, i_2, \ldots i_m]).$$

For $m > 0$ and $n \geq 0$, the following can be verified (with some patience):

$$(x_1+x_2+\ldots+x_m)^n = \sum_{\Psi \in Sig(m,n)} \left(\frac{n!}{i_1!\, i_2!\, \ldots i_m!} \right) x_1^{i_1} x_2^{i_2} \ldots x_m^{i_m} \qquad (11.1)$$

The number

$$\left(\frac{n!}{i_1!\, i_2!\, \ldots i_m!} \right), \qquad (11.2)$$

corresponding to each Ψ, is called the *multinomial coefficient*. Let this number be denoted by $MC(\Psi)$. See the following three cases as illustrative examples:

$$m = 2 \text{ and } n = 2$$

$(x_1 + x_2)^2 = x_1^2 + 2x_1x_2 + x_2^2$		
Ψ	$MC(\Psi)$	strings
$[2,0]$	1	x_1x_1
$[1,1]$	2	$x_1x_2,\ x_2x_1$
$[0,2]$	1	x_2x_2

$$m = 2 \text{ and } n = 3$$

$(x_1 + x_2)^3 = x_1^3 + 3x_1^2x_2 + 3x_1x_2^2 + x_2^3$		
Ψ	$MC(\Psi)$	strings
$[3,0]$	1	$x_1x_1x_1$
$[2,1]$	3	$x_1x_1x_2,\ x_1x_2x_1,\ x_2x_1x_1$
$[1,2]$	3	$x_2x_2x_1,\ x_2x_1x_2,\ x_1x_2x_2$
$[0,3]$	1	$x_2x_2x_2$

$$m = 3 \text{ and } n = 2$$

$(x_1 + x_2 + x_3)^2 = x_1^2 + x_2^2 + x_3^2 + 2x_1x_2 + 2x_2x_3 + 2x_1x_3$		
Ψ	$MC(\Psi)$	strings
$[2,0,0]$	1	x_1x_1
$[0,2,0]$	1	x_2x_2
$[0,0,2]$	1	x_3x_3
$[1,1,0]$	2	$x_1x_2,\ x_2x_1$
$[0,1,1]$	2	$x_2x_3,\ x_3x_2$
$[1,0,1]$	2	$x_1,x_3,\ x_3x_1$

It follows from Equation (11.1), by setting

$$x_1 = x_2 = \ldots = x_m = 1,$$

that for a given m and n, the sum of all the multinomial coefficients is m^n. In other words,

$$m^n = \sum_{\Psi \in Sig(m,n)} \left(\frac{n!}{i_1! \, i_2! \ldots i_m!} \right)$$

$$= \sum_{\Psi \in Sig(m,n)} MC(\Psi).$$

Thus,

$$\sum_{\Psi \in Sig(m,n)} \frac{MC(m,n)}{m^n} = 1. \tag{11.3}$$

Note that the total number of n-mers is also m^n.

11.2.2   Patterns with multiplicities

We pose two problems related to the one in the last section. Let the alphabet be

$$\Sigma = \{\sigma_1, \sigma_2, \ldots, \sigma_l, \ldots, \sigma_m\}.$$

Problem 15 (*Permutations with exact multiplicities*) *Let q be an n-mer generated by a stationary, i.i.d. source which emits $\sigma \in \Sigma$ with probability p_σ. What is the probability that q has exactly i_1 number of σ_1's, i_2 number of σ_2's, ..., i_l number of σ_l's?*

Problem 16 (*Permutations with inexact multiplicities*) *Let q be an n-mer generated by a stationary, iid source which emits $\sigma \in \Sigma$ with probability p_σ. What is the probability that q has at least i_1 number of σ_1's, i_2 number of σ_2's, ..., i_l number of σ_l's?*

In both scenarios,

$$p_{\sigma_1} + p_{\sigma_2} + \ldots + p_{\sigma_l} + \ldots + p_{\sigma_m} = 1,$$

Further, let

$$k = i_1 + i_2 + \ldots + i_l \leq n.$$

Permutations with exact multiplicities.   In the first problem, the characters $\sigma_1, \sigma_2, \ldots, \sigma_l$

1. occur in some k positions on the n-mer with σ_1 occurring i_1 times, σ_2 i_2 times, ..., σ_l occurs i_l times, and

2. do not occur on the remaining $n - k$ positions.

Using Equation (11.2), the number of such distinct occurrences is given by

$$\binom{n}{k} \left(\frac{k!}{i_1! \, i_2! \, \ldots \, i_l!} \right). \tag{11.4}$$

For each distinct occurrence, the probability of its occurrence is given as:

$$(p_{\sigma_1})^{i_1} (p_{\sigma_2})^{i_2} \ldots (p_{\sigma_l})^{i_l} \left((1 - p_{\sigma_1})(1 - p_{\sigma_2}) \ldots (1 - p_{\sigma_l}) \right)^{n-k}. \tag{11.5}$$

For a specific choice, denoted as j, of k out of n locations on the string, let E_j denote the event that $\sigma_1, \sigma_2, \ldots, \sigma_l$ appear only in these k locations satisfying the stated constraints (i.e., exactly i_1 number of σ_1's and so on). Then it can be verified that

$$E_{j_1} \cap E_{j_2} = \emptyset, \tag{11.6}$$

i.e., the events are disjoint for any pair $j_1 \neq j_2$. The proof of this statement is left as Exercise 121 for the reader.

Next, using Equations (11.4) and (11.5), the answer to the first problem, denoted as $P_{i_1+i_2+\ldots+i_l}$, is given as

$$P_{i_1+i_2+\ldots+i_l} = \binom{n}{i_1+i_2+\ldots+i_l} \left(\frac{(i_1+i_2+\ldots+i_l)!}{i_1! \, i_2! \, \ldots \, i_l!} \right)$$
$$(p_{\sigma_1})^{i_1} (p_{\sigma_2})^{i_2} \ldots (p_{\sigma_l})^{i_l} ((1-p_{\sigma_1})(1-p_{\sigma_2})\ldots(1-p_{\sigma_l}))^{n-k}.$$

Permutations with inexact multiplicities. The second problem is a little more complex. We first define a set of l-tuples as follows:

$$C = \left\{ (i_1', i_2', \ldots, i_l') \,\middle|\, \begin{array}{l} i_1' + i_2' + \ldots i_l' = k' \leq n, \\ i_1 \leq i_1', \\ i_2 \leq i_2', \\ \ldots, \\ i_l \leq i_l' \end{array} \right\}.$$

For

$$j = (i_1', i_2', \ldots, i_l') \in C,$$

let E_j denote the event that $\sigma_1, \sigma_2, \ldots, \sigma_l$ occur exactly $i_1', i_2', \ldots, i_l'$ times respectively. Then

$$E_{j_1} \cap E_{j_2} = \emptyset, \tag{11.7}$$

i.e., the events are disjoint for any pair $j_1 \neq j_2 (\in C)$. We leave the proof of this as an exercise for the reader (Exercise 121).

Since the events are disjoint, the answer to the second problem, denoted as $P'_{i_1+i_2+\ldots+i_l}$, is obtained using the solution to Problem 1:

$$P'_{i_1+i_2+\ldots+i_l} = \sum_{(i_1',i_2',\ldots,i_l')\in C} P_{i_1'+i_2'+\ldots+i_l'} \tag{11.8}$$

The reader is also directed to [DS03, HSD05] for results on real data and generalizations to gapped permutation patterns.

11.3 Structured Permutations

The last section dealt with cases where an element of the alphabet occurs significantly many times in the input. But consider a scenario where an element of the alphabet occurs only a few times but the size of the alphabet is fairly large.

We have also seen in Chapter 10 that a permutation pattern can be hierarchically structured as a PQ tree. Thus it is meaningful to ask: *Given a permutation pattern q, where*

$$q = \{\sigma_1, \sigma_2, \ldots, \sigma_l\},$$

that occurs k times in the input, what is the p-value, pr(T, k), of its maximal form given as a PQ tree T ?

What does it mean to compute this probability? We give an exposition based on explicit counting below.

11.3.1   *P*-arrangement

An *arrangement* of size k is defined to be a string (or permutation) of some k consecutive integers

$$i, i+1, i+2, \ldots, i+k-1,$$

and its inversion is obtained by reading the elements from right to left. For example, q_1 and q_2 shown below are arrangements of sizes 5 and 3 respectively.

$$q_1 = 5\,2\,4\,3\,1 \text{ and its inversion is } 1\,3\,4\,2\,5,$$
$$q_2 = 4\,5\,6, \text{ and its inversion is } 6\,5\,4.$$

Recall the following notation from Section 2.8.2:

$$\Pi(q_1[1..5]) = \{1, 2, 3, 4, 5\},$$
$$\Pi(q_2[1..3]) = \{4, 5, 6\}.$$

Let q be an arrangement of size k. Recall from Section 10.5 that

$$[k_1..k_2], \ \ 1 \le k_1 < k_2 \le k,$$

is an *interval* in q if for some integers $i < j$, the following holds:

$$\Pi(q[k_1..k_2]) = \{i, i+1, i+2, \ldots, j\}.$$

Note that

$$[1..k]$$

is always an interval, hence is called the *trivial* interval. Every other interval is *nontrivial*. See the examples below for illustration.

	q_1				q_2		
interval $[k_1..k_2]$	$\Pi(q_1[k_1..k_2])$	*size*					
				interval $[k_1..k_2]$	$\Pi(q_2[k_1..k_2])$	*size*	
$[3..4]$ 5 2 $\boxed{4\,3}$ 1	$\{3,4\}$	2					
$[2..4]$ 5 $\boxed{2\,4\,3}$ 1	$\{2,3,4\}$	3		$[1..2]$	$\boxed{4\,5}$ 6	$\{4,5\}$	2
$[1..4]$ $\boxed{5\,2\,4\,3}$ 1	$\{2,3,4,5\}$	4		$[2..3]$	4 $\boxed{5\,6}$	$\{5,6\}$	2
$[1..5]$ $\boxed{5\,2\,4\,3\,1}$	$\{1,2,3,4,5\}$	5		$[1..3]$	$\boxed{4\,5\,6}$	$\{4,5,6\}$	3

An arrangement of size k is a *P-arrangement* if it has no nontrivial intervals. For example, q_1 and q_2 are not *P*-arrangements but q_3 and q_4 are where

$$q_3 = 1\,2, \quad q_4 = 2\,4\,1\,3.$$

Now, we are ready to state the central problem of the section.

Problem 17 (*P*-arrangement) *What is the number of P-arrangements of size k?*

11.3.2 An incremental method

How does one count such arrangements? Does this have a closed form formula? We give an exposition below that will set the stage for computing this number using an incremental method.

We first clarify the term *position* in an arrangement and the interpretation of an empty symbol, ϕ, in the arrangement. The following example best explains the ideas. For example consider the following:

$$\text{position}\ \ 1\,2\,3\,4\,5\,6\,7$$
$$q = \quad 4\,2\,\phi\,1\,6\,5\,3$$

The nontrivial intervals defined by interval $[2..4]$ and $[5..6]$ in q are shown below:

$$4\ \boxed{2\,\phi\,1}\ 6\,5\,3$$
$$4\,2\,\phi\,1\ \boxed{6\,5}\ 3\,,$$

with

$$\Pi(q[2..4]) = \{1,2\}\ \text{and}$$
$$\Pi(q[5..6]) = \{5,6\}.$$

Base cases. Note that for $k = 1$, the problem is not defined since the size of an interval is at least two. For $k = 2$, the P-arrangements are as follows:

$$1\,2$$

and its inversion

$$2\,1.$$

For $k = 3$, the number of P-arrangements is zero, since every arrangement of the three numbers has at least one nontrivial interval as shown below.

$$1\,\boxed{3\,2}\,,$$
$$\boxed{2\,1}\,3,$$
$$\boxed{3\,2}\,1.$$

For $k = 4$, the P-arrangements are:

$$3\,1\,4\,2$$

and its inversion

$$2\,4\,1\,3.$$

Principle 1. Can we obtain a P-arrangement of size 5 using

$$q_4 = 3\,1\,4\,2,$$

the P-arrangement of size 4? If element 5 is inserted at the start end of q_4 as

$$5\,3\,1\,4\,2,$$

then we have the nontrivial interval as shown below

$$5\,\boxed{3\,1\,4\,2}\,.$$

Similarly, adding element 5 at the other end will give a nontrivial interval. Also, if element 5 is inserted next to 4 as shown below

$$3\,1\,5\,4\,2,$$

we get the nontrivial interval

$$3\,1\,\boxed{5\,4}\,2.$$

Similarly, element 5 inserted to the right of 4 will again give a nontrivial interval. However, the following has no nontrivial intervals

$$q_5 = 3\,5\,1\,4\,2.$$

Thus we can derive a general principle of construction of a P-arrangement of size $k + 1$ from a P-arrangement of size k which is stated as follows.

LEMMA 11.1
Let q be a P-arrangement of
$$1, 2, \ldots, k.$$

Let q' be constructed from from q by inserting element $k+1$ at any of the $k-3$ positions in q that is not an end position and not adjacent to element k. Then q' is a P-arrangement of $1, 2, \ldots, k, k+1$.

Does the converse of Lemma (11.1) hold? In other words, is it true that removing element $k+1$ from any P-arrangement of

$$1, 2, \ldots, k, k+1$$

gives a P-arrangement of $1, 2, \ldots, k$? Consider the following P-arrangement of size 5:

$$4\,2\,5\,1\,3$$

However deleting element 5 gives the following with a nontrivial interval shown boxed:

$$4\;\boxed{2\,1}\;3$$

In other words, this P-arrangement of size 5 could not have been constructed incrementally from one of size 4 and the incremental construction will miss such P-arrangements.

Principle 2[**]. We take a closer look at this arrangement of $1, 2, 3, 4$. and the two nontrivial intervals are as shown below:

$$4\;\boxed{\boxed{2\,1}\;3}$$

It turns out the the smallest nontrivial interval is *nested* in the others, i.e., this interval is a *subset* of the others. An interval $[i_{11} \ldots i_{12}]$ is a subset of $[i_{21} \ldots i_{22}]$, written as

$$[i_{11} \ldots i_{12}] \subset [i_{21} \ldots i_{22}]$$

if and only if the following holds:

$$i_{21} \leq i_{11} < i_{12} \leq i_{22}.$$

In fact, in this example, all the intervals are nested, since otherwise the arrangement of size 5 will not be a P-arrangement. This observation can be generalized as the following lemma.

LEMMA 11.2
Let q be a P-arrangement of

$$1, 2, \ldots, k, k+1.$$

Let q' be obtained from q by removing the element $k+1$ from q. If q' is not a P-arrangement, then the smallest nontrivial interval is nested in every other nontrivial interval.

For example, consider the following P-arrangement

$$q = 9\ 1\ 3\ 6\ 4\ 11\ 7\ 5\ 8\ 2\ 10$$

Removing the element 11, gives the following intervals, that can be arranged as multiple (2 in this example) sequences of nested intervals as shown below:

The sizes of the intervals in q are: 4, 5, 6, 7, 8, 9, 10, and there are two intervals of size 5. We observe the following about nested intervals in an arrangement.

LEMMA 11.3

Let q be an arrangement such that it has r nested nontrivial intervals

$$[i_{11}..i_{12}] \subset [i_{21}..i_{22}] \subset .. \subset [i_{r1}..i_{r2}].$$

Then for each $j = 1, 2, \ldots, r$,

$$q[i_{j1}..i_{j2}]$$

is a P-arrangement of size $i_{j2} - i_{j1} + 1$.

This gives a handle on designing a method for counting (as well as enumerating) arrangements with nested nontrivial intervals. Consider the following arrangement of size 10 with nested intervals as shown:

Note that the smallest interval is a P-arrangement, and when the smallest interval is replaced by its extreme (largest here) element, shown in bold below, the next smallest interval is again a *P*-arrangement. We call this process of replacing an interval by its extreme element as *telescoping*. The next telescoping is shown below.

$$\boxed{8 \ 10 \ \mathbf{7} \ 9}$$

Two more examples of successive telescoping is shown below. Notice that at each stage the smallest interval is a *P*-arrangement.

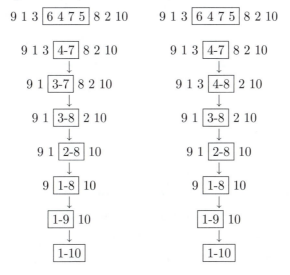

Successive telescoping for the previous example, where an interval is used instead of a single extreme, is shown below:

Putting it all together. We have identified the two properties that an arrangement q of

$$1, 2, \ldots, k$$

must satisfy, so that a *P*-arrangement of

$$1, 2, \ldots, k, k+1$$

may be successfully constructed from q. Now, we are ready to state the following theorem.

THEOREM 11.1

(*P*-arrangement theorem) *Let q be a P-arrangement of size $k+1$. Let q' be obtained by replacing an extreme element (either $k+1$ or 1) from its position j in q, with the empty symbol. Then only one of the following holds.*

1. q' *has no nontrivial intervals, i.e.* q' *is a* P-*arrangement, or,*

2. *every nontrivial interval* $[i_1 \ldots i_2]$ *is such that*

$$i_1 < j < i_2.$$

For convenience, we introduce a new terminology (*telescopic interval size*) that we will use later. Recall that the r nested intervals are given as

$$[i_{11}..i_{12}] \subset [i_{21}..i_{22}] \subset .. \subset [i_{r1}..i_{r2}].$$

Let the size, i_j $(1 \leq j \leq r)$, of each interval be denoted by

$$i_j = i_{j2} - i_{j1} + 1.$$

Then

$$i_1 < i_2 < \ldots < i_{r-1} < i_r.$$

Then the *telescoping sizes*, x_j $(1 \leq j \leq r)$, are defined as:

$$x_1 = i_1,$$
$$x_j = i_j - i_{j-1} + 1,$$

and are written as:

$$x_1 \leftarrow x_2 \leftarrow \ldots \leftarrow x_{r-1} \leftarrow x_r.$$

11.3.3 An upper bound on P-arrangements**

We identify the following functions that can be used to estimate an upper bound on the number of P-arrangements.

1. $Pa(k)$: Let $Pa(k)$ denote the number of P-arrangements of size k.

2. $S(u, l)$: Let $S(u, l)$, $u \geq l$, denote the number of arrangements of size u that has only nested intervals and the size of the smallest interval is l.

3. $Nst(k)$, Nst'(k):

 (a) Let $Nst(k)$ denote the number of arrangements of size k that have only nested intervals. Then $Nst(k)$ can be defined in terms of $S(\cdot, \cdot)$ as follows:

$$Nst(k) = \sum_{l=2}^{k} S(k, l). \tag{11.9}$$

 The summation is used to account for all possible values of l, the size of the smallest interval.

(b) However, we also wish to count the number of potential places in the arrangements where element $k + 1$ can be inserted. The positions are all in the smallest interval of size l. We denote by $Nst'(k)$ the number of positions that the element $k + 1$ can be inserted in.

$$Nst'(k) = \sum_{l=2}^{k} S(k,l)(l-1). \qquad (11.10)$$

Consider $Nst'(k)$ of Equation (11.10). Note that the smallest interval of size l may contain the element k, thus placing the element $k + 1$ in this interval gives a nontrivial interval resulting in an over estimation of the number of P-arrangements. Thus using Theorem (11.1) we get the following:

$$
\begin{aligned}
Pa(k) &\leq (k-4)Pa(k-1) && \text{(using Theorem (11.1)(1))} \\
&+ \quad Nst'(k-1). && \text{(using Theorem (11.1)(2))}
\end{aligned}
$$

Estimating $S(u, l)$. Now we are ready to define $S(u, l)$, $u \geq l > 1$, in terms of $Pa(u')$ where $u' < u$.

First, consider the case when there is exactly one nontrivial interval. The single nontrivial interval is of size l. Then we can consider a P-arrangement of size $u - l + 1$ and each position in this arrangement can be replaced by yet another P-arrangement of the remaining l elements giving the following:

$$S(l,l) = Pa(l)$$

Next consider the case where

$$u \ngeq l.$$

See Exercise 127 to study the simple scenario where the number of nested intervals (r) and the size of each interval is known. However, in the general scenario this number (r) and the size of each interval is not known.

Simple scenario. We then study yet another simple scenario where we compute all arrangements of size u

1. that have only nested intervals,

2. the smallest interval is of size $(l = i_1) > 2$,

3. each successive interval differs from the next in size by at least two and

4. the largest nontrivial interval is of size i_{k_1} or i_{k_2} where

$$l = i_1 < i_{k_1} < i_{k_2} < i_r = u.$$

Note that the telescoping sizes (see Section 11.3.2) are as follows:

$$x_1 = i_1.$$
$$x_{k_1} = i_r - i_{k_1} + 1,$$
$$x_{k_2} = i_r - i_{k_2} + 1.$$

Let the number of such arrangements be N and our task is compute N. Note that we know neither the exact number of nested intervals nor the size of each interval. But this does not matter.

P-arrangement of size > 2. Note that a P-arrangement of size $l > 2$ is such that the largest element is never at the ends of the arrangement. Thus this arrangement (or its inversion) can be inserted within another P-arrangement without producing intervals that are nested.

First, we compute N_1, the number of arrangements with the size of largest nontrivial interval as i_1. Using the principle used in Exercise 127 we obtain

$$N_1 \leq x_{k_1} Pa(x_{k_1}) S(i_{k_1}, l).$$

Similarly, we get

$$N_2 \leq x_{k_2} Pa(x_{k_2}) S(i_{k_2}, l).$$

Thus the required number is

$$
\begin{aligned}
N &= N_1 + N_2 \\
&\leq x_{k_1} Pa(x_{k_1}) S(i_{k_1}, l) + x_{k_2} Pa(x_{k_2}) S(i_{k_2}, l) \\
&= \sum_{j=k_1, k_2} x_j Pa(x_j) S(i_j, l) \\
&= \sum_{\Delta = u - i_{k_1}, u - i_{k_2}} Pa(\Delta + 1)(\Delta + 1) S(u - \Delta, l).
\end{aligned}
$$

Back to computation.   This sets the stage for computing the number of arrangements where the size of the largest nontrivial interval takes *all* possible values. In other words,

$$\Delta = 1, 2, \ldots, (u - l).$$

Thus, in the general case, we get

$$S(u, l) \leq \sum_{\Delta=1}^{u-l} 2(\Delta + 1) Pa(\Delta + 1) S(u - \Delta, l). \tag{11.11}$$

Thus, to summarize,

$$Pa(k) \leq (k - 4) Pa(k - 1) + \sum_{l=2}^{k-2} S(k - 1, l)(l - 1). \tag{11.12}$$

11.3.3.1 A dynamic programming solution

Note that $Pa(\cdot)$ is defined in terms of $Nst(\cdot)$ which is defined in terms of $S(\cdot, \cdot)$ which is again defined in terms of $Pa(\cdot)$. So is this a circular definition or is it possible to successfully compute $Pa(\cdot)$?

Note that in Equation (11.11), $\Delta < u$ holds and $S(u, l)$ is defined in terms of $Pa(k)$ where $k < u$ and thus can be computed using *dynamic programming*.

When the optimal solution to a problem can be obtained from optimal solutions of its subproblems, the problem can be usually solved efficiently by maintaining a table that successively computes the solutions to the subproblems. Thus this table avoids unnecessary re-computations and this approach is called dynamic programming.[3]

Here if $Pa(k)$ can be obtained from $Pa(k')$ where $k' < k$, then it is possible to compute $Pa(k)$ in increasing value of k.

We recall the recursive formulation of the problem of computing $Pa(k)$ the number of P-arrangements of $1, 2, \ldots, k$.

1. For $k > 1$,

$$Pa(2) = 2,$$
$$Pa(3) = 0,$$
$$Pa(4) = 2,$$

$$Pa(k) \le (k-4)Pa(k-1) + \sum_{l=2}^{k-2} S(k-1, l)(l-1).$$

2. For $k \ge l > 1$,

$$S(l, l) = Pa(l),$$

$$S(u, l) \le \sum_{\Delta=1}^{u-l} (\Delta+1)Pa(\Delta+1)S(u-\Delta, l).$$

3. For $k > 1$.

$$Nst'(k) \le \sum_{l=2}^{k} S(k, l)(l-1).$$

Figure 11.1 shows the order in which the functions can be evaluated. For convenience, it has been broken down into four phases as shown. To avoid clutter, the functions $Pa(\cdot)$, $S(\cdot, \cdot)$ and $Nst'(\cdot)$ also refer to the one, two and one dimensional arrays respectively that store the values of the functions as shown in the figure.

---

[3]The 'programming' refers to the particular order in which the tables are filled up and does not refer to 'computer programming'.

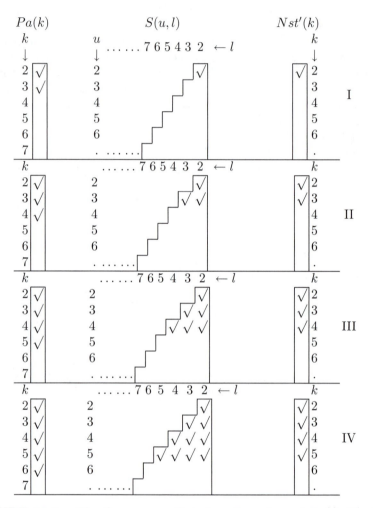

FIGURE 11.1: The three arrays that store the values of $Pa(\cdot)$, $S(\cdot, \cdot)$ and $Nst'(\cdot)$. The order in which the different functions are evaluated and stored in the arrays in a dynamic programming approach is shown above in the first four phases: I, II, III and IV. The check ($\checkmark$) entry indicates that the function has been evaluated at that point. See text for details.

1. Phase I: $Pa(2)$ is evaluated as a base case for function $Pa(\cdot)$.

   (a) Note that $S(2,2) = Pa(2))$.
   (b) Note that $Nst'(2) = S(2,2))$.
   (c) Finally $Pa(3)$ is evaluated, which happens to be a base case for $Pa(\cdot)$.

2. Phase II:

   (a) Note that $S(3,3) = Pa(3)$. Then $S(3,2)$ is evaluated, which depends on $S(2,2)$.
   (b) Next $Nst'(3)$ is evaluated, which depends on row $u = 3$ of the two-dimensional array that stores $S(\cdot,\cdot)$.
   (c) Finally, $Pa(4)$ is evaluated, which depends on $Pa(3)$ and $Nst'(3)$.

3. Phase III:

   (a) Note that $S(4,4) = Pa(4)$. Then $S(4,3)$ is evaluated, which depends on column $l = 3$ of array $S(\cdot,\cdot)$ array. Then $S(4,2)$ is evaluated, which depends on column $l = 2$ of array $S(\cdot,\cdot)$.
   (b) Next $Nst'(4)$ is evaluated, which depends on row $u = 4$ of array $S(\cdot,\cdot)$.
   (c) Finally $Pa(5)$ is evaluated, which depends on $Pa(4)$ and $Nst'(4))$.

4. Phase K $(K > 1)$: We can now generalize the evaluations in phase K.

   (a) Note that $S(K + 1, K + 1) = Pa(K + 1)$. Then $S(K + 1, j)$, as j takes values from K down to 2, is evaluated. At each value of j, $S(K+1,j)$ depends on column j entries evaluated up to this point in the array $S(\cdot,\cdot)$.
   (b) Next $Nst'(K + 1)$ is evaluated whose values depend on the entries in row $u = K + 1$ of the array $S(\cdot,\cdot)$.
   (c) Finally, $Pa(K + 2)$ is evaluated whose value depends on $Pa(K+1)$ and $Nst'(K + 1)$.

11.3.4 A lower bound on P-arrangements

We can obtain an easy lower bound on $Pa(k)$ (which we will see is already quite high), or an underestimate of the number of P-arrangements by only using Principle 1 of Theorem (11.1). We achieve this using the following recurrence equations (see Section 2.8):

$$Pa(2) = 2,$$
$$Pa(3) = 0,$$
$$Pa(4) = 2,$$
$$Pa(k) \geq (k - 4)Pa(k - 1), \quad \text{for } k > 4.$$

Solving this recurrence form (see Section 2.8) gives

$$Pa(k) \geq 2(k-4)!, \text{ for } k \geq 4. \tag{11.13}$$

The over- and underestimates on the number of P-arrangements are given by Equations (11.12) and (11.13) respectively.[4]

11.3.5   Estimating the number of frontiers

Let T be a PQ tree (see Section 10.3.1) with k leaf nodes labeled by k integers

$$\Sigma = 1, 2, \ldots, k.$$

Recall that this tree T

1. has k leaf nodes, and,

2. has N internal nodes (some P nodes and some Q nodes), with

$$N < k.$$

Recall from Section 10.6.2 that a PQ tree T encodes $\mathcal{I}(T)$, a collection of subsets of Σ as shown in Equation (10.5). Let an arrangement q of elements of Σ be a *frontier* of T if for every set

$$I \in \mathcal{I}(T),$$

there exists

$$1 \leq i_1 < i_2 \leq k$$

such that

$$\Pi(q[i_1..i_2]) = I.$$

Let

$$Fr(T) = \{q \mid q \text{ is a frontier } T\}. \tag{11.14}$$

Two PQ trees T and T' are *equivalent*, denoted

$$T \equiv T',$$

if one can be obtained from the other by applying a sequence of the following transformation rules:

1. Arbitrarily permute the children of a P-node, and

2. Reverse the children of a Q-node.

There is yet another view to frontiers of a PQ tree. The frontier of a tree T, denoted by $F(T)$, is the arrangement obtained by reading the labels of the leaves from left to right. An alternative definition for $Fr(T)$ is as follows:

$$Fr(T) = \{F(T') \mid T' \equiv T\}. \tag{11.15}$$

---

[4]However, an easy upper bound is $Pa(k) < k!$.

What is the size of $Fr(T)$? The burning question of this section is: *Given a PQ tree T with k leaf nodes, what is the size of $Fr(T)$?*

In other words, what is the number of arrangements that encode exactly the same subsets of Σ as T?

We define $\#(A)$, for each node A of T as follows. Let node A in the PQ tree T have c children $A_1, A_2, \ldots, A_c$. Then

$$
\#(A) = \begin{cases} 1 & \text{if } A \text{ is a leaf node,} \\ 2 \prod_{j=1}^{c} \#(A_j) & \text{if } A \text{ is a } Q \text{ node,} \\ Pa(c) \prod_{j=1}^{c} \#(A_j) & \text{if } A \text{ is a } P \text{ node.} \end{cases} \tag{11.16}
$$

Recall that $Pa(c)$ is the number of P-arrangements of size $1, 2, \ldots, c$. We next claim the following.

$$
\#(Root(T)) = |Fr(T)|,
$$

where $Root(T)$ is the root node of the PQ tree T. This is best explained through a concrete example shown in Figure 11.3.

Note that the number of leaf nodes is 7 in this example. We begin by first relabeling the nodes in the left to right from 1 to 7 as shown. The arrangements

$$1\,2\,3\,4\,5\,6\,7$$

and its inversion

$$7\,6\,5\,4\,3\,2\,1,$$

are clearly frontiers. The others are computed as follows. For each P node of T with c children, $Pa(c)$ gives the number of possible arrangements of the children. For each Q node, the number of possible arrangements is only two as illustrated in the figure.

See Example (133) for another example and $\#(A)$ for each node A is computed as shown in Figure 11.2 for this T.

A practical solution. In the last section we derived lower and upper bounds for $Pa(c)$. Let the under and over bound of $Pa(c)$ be given as

$$
L(c) \le Pa(c) \le U(c).
$$

Then Equation (11.16) can be re-written as:

$$
\#(A) \begin{cases} = & 1 & \text{if } A \text{ is a leaf node,} \\ = & 2 \prod_{j=1}^{c} \#(A_j) & \text{if } A \text{ is a } Q \text{ node,} \\ \ge & L(c) \prod_{j=1}^{c} \#(A_j) \Big] \\ < & U(c) \prod_{j=1}^{c} \#(A_j) \Big] & \text{if } A \text{ is a } P \text{ node.} \end{cases} \tag{11.17}
$$

See Figure 11.4 for an example.

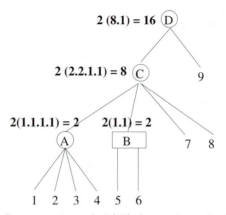

FIGURE 11.2: Computation of $\#(X)$ for each node X in the PQ tree using Equation (11.16). Note that the internal nodes are labeled $A, \ldots, D$ and $\#(A) = \#(B) = 2$, $\#(C) = 8$, and $\#(D) = 16$.

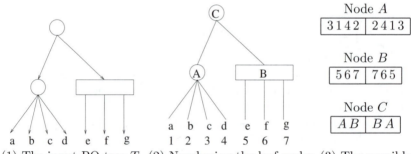

(1) The input PQ tree T. (2) Numbering the leaf nodes (3) The possible
                    & labeling the internal nodes.   arrangmeents.

1234567		abcdefg	
3142567	765 2413	cadbefg	gfe bdac
3142765	5672 413	cadbgfe	efgb dac
2413567	7653142	bdacefg	gfecadb
2413765	5673142	bdacgfe	efgcadb

7654321   gfedcba

(4) The 10 possible arrangements. (5) Arrangements in the input alpahabet.

FIGURE 11.3: Different steps involved in computing $|Fr(T)|$ are shown in (1), (2) and (3). The different arrangements are shown in (4) and (5) above.

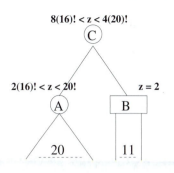

FIGURE 11.4: The P node A has 20 children and the Q node B has 11 children. $\#(X)$ for node X is given by z. The lower and upper estimates have been used for each node.

11.3.6 Combinatorics to probabilities

We connect the final dot by computing the probabilities from the functions that we have evaluated so far in the preceding sections.

To understand this we make a comparison to DNA patterns. Consider the following question: *What is the probability, $pr(x)$, of seeing a pattern x which has n_1 purines and $n - n_1$ pyramidines amongst all n length patterns?*

For this we compute N_x which is the number of n length patterns with n_1 purines and $n - n_1$ pyramidines. The total number of patterns of length n is

$$2^n.$$

Then, the probability is given as

$$pr(x) = \frac{N_x}{2^n}.$$

First we define *compatibility* as follows. Let q be a permutation of some finite set Σ and let T be a PQ tree with its leaf nodes labeled by elements of Σ. Further, let q be of size $|\Sigma|$ and let T have $|\Sigma|$ leaf nodes. Then T is compatible with q and vice-versa, if and only if the following holds.

$$q \in Fr(T).$$

We are now ready to pose our original question along the lines of the earlier one: *What is the probability, $pr(T)$, of a PQ tree T which has n leaf nodes, labeled by integers $1, 2, \ldots, n$, being compatible with a random permutation of $1, 2, \ldots, n$?*

For this we compute N_T which is the number of permutations that are compatible with T. Note that

$$N_T = |Fr(T)|.$$

The total number of permutations of length n is

$$n!$$

Then, the probability is given as

$$pr(T) = \frac{N_T}{n!} = \frac{|Fr(T)|}{n!}. \tag{11.18}$$

An alternative view. Let T be a tree with n leaf nodes. We label the leaf nodes by integers

$$1, 2, \ldots, n$$

in the left to right order.[5] Let q be a random permutation of integers

$$1, 2, \ldots, n.$$

See Section 5.2.3 for a definition of random permutation. Then the probability, $pr(T)$, of the occurrence of the event

$$q \in Fr(T)$$

is given by

$$pr(T) = \frac{|Fr(T)|}{n!}. \tag{11.19}$$

Generalization. We have computed the probability of seeing a structured permutation pattern T two times ($K = 2$) in Equations (11.18) and (11.19). This can be generalized to any K as

$$(pr(T))^{K-1}. \tag{11.20}$$

11.4   Exercises

For the problems below, see Section 5.2.4 for a definition of *random strings* and Section 5.2.3 for a definition of *random permutations*.

Exercise 121 (Disjoint events) *Consider the discussion in Section 11.2.2. Show that the events are disjoint in each of that cases below. In other words,*

  1. show that Equation (11.6) holds and

---

[5]In fact, the leaves could be labeled in any order (as long as it is a bijective mapping) and the arguments still hold.

2. show that Equation (11.7) holds.

Hint: Note that the event is defined as one with i_1 number of σ_1's *and* i_2 number of σ_2's, *and* i_3 number of σ_3's, ..., *and* i_l number of σ_l's. However, if the event is defined as one with i_1 number of σ_1's *or* i_2 number of σ_2's, *or* i_3 number of σ_3's, ..., *or* i_l number of σ_l's, then would the equations hold?

Exercise 122 (Effect of boundary) *Let*

$$|\Sigma| = n.$$

If two input strings s_1 and s_2, with no multiplicities, defined on Σ, of length n are circular, then show that for $1 \le k < n$,

$$p_k = p_{n-k},$$

where p_k is the probability of seeing a set of size k appear together is s_1 and s_2.

Exercise 123 (P-arrangement) *For $k > 1$, let $Pa(k)$ denote the number of P-arrangements of size k. Then what is $S(k)$, the number of permutations (arrangements) of size k that has exactly one nontrivial interval of size $l < k$?*

Hint: An interval of size l can be treated as a single number, that can then be expanded to admit its own P-arrangement. Then does the following hold?

$$S(k) = Pa(l)Pa(k - l + 1).$$

Exercise 124 (P-arrangement structure) *Let q be a P-arrangement of integers*

$$i, i + 1, \ldots, j - 1, j.$$

1. *Which is the P-arrangement for which i and/or j occur(s) at the end position(s)?*

2. *Show that the elements i and j do not occur at the end positions of q when $j - 1 > 1$.*

Exercise 125 *Prove statements (1) and (2) of Theorem (11.1).*

Hint: (1) Use proof by contradiction. (2) If the nontrivial intervals are not nested, is it possible that q also has a nontrivial interval?

Exercise 126 (Enumeration) *Construct all arrangements of elements* $1, 2, 3, 4, 5$ *such that each has*

1. *exactly one nontrivial interval and this interval is of size 4.*

2. *exactly one nontrivial interval and this interval is of size 2.*

3. *the smallest nontrivial interval is of size 2.*

Hint: Invert the telescoping and do an appropriate renumbering. In the tables below, row marked (1a) shows the position, as underlined, of the arrangement in row (1) that is expanded (inverse of telescoping) in the next row, along with appropriate renumbering. Row (2a) has a similar interpretation.

1. Row (1) shows the P-arrangements of $1, 2$ and row(2) shows its inversion.

2 1				(1)

2̲ 1		2 1̲		(1a)
4 2 5 3 | 1	3 5 2 4 | 1	5 | 3 1 4 2	5 | 2 4 1 3	

(inversion)

1 2				(2)

1̲ 2		1 2̲		(2a)
3 1 4 2 | 5	2 4 1 3 | 5	1 | 4 2 5 3	1 | 3 5 2 4	

2. Row (1) shows the P-arrangements of $1..4$ and row(2) shows its inversion.

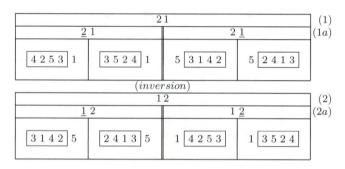

3. When $l = 4$, then the only possible nested intervals are size 4 and 5. When $l = 3$, the number is zero since $Pa(3) = 0$. When $l = 2$, let r be the number of intervals, then for each case the possible (nested) interval sizes are given in the following table. For example when $r = 3$, $2 < 3 < 5$ is a possible configuration of the interval sizes and in the

inverted telescoping process the P-arrangement sizes are 3, 2 and 2.

$$l = 2$$

$r = 2$	$r = 3$	$r = 4$
$2 < 5$	$2 < 3 < 5$	$2 < 3 < 4 < 5$
$2 \leftarrow 4$	$2 \leftarrow 2 \leftarrow 3$	$2 \leftarrow 2 \leftarrow 2 \leftarrow 2$
	$2 < 4 < 5$	
	$2 \leftarrow 3 \leftarrow 2$	

The first column (2 intervals) is already answered in part 1. The second column gives zero P-arrangements since p-arrangement of size 3 is involved. See case $r = 4$ below.

Exercise 127 *Let $S'(i_r, i_1)$ be the number of arrangements of size i_r such that each has r nested intervals of sizes*

$$(l = i_1) < i_2 < \ldots < i_{r-1} < (i_r = u),$$

and for $1 < j < r$,

$$x_j > 2,$$

where

$$x_1 = i_1,$$
$$x_j = i_j - i_{j-1} + 1, \ \ for \ r \geq j > 1.$$

1. *Show the following:*

$$S'(i_r, i_1) \leq \prod_{j=1}^{r} x_j Pa(x_j).$$

2. *If $x_j > 1$, for all j, what is $S'(i_r, i_1)$?*

Hint: 1. Use the ideas of Exercise 126. 2. Consider the scenario when x_j and x_{j+1} are both of size 2. See also Exercise 126(3) for an illustration.

Exercise 128 *Enumerate all the P-arrangements of elements $1\ldots 5$.*

Hint: Use Theorem (11.1).

Principle 1	Principle 2		
2 4 1 3	3 $\boxed{1\ 2}$ 4		
			inversions
2 4 1 5 3	3 1 5 2 4	3 5 1 4 2	4 2 5 1 3

Exercise 129 (Rearrangements) *Let q be an arrangement of k elements that has $r > 1$ nested intervals and the size of the smallest one is l.*

1. *Then the elements 1 and k do not occur together in the smallest interval.*

2. *Each element i in q is replaced by $i + 1$ to obtain an arrangement q' of elements $2, 3, \ldots, k, k + 1$. Show that*

$$[i_{11} \ldots i_{12}], [i_{21} \ldots i_{22}], \ldots, [i_{r1} \ldots i_{r2}]$$

are the r intervals in both q and q'.

Prove the two statements.

Hint: (1) Observe that $l \lesssim k$. (2) Use proof by contradiction.

Exercise 130 (Over-counting) *Recall the following from Section 11.3.1:*

$$N st'(k) \leq \sum_{l=2}^{k-2} S(k-1,l)(l-1).$$

Consider the following line of argument.
Recall that the smallest interval is of size l. Using statement (1) of Exercise 129, both elements 1 and $k-1$ do not occur together in the smallest nested interval. Then only one of the following holds.

1. *(Case 1): If element $k-1$ does not occur, element k can be inserted in any of the $l-1$ positions in the smallest interval.*

2. *(Case 2): If element $k-1$ does occur, 1 does not occur and using statement (2) of Exercise 129, element 0 can be inserted in any of the $l-1$ positions. The arrangement of elements $0, 1, \ldots, k-1$ is simply renumbered to elements $1, 2, \ldots, k-1, k$.*

Thus

$$N st'(k) = \sum_{l=2}^{k-2} S(k-1,l)(l-1).$$

What is incorrect with this line of argument?

Hint: In Case 2, is it possible that this arrangement has already been accounted for? Are all the nested interval sequences accounted for?

Exercise 131 (Correction factors) *In Section 11.3.2, an overestimate of $Pa(k)$ is computed. What are the cases that need to be handled to get an exact estimate of $Pa(k)$?*

Hint: How is the counting done if two successive telescopic sizes of the intervals are 2 each? In the arrangements that $Nst(k)$ counts, how many are such that element k occurs in the smallest interval? How are the intervals that are not strictly nested taken care of?

Exercise 132 (Frontier equivalence) *Show that $Fr(T)$ in Equations (11.14) and (11.15) describe the same set of arrangements.*

Exercise 133 *Enumerate the frontiers of the PQ tree T shown below. What is $|Fr(T)|$?*

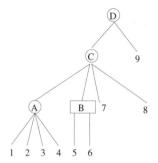

Hint: Notice that the leaf nodes are already labeled in the left to right order. Each internal node and the possible arrangements of the children are shown below.

Exercise 134 *Let T be a tree with n leaf nodes which are labeled by integers $1, 2, \ldots, n$ in the left to right order and let q be a random permutation of integers $1, 2, \ldots, n$.*

1. *If T has exactly one Q node and no P nodes, then what is the probability of the following event:*

$$q \in Fr(T)?$$

2. *If T has exactly one P node and no Q nodes, then what is the probability of the following event:*

$$q \in Fr(T)?$$

Exercise 135 ** **(Human-rat data)** *A common cluster of 380 genes in human chromosome 11 and rat chromosome 1 is shown below with the leaf nodes being labeled with the genes.*

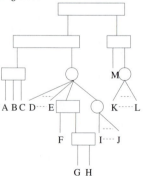

The labeled nodes of T are to be interpreted as follows.

A:	Q node with 146 genes,
B:	Q node with 4 genes,
C:	Q node with 64 genes,
D...E:	six nodes with 1 gene each, four Q nodes with 2 genes each, one Q node with 82 genes,
F:	1 gene,
G:	Q node with 5 genes,
H:	Q node with 2 genes,
I....J:	one Q node with 2 genes, 11 nodes with 1 gene each,
K...L:	twelve with 1 gene each, three Q nodes of 2 genes each, one Q node with 3 genes, one Q node with 4 genes,
M:	Q node with 24 genes.

Compute pr(T).

Hint: Use Equation (11.17) to get the under- and overestimates of $|Fr(T)|$.

Exercise 136 *Argue that Equation (11.20) is the probability of $K > 1$ occurrences of a structured permutation pattern (PQ tree) T.*

Hint: How is the probability space defined for K occurrences?

Chapter 12

Topological Motifs

Through science we prove,
through intuition we discover.
- attributed to J. H. Poincare

12.1 Introduction

Due to some unknown reason,[1] nature has organized the blue-print of a living organism along a line. Thus nucleotides on a strand of DNA or amino acids on a protein sequence (the primary structure) or genes on a chromosome are linearly arranged and the study of strings has been an important component in the general area of bioinformatics.

But sometimes, there is a deviation from this clean organizational simplicity, for instance, a cell's *metabolic network*, as we understand it. A *metabolic pathway* is a series of chemical reactions that occur within a cell, usually catalyzed by enzymes, resulting in the synthesis of a metabolic product that is stored in the cell. Sometimes, instead of creating such a product, the pathway may simply initiate another series of reactions (yet another pathway). Various such metabolic pathways within a cell have a large number of common components and thus form the cell's metabolic network. Figure 12.1 shows an example.

To study and gain an understanding in a domain such as this, one abstracts the organization of this information as a graph. A graph captures this kind of complex interrelationships of the entities. Continuing the theme of this book, we seek the recurring structures in this data.

12.1.1 Graph notation

We enhance the graph notation of Section 2.2 here. A graph G is defined by a quadruplet (V, E, A_V, A_E) as follows:

1. V is the set of vertices.

---

[1]However, speculations abound and there are almost as many theories as there are scientists.

2. $E \, (\subseteq V \times V)$ is the set of edges. Each element $e \in E$ is usually written as $v_i v_j$ (where $v_i, v_j \in V$). When the graph is directed, the edge $e = v_i v_j$ is to be interpreted as an edge from v_i to v_j.

3. Let $\mathcal{A}_V$ be the set of all possible attributes that can be assigned to a vertex. Then

$$A_V : V \to \mathcal{A}_V.$$

Thus the attribute of $v \in V$ is written as $A_V(v)$ or simply $att(v)$.

4. Let $\mathcal{A}_E$ be the set of all possible attributes that can be assigned to an edge. Then

$$A_E : E \to \mathcal{A}_E.$$

Thus the attribute of $e \in E$ is written as $A_E(e)$ or simply $att(e)$ (or $att(v_i v_j)$ where $e = v_i v_j$).

Usually $\mathcal{A}_V \cap \mathcal{A}_E$ is assumed to be empty. In other words the attributes of a vertex and that of an edge do not 'mix'.

The size of G will be denoted by N_G and is defined to be

$$N_G = |V| + |E|.$$

However, for convenience we denote a graph as $G(V, E)$ with vertex set V and edge set E. The vertex and edge attribute mappings A_V, A_E are assumed to be implicit.

12.2 What Are Topological Motifs?

In this section we give an informal, intuitive introduction to topological motifs.[2] Consider the graph in Figure 12.2 with seven connected components numbered for convenience from (1) to (7). The vertex attributes are represented by the color of the vertex in the figure. Thus a vertex attribute can be red, blue, green or black. The edges are directed and the attribute is denoted by the type of edge (solid or dashed) in the figure.

What are the recurring structures in this graph? For example, a vertex with an attribute red occurs seven times in the graph. Is the occurrence of a red vertex always accompanied by another vertex (of a specific attribute) along with an edge (of a specific attribute)? So we use the natural notion of *maximality* when we enumerate these common structures or motifs. A formal definition of *maximal motifs* is given in the later sections.

---

[2]In literature they are also called *network motifs*.

Figure 12.3 gives the exhaustive list of *maximal* and *connected* structures (subgraphs or *motifs*) that occurs at least two times in the input. Here connected is defined as follows:

> For any pair of vertices v and v' there is a path that can be obtained by ignoring the direction of the edge from v to v'.

In this example, each occurrence of the motif is in a distinct connected component of the graph. There are seventeen such motifs.

12.2.1 Combinatorics in topologies

We next make the problem a little harder by making all the edges undirected and having the same attribute. However, we maintain the following simplifying property:

> No two adjacent vertices have the same attribute and no two vertices adjacent to one vertex have the same attribute.

Consider the graph shown in Figure 12.4. Again, the color of the vertex denotes its associated attribute and the graph has seven connected components numbered, for convenience, from (1) to (7).

The exhaustive list of maximal common structures (motifs) on this input graph is 63 and is shown in Figure 12.5.

Estimating the output size. We spell out the underlying combinatorics in terms of common structures, in this extreme example. It is not difficult but tedious. But it is a good exercise that shows the enormity of the task involved, in a worst case scenario.

Assuming the output is the list of all the maximal motifs and their occurrences in the graph, we do a simple exercise of estimating the size of the output.

Figure 12.7 gives the input size and the details involved in calculating the output size. This shows that the size of the input (which is simply the sum of the number of vertices and number of edges) is only 144 whereas the size of the output is 1448. The size of the output also includes the details of the occurrences of each motif.

The occurrence of a motif is defined by a subset of vertices and edges in the input graph. Thus each vertex in the motif defines a list of vertices and each edge defines a list of edges that correspond to the occurrence of this motif. Figure 12.8 gives the number of these vertex and edge lists in the graph that correspond to all the maximal motifs.

12.2.2   Input with self-isomorphisms

A graph *isomorphism* is a bijection f, i.e., a one-to-one and onto mapping, between the vertices of two graphs $G_1(V_1, E_1)$ and $G_2(V_2, E_2)$,

$$f : V_1 \rightarrow V_2,$$

with the property that for $v_1, v_2 \in V$, if

$$(v_1 v_2) \in E_1$$

then

$$(f(v_1)f(v_2)) \in E_2.$$

Two graphs are *isomorphic* if such a bijection f exists.

A graph $G(V, E)$ is *self-isomorphic* if f is a bijection as above and there exists some v such that

$$v \neq f(v).$$

We abuse notation here and say that given $l > 0$, a graph $G(V, E)$ displays self-isomorphism, if for some nonempty sets

$$V_1 \neq V_2 \subset V$$

and $|V_1| > l$ there exists a bijection

$$f : V_1 \rightarrow V_2,$$

with the property that for $v_1, v_2 \in V_1$, if

$$(v_1 v_2) \in E$$

then

$$(f(v_1)f(v_2)) \in E.$$

Self-isomorphism is an important property of a graph, since it can be particularly confounding to methods that attempt to recognize common structures in the graph.

Concrete example.   We next modify the graph under study by changing the attributes of some of the vertices to obtain the graph shown in Figure 12.9. Now the vertices have only two types of attributes.

Now the graph becomes highly self-isomorphic. Figure 12.10 shows the isomorphisms in the different connected components of the graph. Components (1), (4) and (5) are isomorphic to each other and components (2), (3) and (6) are isomorphic to each other. Thus, for all practical purposes, the graph has three connected components number (5), (6) and (7).

Figure 12.11 gives an exhaustive list of maximal motifs that occur at least two times in the graph. Note that a motif can have two occurrences that overlap. We consider two occurrences to be *distinct*,

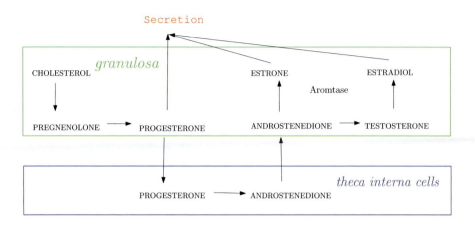

FIGURE 12.1: Synthesis of estrogens in the ovary: Estrogens (estradiol, estrone) are synthesized via androgen intermediates which takes place in the *granulosa* and the *theca interna cells* of the follicle, each of which is under the control of a different hormone.

if in the two occurrences there is at least one edge that is not present in both the occurrences.

In the figure, we count only distinct occurrences. Thus an occurrence of $5(2), 6, 7(3)$ is to be interpreted as having two distinct occurrences in component (5), one occurrence in component (6) and three distinct occurrences in component (7).

In this scenario, when is a motif maximal? Usually, a subgraph M' of a motif M is considered nonmaximal with respect to (w.r.t.) M. We use the definition that takes the occurrences as well into account. Thus the occurrences may actually determine if a M' is maximal or not:

1. If the number of distinct occurrences of M and M' differ, then M' is maximal w.r.t. M.

2. However if the number of distinct occurrences are the same then M' is nonmaximal w.r.t. M.

12.3 The Topological Motif

In the remainder of the discussion we deal with graphs with undirected edges and attributes defined only on the vertices. See Exercises 141 and 142 for a

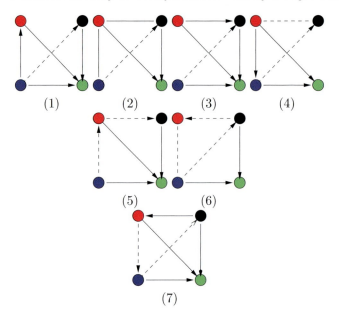

FIGURE 12.2:   The input graph with 7 connected components where the edges may be directed and have different attributes. The two attributes are shown as solid and dashed edges.

systematic reduction of a directed graph with attributes on both edges and vertices to an undirected graph with attributes only on the vertices. However this conversion results in a graph with a larger vertex set. But, the methods presented here can be easily adapted to directed graphs (with edge attributes) and in fact simpler than the scenario discussed here.

Informally, as we have seen in the last sections, a topological motif is a 'graph' that occurs multiple times in a given graph G (or a collection of graphs). Usually the interest is in a motif (say, M) that *occurs* at least K times in a input graph. This K is usually referred to as a *quorum* constraint. This is so called because if a subgraph occurs less than K times, it may not be of interest. Formally a topological motif and its occurrence is defined as follows.

DEFINITION 12.1   *(topological motif, occurrence, mappings F_i, location list) Given a graph*

$$G(V, E),$$

a topological motif *is a connected graph*

$$M(V_M, E_M),$$

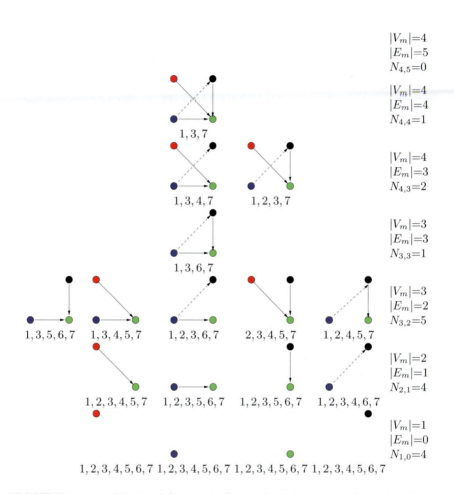

$$|V_m|=4$$
$$|E_m|=5$$
$$N_{4,5}=0$$

$$|V_m|=4$$
$$|E_m|=4$$
$$N_{4,4}=1$$

1, 3, 7

$$|V_m|=4$$
$$|E_m|=3$$
$$N_{4,3}=2$$

1, 3, 4, 7   1, 2, 3, 7

$$|V_m|=3$$
$$|E_m|=3$$
$$N_{3,3}=1$$

1, 3, 6, 7

$$|V_m|=3$$
$$|E_m|=2$$
$$N_{3,2}=5$$

1, 3, 5, 6, 7   1, 3, 4, 5, 7   1, 2, 3, 6, 7   2, 3, 4, 5, 7   1, 2, 4, 5, 7

$$|V_m|=2$$
$$|E_m|=1$$
$$N_{2,1}=4$$

1, 2, 3, 4, 5, 7   1, 2, 3, 5, 6, 7   1, 2, 3, 5, 6, 7   1, 2, 3, 4, 6, 7

$$|V_m|=1$$
$$|E_m|=0$$
$$N_{1,0}=4$$

1, 2, 3, 4, 5, 6, 7 1, 2, 3, 4, 5, 6, 7 1, 2, 3, 4, 5, 6, 7 1, 2, 3, 4, 5, 6, 7

FIGURE 12.3:   Maximal (connected) motifs that occur at least two times in the input graph of Figure 12.2. $N_{x,y}$ denotes the number of motifs with x vertices and y edges and each row shows motifs with x number of vertices and y number of edges. For example Row 2 shows the maximal motifs with 5 vertices and 4 edges each. The list of numbers below the motif gives the occurrence list, for example $1, 3, 7$ indicates that the motif occurs in components numbered (1), (3) and (7).

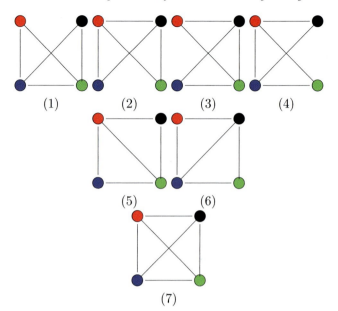

FIGURE 12.4: The input graph with 7 connected components numbered (1) to (7). All the edges are undirected and have the same attribute. The attributes of the vertices are given by their color.

where

$$V_M = \{u_1, u_2, \ldots, u_p\}, \ p \geq 1$$

and is said to occur on

$$O_i = \{v_{i1}, v_{i2}, \ldots, v_{ip}\} \subseteq V,$$

if and only if there is a mapping

$$F_i : V_M \to O_i,$$

such that,

1. *for each $u \in V_M$,*

$$att(u) = att(F_i(u)), \ and$$

2. *for each $(u_1 u_2) \in E_M$,*

$$(u_1 u_2) \in E_M \Rightarrow (F_i(u_1)F_i(u_2)) \in E.$$

Let the number of such distinct mappings (F_i's) be K'. Then the number of occurrences is K' and the occurrence lists are

$$O_1, O_2, \ldots, O_{K'}.$$

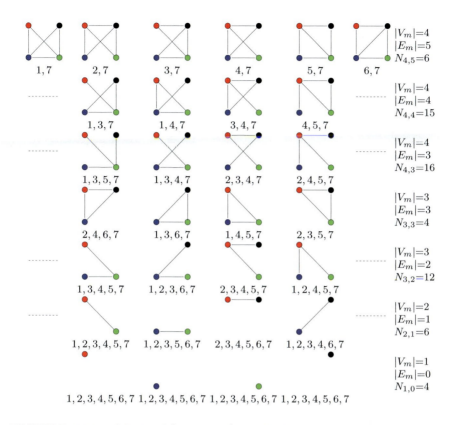

FIGURE 12.5: Maximal (connected) motifs that occur at least two times in the input graph of Figure 12.4. In this example each occurrence of the motif is in a distinct component of the graph. $N_{x,y}$ denotes the number of motifs with x vertices and y edges. Each row shows motifs with x number of vertices and y number of edges. For example Row 1 shows some maximal motifs with 4 vertices and 5 edges each.

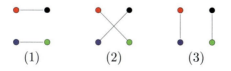

FIGURE 12.6: Continuing the example of Figure 12.4. Any two edges many not be adjacent, leading to disconnected structures, hence the three structures are not motifs.

Input				
vertices $\|V\|$	edges $\|E\|$		-	input size $\|V\| + \|E\|$
28	36		-	64

Output				
vertices $\|V_m\|$	edges $\|E_m\|$	number l	occurrences K'	output size $lK'(\|V_m\| + \|E_m\|)$
4	5	6	2	108
4	4	15	3	360
4	3	16	4	448
3	3	4	4	96
3	2	12	5	300
2	1	6	6	108
1	0	4	7	28
		63		1448

FIGURE 12.7:   The number of distinct maximal (connected) motifs that occur at least two times in the graph of Figure 12.4 is 63.

Vertex Lists (n_V)					Edge Lists			
$\|V_m\|$	l	K'	$n_V = l\|V_m\|$	$K'n_V$	$\|E_m\|$	l	K'	$l\|E_m\|$
4	6	2	24	48	5	6	2	30
4	15	3	60	180	4	15	3	60
4	16	4	64	256	3	16	4	48
3	4	4	12	48	3	4	4	12
3	12	5	36	180	2	12	5	24
2	6	6	12	72	1	6	6	6
1	4	7	4	28	0	4	7	0
	63		212	812		63		180

FIGURE 12.8:   Number of distinct location lists of the maximal motifs in Figure 12.7.

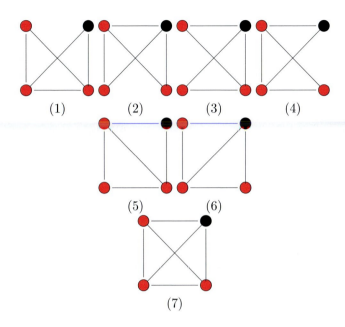

FIGURE 12.9:   The input graph with with only two (vertex) attributes.

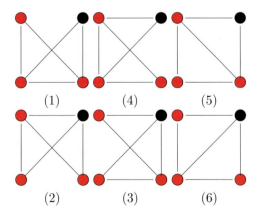

FIGURE 12.10:   Consider the graph of Figure 12.9. The top 3 components are isomorphic to each other and so are the bottom three components.

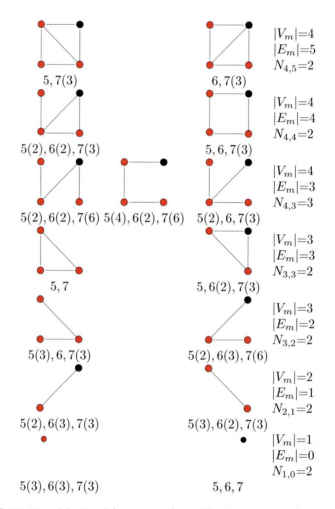

FIGURE 12.11:   Maximal (connected) motifs that occur at least two times in the input graph of Figure 12.9. Since the components (1), (4) and (5) are isomorphic (or identical) and so are components (2), (3) and (6), the occurrences are listed only for components (5), (6) and (7) for each motif.

For a set of vertices $U \subseteq V_M$, let

$$F_i(U) = \{F_i(v) \mid v \in U \, (\subseteq V_M)\}.$$

The location list of U, $\mathcal{L}_U$, is defined as

$$\mathcal{L}_U = \{F_i(U) \mid 1 \leq i \leq K'\}.$$

If U is a singleton set,
$$U = \{u_j\},$$

then its location list may also be written as

$$\mathcal{L}_{u_j}.$$

$\mathcal{L}_U$, *the location list of U is given by the following*

$$\mathcal{L}_U = \{F_1(U), F_2(U), \ldots, F_{K'}(U)\}.$$

The graph induced by the vertices $F_i(u)$, $u \in V_m$, is the ith occurrence subgraph of motif $M(V_M, E_M)$ on the input graph $G(V, E)$.

Notice that a mapping (called F in the definition) is required to unambiguously define an occurrence of a motif. Also when U is not a singleton set it is usually a collection of vertices that have the same attribute and $\mathcal{L}_U$ is a multi-set or a set of sets of vertices.

12.3.1 Maximality

Consider the input graph with two connected components shown in Figure 12.12(a). Let the quorum be 2. A motif

$$M(V_M, E_M)$$

with

$$V_M = \{u_1, u_2, u_3\}$$

is shown in Figure 12.12(b). The two occurrences of the motif with

1. $att(u_1) = $ blue,

2. $att(u_2) = $ green,

3. $att(u_3) = $ red,

are given as follows:

1. $O_1 = \{v_2, v_3, v_1\}$, with
   $F_1(u_1) = v_2$, $F_1(u_2) = v_3$, $F_1(u_3) = v_1$ and

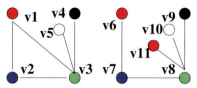

(a) Input graph with two connected components.

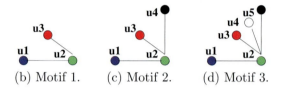

(b) Motif 1.   (c) Motif 2.   (d) Motif 3.

FIGURE 12.12: (a) The different attributes of the nodes (vertices) are shown in different colors. v_1 to v_5 form one connected component and the other is formed by v_6 to v_{11}. (b) and (c) show two motifs that occur once in each connected component of input graph. (d) The maximal version of these motifs, i.e., no more vertices or edges can be added to this motif.

2. $O_2 = \{v_7, v_8, v_{11}\}$, with
   $F_2(u_1) = v_7$, $F_2(u_2) = v_8$ and $F_2(u_3) = v_{11}$.

The location list of the vertices of the motif are:

1. $\mathcal{L}_{u_1} = \{v_2, v_7\}$,

2. $\mathcal{L}_{u_2} = \{v_3, v_8\}$, and

3. $\mathcal{L}_{u_3} = \{v_1, v_{11}\}$.

   Notice that Motif 1 is a subgraph of Motif 2 and Motif 2 is a subgraph of Motif 3. Each of them occurs exactly two times in the input graph. Thus all the information about Motifs 1 and 2 is already contained in Motif 3. This calls for a notion of *maximality* of motifs, which we formally define below.

DEFINITION 12.2   *(maximal motif, edge-maximal motif, vertex-maximal motif) Given $G(V, E)$, let*

$$M(V_m, E_m)$$

be a topological motif with its complete occurrence list

$$O_1, O_2, \ldots, O_l,$$

with the mappings for $1 \leq i \leq l$ as

$$F_i : V_M \to O_i.$$

- Edge-maximal: *The motif $M(V_m, E_m)$ is edge-maximal when for all pairs $u_1, u_2 \in V_m$,*

$$\text{if } (F_i(u_1)F_i(u_2)) \in E \text{ for all } i, \text{ then } (u_1 u_2) \in E_m \text{ holds.}$$

- Vertex-maximal: *The motif $M(V_m, E_m)$ is vertex-maximal when there do not exist vertices*

$$v_1, v_1', v_2, v_2', \ldots, v_l, v_l' \in V$$

such that for some $u \in V_m$,

$$F_i(u) = v_i', \text{ for each } i,$$

and, the following hold for all i:

1. *$v_i \notin O_i$, $v_i' \in O_i$,*
2. *$att(v_i) = a$, for some attribute a, and,*
3. *$(v_i v_i') \in E$.*

The motif is maximal *if both edge-maximality and vertex-maximality hold.*

In other words, edge-maximality ensures that no more edges can be added to the motif and vertex-maximality ensures that no more vertices can be added to the motif without altering the occurrence list. Continuing the example of Figure 12.12, motifs 1 and 2 are nonmaximal. Since no more edges or vertices can be added to motif 3, it is maximal.

12.4 Compact Topological Motifs

We first discuss an important issue with counting the number of occurrences of a motif: we call this the *combinatorial explosion due to occurrence-isomorphisms*.

12.4.1 Occurrence-isomorphisms

We next consider a slightly modified input graph shown in Figure 12.13(a). Here the vertex attributes black and white have both been replaced by red. How does the problem scenario change?

Motif 1 of Figure 12.13(b). This motif is given by $M(V_m, E_m)$ where $V_m = \{u_1, u_2, u_3, u_4, u_5\}$ and

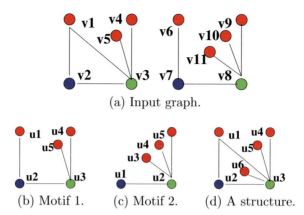

(a) Input graph.

(b) Motif 1.   (c) Motif 2.   (d) A structure.

FIGURE 12.13:   (a) The input graph with two connected components. (b) and (c) show motifs that occur at least twice on the graph. (d) A structure that is not a motif in (a).

$$att(u_1) = att(u_4) = att(u_5) = \text{red},$$

$$att(u_2) = \text{blue and}$$

$$att(u_3) = \text{green}.$$

The eight occurrences of this motif are given as follows:

- First connected component of the input graph:

  1. $O_1 = \{v_1, v_2, v_3, v_4, v_5\}$, with
     $F_1(u_1) = v_1$, $F_1(u_2) = v_2$, $F_1(u_3) = v_3$, $F_1(u_4) = v_4$, $F_1(u_5) = v_5$.

  2. $O_2 = \{v_1, v_2, v_3, v_4, v_5\}$, with
     $F_1(u_1) = v_1$, $F_1(u_2) = v_2$, $F_1(u_3) = v_3$, $F_1(u_4) = v_5$, $F_1(u_5) = v_4$.

- Second connected component of the input graph:

  3. $O_3 = \{v_6, v_7, v_8, v_9, v_{10}\}$, with
     $F_1(u_1) = v_6$, $F_1(u_2) = v_7$, $F_1(u_3) = v_8$, $F_1(u_4) = v_9$, $F_1(u_5) = v_{10}$.

  4. $O_4 = \{v_6, v_7, v_8, v_{10}, v_9\}$, with
     $F_1(u_1) = v_6$, $F_1(u_2) = v_7$, $F_1(u_3) = v_8$, $F_1(u_4) = v_{10}$, $F_1(u_5) = v_9$.

  5. $O_5 = \{v_6, v_7, v_8, v_{10}, v_{11}\}$, with
     $F_1(u_1) = v_6$, $F_1(u_2) = v_7$, $F_1(u_3) = v_8$, $F_1(u_4) = v_{10}$, $F_1(u_5) = v_{11}$.

  6. $O_6 = \{v_6, v_7, v_8, v_{11}, v_{10}\}$, with
     $F_1(u_1) = v_6$, $F_1(u_2) = v_7$, $F_1(u_3) = v_8$, $F_1(u_4) = v_{11}$, $F_1(u_5) = v_{10}$.

7. $O_7 = \{v_6, v_7, v_8, v_9, v_{11}\}$, with
   $F_1(u_1) = v_6, F_1(u_2) = v_7, F_1(u_3) = v_8, F_1(u_4) = v_9, F_1(u_5) = v_{11}$.

8. $O_8 = \{v_6, v_7, v_8, v_{11}, v_9\}$. with
   $F_1(u_1) = v_6, F_1(u_2) = v_7, F_1(u_3) = v_8, F_1(u_4) = v_{11}, F_1(u_5) = v_9$.

When attributes of two or more vertices of the motif are identical, sometimes they can be mapped to a fixed set of vertices of the input graph in combinatorially all possible ways. For example u_4 and u_5 of the motif are mapped onto the pair v_4 and v_5 in two possible ways (given by, F_2 and F_3). Similarly, u_4 and u_5 are mapped to any two of v_9, v_{10} and v_{11} in six possible ways.

We term this explosion in the number of distinct mappings as *combinatorial explosion due to occurrence-isomorphism.*

Motif 2 of Figure 12.13(b). This motif is given by $M(V_m, E_m)$ where $V_m = \{u_1, u_2, u_3, u_4, u_5\}$ and

$att(u_1) = $ blue,

$att(u_2) = $ green, and

$att(u_3) = att(u_4) = att(u_5) = $ red.

The twelve occurrences of this motif are given as follows:

- First connected component of the input graph:

    1. $O_1 = \{v_2, v_3, v_4, v_5, v_1\}$, with
       $F_1(u_1) = v_2, F_1(u_2) = v_3, F_1(u_3) = v_4, F_1(u_4) = v_5, F_1(u_5) = v_1$.
    2. $O_2 = \{v_2, v_3, v_4, v_5, v_1\}$, with
       $F_1(u_1) = v_2, F_1(u_2) = v_3, F_1(u_3) = v_4, F_1(u_4) = v_1, F_1(u_5) = v_5$.
    3. $O_3 = \{v_2, v_3, v_4, v_5, v_1\}$, with
       $F_1(u_1) = v_2, F_1(u_2) = v_3, F_1(u_3) = v_1, F_1(u_4) = v_4, F_1(u_5) = v_5$.
    4. $O_4 = \{v_2, v_3, v_4, v_5, v_1\}$, with
       $F_1(u_1) = v_2, F_1(u_2) = v_3, F_1(u_3) = v_1, F_1(u_4) = v_5, F_1(u_5) = v_4$.
    5. $O_5 = \{v_2, v_3, v_4, v_5, v_1\}$, with
       $F_1(u_1) = v_2, F_1(u_2) = v_3, F_1(u_3) = v_5, F_1(u_4) = v_1, F_1(u_5) = v_4$.
    6. $O_6 = \{v_2, v_3, v_4, v_5, v_1\}$, with
       $F_1(u_1) = v_2, F_1(u_2) = v_3, F_1(u_3) = v_5, F_1(u_4) = v_4, F_1(u_5) = v_1$.

- Second connected component of the input graph:

    7. $O_7 = \{v_7, v_8, v_9, v_{10}, v_{11}\}$, with
       $F_1(u_1) = v_7, F_1(u_2) = v_8, F_1(u_3) = v_9, F_1(u_4) = v_{10}, F_1(u_5) = v_{11}$.

8. $O_8 = \{v_7, v_8, v_9, v_{10}, v_{11}\}$, with
   $F_1(u_1) = v_7$, $F_1(u_2) = v_8$, $F_1(u_3) = v_9$, $F_1(u_4) = v_{11}$, $F_1(u_5) = v_{10}$.

9. $O_9 = \{v_7, v_8, v_9, v_{10}, v_{11}\}$, with
   $F_1(u_1) = v_7$, $F_1(u_2) = v_8$, $F_1(u_3) = v_{10}$, $F_1(u_4) = v_{11}$, $F_1(u_5) = v_9$.

10. $O_{10} = \{v_7, v_8, v_9, v_{10}, v_{11}\}$, with
    $F_1(u_1) = v_7$, $F_1(u_2) = v_8$, $F_1(u_3) = v_{10}$, $F_1(u_4) = v_9$, $F_1(u_5) = v_{11}$.

11. $O_{11} = \{v_7, v_8, v_9, v_{10}, v_{11}\}$, with
    $F_1(u_1) = v_7$, $F_1(u_2) = v_8$, $F_1(u_3) = v_{11}$, $F_1(u_4) = v_9$, $F_1(u_5) = v_{10}$.

12. $O_{12} = \{v_7, v_8, v_9, v_{10}, v_{11}\}$, with
    $F_1(u_1) = v_7$, $F_1(u_2) = v_8$, $F_1(u_3) = v_{11}$, $F_1(u_4) = v_{10}$, $F_1(u_5) = v_9$.

In this example, the *combinatorial explosion due to occurrence-isomorphism* is from the motif vertices u_3, u_4, u_5 being mapped in all possible ways to v_4, v_5, v_1 in the first connected component and to v_9, v_{10}, v_{11} in the second connected component of the input graph.

12.4.2   Vertex indistinguishability

Is it possible to count and describe the distinct occurrences without the combinatorial explosion as seen in the last two examples? For this we need to first recognize *indistinguishable* vertices, which are defined below. Given a graph [3]

$$M(V_M, E_M)$$

the vertices in

$$U_1 \subseteq V_M$$

are *indistinguishable* w.r.t. (w.r.t.) $U_2 \subseteq V_M$ if and only if

(1) $att(u_i) = a_1$, and $att(u_j) = a_2$, for all $u_i \in U_1$, $u_j \in U_2$, for some attributes a_1 and a_2, and,

(2) there is an edge $(u_i u_j) \in E_M$, for each $u_i \in U_1$ and $u_j \in U_2$.

Vertices in U_1 are said to be indistinguishable from each other w.r.t. U_2. Further, if there exists no

$$U_1' \supseteq U_1$$

---

[3]Although usually a graph is written as $G(V, E)$, here we use a motif notation $M(V_M, E_M)$ for the graph to emphasize the fact that *indistinguishability* of vertices is primarily associated with motifs (but evidenced in the input graph).

such that U_1' is indistinguishable w.r.t. U_2, then the set U_1 is *maximally indistinguishable* w.r.t. U_2.

12.4.3 Compact list

To ease handling of the combinatorial explosion due to occurrence isomorphisms, we introduce the compact list notation for the location lists of topological motifs [Par07a]. It is often possible to represent $\mathcal{L}_U$ in a much more compact way taking into account the fact that the vertices in U are indistinguishable.

For instance, in the motif in Figure 12.13(b),

$$U = \{u_4, u_5\}$$

is a maximal set of indistinguishable vertices (w.r.t $\{u_3\}$) and

$$\mathcal{L}_U = \{\{v_4, v_5\}, \{v_9, v_{10}\}, \{v_{11}, v_{10}\}, \{v_9, v_{11}\}\}.$$

However, a more compact way to represent $\mathcal{L}_U$ is by the set,

$$\mathcal{L}_U^c = \{\{v_4, v_5\}, \{v_9, v_{10}, v_{11}\}\}.$$

and one recovers $\mathcal{L}_U$ from $\mathcal{L}_U^c$ by taking all two (the smallest cardinality of the sets in $\mathcal{L}_U$) element subsets of the sets in $\mathcal{L}_U^c$. We call

1. $\mathcal{L}_U$ the expansion of the set $\mathcal{L}_U^c$, and

2. $\mathcal{L}_U^c$ a compact form of $\mathcal{L}_U$.

In the rest of the discussion, we denote a compact list $\mathcal{L}_U^c$ simply by $\mathcal{L}_U$ and it should be clear from the context what we mean.

12.4.4 Compact vertex, edge & motif

Next, we define a *compact vertex* and a *compact edge* which is a natural next step after recognizing indistinguishable vertices and compact location lists. Given a maximal motif

$$M(V_M, E_M),$$

if

1. $U_1 (\subset V_M)$ is maximally indistinguishable w.r.t. $U_2 (\subset V_M)$ and

2. U_2 is maximally indistinguishable w.r.t. U_1,

then U_1 and U_2 are called *compact vertices*. Further, each of the following holds.

1. Each vertex in U_i has the same attribute, for $i = 1, 2$.

2. $U_1 \cap U_2 = \emptyset$.

3. Since there is an edge from each vertex $u_1 \in U_1$ to each vertex $u_2 \in U_2$, there are

$$|U_1| \times |U_2|$$

edges between U_1 and U_2. We represent the $|U_1| \times |U_2|$ edges by a single edge, written as

$$(U_1 U_2).$$

This is also called a *compact edge*. In other words, there is a compact edge between two compact vertices.

This naturally leads to the compact notation for the motif where the vertex set is a collection of compact vertices and compact edges defined on them. For convenience, we represent a compact motif as C with the following notation

$$C(V_C, E_C) \equiv M(V_M, E_M),$$

where U_C is the set of compact vertices and E_C the set of compact edges.

It is important to note that two compact vertices may have a nonempty intersection. The second example shown in Figure 12.14 illustrates such a nonempty intersection.

12.4.5   Maximal compact lists

We have defined compact lists, but what is a *maximal* compact list? As we have seen earlier a compact list is inalienably associated with a compact vertex. The maximality of a compact list is 'inherited' from the associated compact vertex.

Given a graph $G(V, E)$, we say a compact list $\mathcal{L}$ is a *maximal* compact list, if there exists some maximal compact motif $C(V_C, E_C)$ such that $\mathcal{L} = \mathcal{L}_U$ with $U \in V_C$.

The maximality property of the compact list is central to the discovery method of [Par07a]. Great care is taken to ensure that a new compact list that is generated is maximal. With this in mind, we discuss some operations that maintain this maximality of the list.

12.4.6   Conjugates of compact lists

In the following discussion let D be the number of distinct attributes in the given graph $G(V, E)$. Also for an attribute x, let

$$V_x = \{v \in V \mid att(v) = x\}.$$

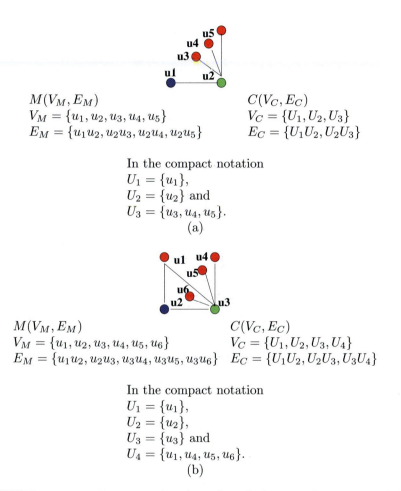

$M(V_M, E_M)$ $C(V_C, E_C)$

$V_M = \{u_1, u_2, u_3, u_4, u_5\}$ $V_C = \{U_1, U_2, U_3\}$

$E_M = \{u_1u_2, u_2u_3, u_2u_4, u_2u_5\}$ $E_C = \{U_1U_2, U_2U_3\}$

In the compact notation
$U_1 = \{u_1\}$,
$U_2 = \{u_2\}$ and
$U_3 = \{u_3, u_4, u_5\}$.
(a)

$M(V_M, E_M)$ $C(V_C, E_C)$

$V_M = \{u_1, u_2, u_3, u_4, u_5, u_6\}$ $V_C = \{U_1, U_2, U_3, U_4\}$

$E_M = \{u_1u_2, u_2u_3, u_3u_4, u_3u_5, u_3u_6\}$ $E_C = \{U_1U_2, U_2U_3, U_3U_4\}$

In the compact notation
$U_1 = \{u_1\}$,
$U_2 = \{u_2\}$,
$U_3 = \{u_3\}$ and
$U_4 = \{u_1, u_4, u_5, u_6\}$.
(b)

FIGURE 12.14:   Two examples of motifs with their usual notation and the corresponding compact notation. Notice that in the motif in (b), two compact vertices U_1 and U_4 have a nonempty intersection, i.e., $U_1 \cap U_4 = \{u_1\}$.

Conjugate of compact vertices. If edge $(v_1v_2) \in E$, then v_2 is an immediate neighbor of v_1 and v_1 is an immediate neighbor of v_2. Just as a vertex has an immediate neighbor in a graph, we define such a notion for a compact vertex and call it the *conjugate*.

For attributes a and x, let $L \subset V_a$. Then

$$conj_x(L) = \{v \in V_x \mid \text{ for each } v' \in L \ (vv') \in E\}. \tag{12.1}$$

Thus

1. $conj_b(L)$ is a conjugate if L and

2. L is a conjugate of $conj_b(L)$.

Further, L has at most D nonempty conjugates.

The conjugates can also be seen from the perspective of a compact motif as follows. Let U_1 and U_2 be compact vertices in a compact motif

$$C(V_C, E_C),$$

with

$$(U_1U_2) \in E_C.$$

Then we call

1. U_1 to be a conjugate of U_2 and

2. similarly U_2 to be a conjugate of U_1.

This leads to the following:

For each $L_1 \in \mathcal{L}_{U_1}$ there is at least one $L_2 \in \mathcal{L}_{U_2}$ such that for each pair $v_1 \in L_1$, $v_2 \in L_2$, $(v_1v_2) \in E$.

Similarly for each $L_2 \in \mathcal{L}_{U_2}$ there is at least one $L_1 \in \mathcal{L}_{U_1}$ such that for each pair $v_1 \in L_1$, $v_2 \in L_2$, $(v_1v_2) \in E$.

Then L_1 is a conjugate of L_2 and L_2 is a conjugate of L_1.

Conjugate of location lists. For an attribute x, the *conjugate* list of a given list $\mathcal{L}$ is defined as follows (using Equation (12.1)):

$$Conj_x(\mathcal{L}) = \{conj_x(L) \mid L \in \mathcal{L}\}. \tag{12.2}$$

Note that $\mathcal{L}$ has at most D nonempty conjugates.

The following is a critical property of the conjugate list that we exploit in the algorithm to discover the compact (maximal) motifs. This fact can be verified (using proof by contradiction) and we leave the proof as an exercise for the reader.

THEOREM 12.1
(Conjugate maximality theorem) *The conjugate of a maximal location list is maximal.*

Continuing example. Consider motif 2 of the example in Figure 12.13.

$$Conj_{green}(\mathcal{L}_U) = \{\{v_3\}, \{v_8\}, \{v_8\}, \{v_8\}\}$$
$$= \{\{v_3\}, \{v_8\}\}$$
$$= \{v_3, v_8\}. \tag{12.3}$$

The conjugate of the compact list is shown as below,

$$Conj_{green}(\mathcal{L}_U) = \{\{v_3\}, \{v_8\}\}$$
$$= \{v_3, v_8\}.$$

Conjugate notation. In an implementation, the conjugate relation is stored as pointers. However in the description here, we show the conjugate relation of the list as $\Updownarrow$, and that of each of its elements as $\Uparrow$. For example,

$$
\begin{array}{lccc}
\mathcal{L}_U = & \{\{v_4, v_5\}, & \{v_9, v_{10}, v_{11}\}\} \\
\Updownarrow & \Uparrow & \Uparrow \\
\mathcal{L}_{u_2} = & \{v_3, & v_8\} \\
\Updownarrow & \Uparrow & \Uparrow \\
\mathcal{L}_{u_1} = & \{v_2, & v_7\}
\end{array}
\tag{12.4}
$$

Note that in this example the following hold:

1. $Conj_{red}(\mathcal{L}_{u_2}) = \mathcal{L}_U$ and $Conj_{blue}(\mathcal{L}_{u_2}) = \mathcal{L}_{u_1}$,

2. $Conj_{green}(\mathcal{L}_U) = \mathcal{L}_{u_2}$, and

3. $Conj_{blue}(\mathcal{L}_{u_1}) = \mathcal{L}_{u_2}$.

Multiplicity in (compact) location lists. Consider the location list given in Equation (12.3). Here we have replaced three instances of v_8, as determined by the conjugates, with just one. In the rest of the treatment, we ignore such multiplicities of the elements of the location list. Note that there is no loss of information. For instance, consider the first two lists in Equation (12.4) along with the conjugacy relations:

$$
\begin{array}{lcc}
\mathcal{L}_U = & \{\{v_4, v_5\}, & \{v_9, v_{10}, v_{11}\}\} \\
\Updownarrow & \Uparrow & \Uparrow \\
\mathcal{L}_{u_2} = & \{v_3, & v_8\}
\end{array}
$$

This implicity implies the following conjugacies:

$$
\begin{array}{lcccc}
\mathcal{L}_U = & \{\{v_4, v_5\}, & \{v_9, v_{10}\}, & \{v_9, v_{11}\}, & \{v_{10}, v_{11}\}\} \\
\Updownarrow & \Uparrow & \Uparrow & \Uparrow & \Uparrow \\
\mathcal{L}_{u_2} = & \{v_3, & v_8, & v_8, & v_8\}
\end{array}
$$

12.4.7 Characteristics of compact lists

Given a graph $G(V, E)$ and a maximal motif

$$M(V_M, E_M)$$

on this graph, let

$$U(\subseteq V_M)$$

be a set of maximal indistinguishable vertices (w.r.t. some $U' \subset V_M$), the the compact location list of U is written as: [4]

$$\mathcal{L}_U = \{L_1, L_2, \ldots, L_\ell\} \subset 2^V.$$

Then the five characteristics of $\mathcal{L}_U$ are as follows:

1. Let

$$d = \min_{i=1}^{\ell} |L_i|.$$

   If for some $L' \in \mathcal{L}$,

$$|L'| = d,$$

   then L' is called a *discriminant* of $\mathcal{L}$. We write

$$d = discSz(\mathcal{L}).$$

2.

$$flat(\mathcal{L}) = \bigcup_{L \in \mathcal{L}} L.$$

3. *Expansion* of $\mathcal{L}$, $Exp(\mathcal{L}) \subset 2^V$ is given as:

   $Exp(\mathcal{L}) = \{L \in 2^V \mid \text{ there exists some } L_i \in \mathcal{L} \text{ with } L \subset L_i \text{ and } |L| = d\}.$

4. All the vertices in $flat(\mathcal{L})$ have the same attribute given by $att(\mathcal{L})$.

5. For $1 \leq l \leq \ell$, let

$$E_{L_l} = \{(v_1 v_2) \in E \mid v_1, v_2 \in L_l\}.$$

   Then

$$G_{L_l}(L_l, E_{L_l})$$

   is called the *induced subgraph* on L_l. Further, if for each $u_1, u_2 \in L_l$, $(u_1 u_2) \in E_{L_l}$, then the graph G_{L_l} is called a *clique*.

   $clq(\mathcal{L})$ is an indicator that is set 1 if *all* the ℓ induced subgraphs are cliques. Formally,

$$clq(\mathcal{L}) = \begin{cases} 1, & \text{if for each } L \in \mathcal{L}, G_L(L, E_L) \text{ is a clique,} \\ 0, & \text{otherwise.} \end{cases}$$

---

[4]For example, compact list $\mathcal{L}_U = \{\{v_1, v_2, v_3\}, \{v_4, v_5\}\}$ is written as $\mathcal{L}_U = \{L_1, L_2\}$, where $L_1 = \{v_1, v_2, v_3\}$ and $L_2 = \{v_4, v_5\}$.

Continuing example. We compute the five characteristics for the example of the motif in Figure 12.13(b). Note that

$$U_1 = \{u_4, u_5\}$$

is a maximal set of indistinguishable vertices w.r.t.

$$U_2 = \{u_3\}$$

and vice-versa. The compact location lists of U_1 and U_2 along with their characteristics is given below.

(a) $\mathcal{L}_{U_1} = \{\{v_4, v_5\}, \{v_9, v_{10}, v_{11}\}\}$.

     1. Discriminant of $\mathcal{L}_{U_1}$ is $\{v_4, v_5\}$ and thus $discSz(\mathcal{L}_{U_1}) = 2$.

     2. $flat(\mathcal{L}_{U_1}) = \{v_4, v_5, v_9, v_{10}, v_{11}\}$.

     3. $Exp(\mathcal{L}_{U_1}) = \{\{v_4, v_5\}, \{v_9, v_{10}\}, \{v_{10}, v_{11}\}, \{v_9, v_{11}\}\}$.

     4. $att(\mathcal{L}_{U_1}) =$ red,

     5. $clq(\mathcal{L}_{U_1}) = 0$.

(b) $\mathcal{L}_{U_2} = \{\{v_3\}, \{v_8\}\}$ or simply $\mathcal{L}_{u_3} = \{v_3, v_8\}$ with $att(\mathcal{L}_{U_2}) =$ green.

Next consider the motif in Figure 12.13(c).

$$U_3 = \{u_3, u_4, u_5\}$$

is a maximal set of indistinguishable vertices w.r.t.

$$U_4 = \{u_2\}$$

and vice-versa. The compact location lists of U_3 and U_4 is given as:

(a) $\mathcal{L}_{U_3} = \{\{v_1, v_4, v_5\}, \{v_9, v_{10}, v_{11}\}\}$.

     1. Discriminant of $\mathcal{L}_{U_3}$ is $\{v_1, v_4, v_5\}$ and so is $\{v_9, v_{10}, v_{11}\}$. Thus $discSz(\mathcal{L}_{U_3}) = 3$.

     2. $flat(\mathcal{L}_{U_3}) = \{v_1, v_4, v_5, v_9, v_{10}, v_{11}\}$.

     3. $Exp(\mathcal{L}_{U_3}) = \mathcal{L}_{U_3}$.

     4. $att(\mathcal{L}_{U_3}) =$ red,

     5. $clq(\mathcal{L}_{U_3}) = 0$.

(b) $\mathcal{L}_{U_4} = \{\{v_3\}, \{v_8\}\}$ or simply $\mathcal{L}_{u_2} = \{v_3, v_8\}$ with $att(\mathcal{L}_{U_4}) =$ green.

12.4.8   Maximal operations on compact lists

Once we recognize that a compact list is merely a concise notation for its expansion, the following is easy to see.

1. $\mathcal{L}_1 =_c \mathcal{L}_2$ if and only if $Exp(\mathcal{L}_1) = Exp(\mathcal{L}_2)$.

   In the rest of the chapter we use '=' also as an assignment symbol. Thus when we say $\mathcal{L}_1 = \mathcal{L}_2$, we mean that list $\mathcal{L}_1$ is assigned to be the list $\mathcal{L}_2$.

2. $\mathcal{L}_1 \subset_c \mathcal{L}_2$ if and only if
   for each $L_1 \in \mathcal{L}_1$, there exists some $L_2 \in \mathcal{L}_2$ such that $L_1 \subset L_2$.

3. The intersection of two compact lists is written as

$$\mathcal{L}_3 = \mathcal{L}_1 \cap_c \mathcal{L}_2,$$

   and is defined as follows:

   $v_1, v_2 \in L_3 (\in \mathcal{L}_3)$
   if and only if $v_1, v_2 \in L_1, L_2$ for some $L_1 \in \mathcal{L}_1$ and $L_2 \in \mathcal{L}_2$.

   The intersection of p compact lists,

$$\mathcal{L}_1 \cap_c \mathcal{L}_2 \cap_c \ldots \cap_c \mathcal{L}_p,$$

   can be generalized from intersection of two lists.

   The following can be verified and is left as an exercise for the reader.

$$flat(\mathcal{L}_3) = flat(\mathcal{L}_1) \cap flat(\mathcal{L}_2).$$

4. The union of two compact lists is written as

$$\mathcal{L}_3 = \mathcal{L}_1 \cup_c \mathcal{L}_2,$$

   and is defined as follows:

$$\mathcal{L}_3 = \left\{ L_1 \cup L_2 \;\middle|\; \begin{array}{ll} L_1 \in \mathcal{L}_1 & \text{and} \quad L_2 \in \mathcal{L}_2 \quad \text{and} \\ L_1 \subset L_2 & \text{or} \quad L_2 \subset L_1 \end{array} \right\}.$$

   The union of p compact lists,

$$\mathcal{L}_1 \cup_c \mathcal{L}_2 \cup_c \ldots \cup_c \mathcal{L}_p,$$

   can be generalized from union of two lists.

   The following can be verified and is left as an exercise for the reader.

$$flat(\mathcal{L}_3) = flat(\mathcal{L}_1) \cup flat(\mathcal{L}_2).$$

5. The difference of two compact lists is written as

$$\mathcal{L}_3 = \mathcal{L}_1 \setminus_c \mathcal{L}_2,$$

and is defined as follows:

$v_1 \in L_3 (\in \mathcal{L}_3)$ if and only if

1. $v_1, v_2 \in L_1$ and
2. $v_2 \in L_2$, but $v_1 \notin L_2$,

for some $L_1 \in \mathcal{L}_1$ and $L_2 \in \mathcal{L}_2$.

However, when

$$\mathcal{L}_2 = \mathcal{L}_1 \cap_c \mathcal{L}_1',$$

for some $\mathcal{L}_1'$, then,

$$
\begin{aligned}
\mathcal{L}_3 &= \mathcal{L}_1 \setminus_c \mathcal{L}_2 \\
&= \mathcal{L}_1 \setminus_c (\mathcal{L}_1 \cap_c \mathcal{L}_1') \\
&= \{L_1 \setminus L_2 \mid L_1 \in \mathcal{L}_1, L_2 \in \mathcal{L}_2 \text{ and } L_1 \cap L_2 \neq \emptyset \}.
\end{aligned}
$$

Note that in general, given maximal compact lists $\mathcal{L}_1$ and $\mathcal{L}_2$,

$$\mathcal{L}_1 \setminus_c \mathcal{L}_2$$

is not necessarily maximal.

12.4.9 Maximal subsets of location lists

A location list can have a large number of subsets, but which subsets are maximal? Our interest is only in these specific subsets. We define two ways of generating new lists given one location list $\mathcal{L}$.

1. Let

$$
\begin{aligned}
d &= discSz(\mathcal{L}) \quad \text{and} \\
d_{\max} &= \max_{L_i \in \mathcal{L}} (|L_i|).
\end{aligned}
$$

$Enrich(\mathcal{L})$ is a set of lists defined as

$$Enrich(\mathcal{L}) = \left\{ \mathcal{L}(p) \;\middle|\; \begin{array}{l} p = |L_i| \text{ for some } L_i \in \mathcal{L}, \text{ and} \\ d \le p \le d_{\max}. \end{array} \right\},$$

where

$$\mathcal{L}(p) = \{L_i \in \mathcal{L} \mid |L_i| \ge p\}.$$

Thus new lists are generated only when

$$d_{\max} > d > 1,$$

and for each such newly generated list $\mathcal{L} \in Enrich(\mathcal{L})$,

$$discSz(\mathcal{L}) > d.$$

2. Let

$$L_i(\subset V),$$

where all the vertices have the same attribute. If

$$U \subset L_i$$

induces a clique,[5] then it is also a *maximal clique* if there is no U' such that

$$U \subsetneq U' \subseteq L_i$$

that also induces a clique on the input graph.

Let

$$maxClk(L_i) = \{L_{1_i}, L_{2_i}, \ldots, L_{k_i}\}$$

where each L_j, $1_i \leq j \leq k_i$ induces a maximal clique. Then $clique(\mathcal{L})$ is a list defined as:

$$clique(\mathcal{L}) = \{L_{j_i} \mid L_{j_i} \in maxClk(L_i) \text{ and } L_i \in \mathcal{L}\}.$$

Clearly

$$clq(clique(\mathcal{L})) = 1.$$

Note that $Enrich(\mathcal{L})$ results in possibly more than one new list whereas $clique(\mathcal{L})$ results in at most one new list.

We next make another observation that is important for the algorithm that is discussed in the later sections. It states that conjugate lists can be computed by simply taking appropriate subsets of other known compact lists rather than using the original graph $G(V, E)$ as in Equation (12.1).

LEMMA 12.1
(Intersection conjugate lemma) *For lists*

$$\mathcal{L}_1, \mathcal{L}_2, \ldots, \mathcal{L}_p$$

let the conjugates for an attribute a be

$$\mathcal{L}_{1_a}, \mathcal{L}_{2_a}, \ldots, \mathcal{L}_{p_a}.$$

For

$$1 \leq j \leq p,$$

---

[5]See Section 12.4.7 for the definition of clique.

define relations R_j as follows:

$$(L_j, L_{a_j}) \in R_j \Leftrightarrow \begin{cases} L_j \text{ is a conjugate of } L_{a_j} \\ \text{for } L_j \in \mathcal{L}_j \text{ and } L_{a_j} \in \mathcal{L}_{a_j}. \end{cases}$$

Next, let

$$\mathcal{L}' = \mathcal{L}_1 \cap_c \mathcal{L}_2 \cap_c \ldots \cap_c \mathcal{L}_p.$$

Then the conjugate, L'_a, of each element

$$L' \in \mathcal{L}'$$

is given as follows: [6]

$$L'_a = \bigcup_j \left(\bigcup_i L_{a_{j_i}} \right) \quad \text{where} \quad \begin{cases} L' \subset L_{j_i} \in \mathcal{L}_j \text{ and} \\ (L_{j_i}, L_{a_{j_i}}) \in R_j. \end{cases}$$

Further, if each of the following is maximal (then $\mathcal{L}_c$ is maximal):

$$\mathcal{L}_1, \mathcal{L}_2, \ldots, \mathcal{L}_p,$$
$$\mathcal{L}_{1_a}, \mathcal{L}_{2_a}, \ldots, \mathcal{L}_{p_a},$$

then this conjugate, L'_a, is the same set given by

$$conj_a(L') \quad (\text{of Equation (12.1)})$$

that is obtained directly using the input graph $G(V, E)$.

COROLLARY 12.1

(Subset conjugate lemma) *For an attribute a, let $\mathcal{L}_a$ be a conjugate of $\mathcal{L}$. Define a relation R as follows:*

$$(L, L_a) \in R_j \Leftrightarrow \begin{cases} L \text{ is a conjugate of } L_a \\ \text{for } L \in \mathcal{L} \text{ and } L_a \in \mathcal{L}_a. \end{cases}$$

Let

$$\mathcal{L}' \subset_c \mathcal{L}.$$

Then the conjugate, L'_a, of each element $L' \in \mathcal{L}'$ is given as follows.

$$L'_a = \bigcup_i L_{a_i}, \quad \text{where} \quad \begin{cases} L' \subset L_i \in \mathcal{L} \text{ and} \\ (L_i, L_{a_i}) \in R. \end{cases}$$

Further, if $\mathcal{L}$, $\mathcal{L}_a$ and $\mathcal{L}'$ are maximal, then this conjugate, L'_a, is the same set given by

$$conj_a(L') \quad (\text{of Equation (12.1)})$$

that is obtained directly using the input graph $G(V, E)$.

---

[6]The two 'unions' here are due to the fact that in the intersection set $\mathcal{L}'$, more than one element (denoted here as L_{j_i}) of $\mathcal{L}_j$ may be involved.

12.4.10   Binary relations on compact lists

We study some binary relations on compact lists that will be used in the following section to construct the compact motifs. Recall that binary relations can be captured by a graph.

1. (complete intersection) An intersection of $\mathcal{L}_1$ of $\mathcal{L}_2$ is called a *complete intersection* if

   $\mathcal{L}_1 \cap_c \mathcal{L}_2$ is a *complete subset* of $\mathcal{L}_1$ and $\mathcal{L}_2$.

   Note that $\mathcal{L}' \subset_c \mathcal{L}$ is a *complete subset* if

   for each $L \in \mathcal{L}$ there is $L' \in \mathcal{L}'$ such that $L' \subset L$.

2. (compatible) $\mathcal{L}_1$ and $\mathcal{L}_2$ are *compatible* if

   (a) $\mathcal{L}_1 \cap_c \mathcal{L}_2 = \emptyset$, or,

   (b) if $\mathcal{L}_1 \cap_c \mathcal{L}_2 \neq \emptyset$, then the intersection is complete.

3. (incompatible) If $\mathcal{L}_1$ and $\mathcal{L}_2$ are not compatible they are called *incompatible*.

4. (conjugate) We have already seen the condition when $\mathcal{L}_1$ is a conjugate of $\mathcal{L}_2$ (see Equation (12.2)).

12.4.11   Compact motifs from compact lists

How are compact motifs computed from compact lists? We define a *meta-graph*

$$\mathbf{G(L, E)}$$

on a collection of compact lists $\mathbf{L}$ where the labeled edges are defined as follows. Figure 12.15 shows a concrete example.

1. With a slight abuse of notation, we call a compact list

   $$\mathcal{L} \in \mathbf{L}$$

   a vertex in this meta-graph.

2. The edges in the meta-graph are of three types:[7]

   (a) If $\mathcal{L}_2$ is a conjugate of $\mathcal{L}_1$ then

      $$(\mathcal{L}_1 \mathcal{L}_2) \in \mathbf{E},$$

      and the edge type is 'link'. In Figure 12.15(a) this is shown as a regular solid edge.

---

[7]To avoid confusion with attributes on vertices we call this the 'edge type' instead of 'edge attribute'.

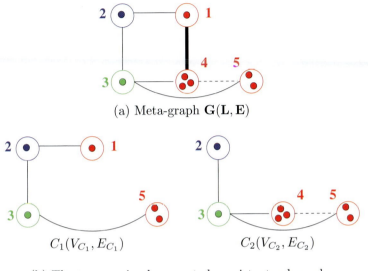

(a) Meta-graph $\mathbf{G(L, E)}$

$C_1(V_{C_1}, E_{C_1})$ $C_2(V_{C_2}, E_{C_2})$

(b) The two maximal connected consistent subgraphs.

$M_1(V_{M_1}, E_{M_1})$ $M_2(V_{M_2}, E_{M_2})$

(c) Maximal motifs.

FIGURE 12.15: (a) A meta-graph where the the 'forbidden' edge is shown in bold (connects an *inconsistent* pair), the 'subsume' edge is shown dashed (connects a *complete intersection* pair) and the remaining edges are 'link' edges (denoting the conjugacy relation). The correspondence between the node numbers shown here to the location lists shown in Figure 12.17 are as follows: $1 \leftrightarrow \mathcal{L}_1$, $2 \leftrightarrow \mathcal{L}_2$, $3 \leftrightarrow \mathcal{L}_3$, $4 \leftrightarrow \mathcal{L}_4$, $5 \leftrightarrow \mathcal{L}_4 \setminus_c \mathcal{L}_1$. The singleton lists are not shown here. (b) and (c) show the MCCSs and the maximal motifs respectively.

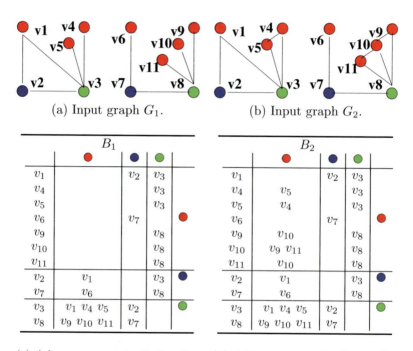

(a) Input graph G_1.   (b) Input graph G_2.

(c) Adjacency matrix B_1 for G_1.   (d) Adjacency matrix B_2 for G_2.

FIGURE 12.16:   The two running examples G_1 and G_2 with their adjacency matrices. Each graph has two connected components.

Initialization ($\mathbf{L}_{init}$)

$$\mathcal{L}_1 \quad = \quad \{v_1, \qquad\qquad v_6\}$$
$$\Updownarrow \qquad\qquad\qquad \updownarrow \qquad\qquad \updownarrow$$
$$\mathcal{L}_2 \quad = \quad \{v_2, \qquad\qquad v_7\}$$
$$\Updownarrow \qquad\qquad\qquad \updownarrow \qquad\qquad \updownarrow$$
$$\mathcal{L}_3 \quad = \quad \{v_3, \qquad\qquad v_8\}$$
$$\Updownarrow \qquad\qquad\qquad \updownarrow \qquad\qquad \updownarrow$$
$$\mathcal{L}_4 \quad = \quad \{\{v_1, v_4, v_5\}, \quad \{v_9, v_{10}, v_{11}\}\}$$

$\mathcal{L}_1$ and $\mathcal{L}_4$ are *inconsistent*.

Iterative Step

$$\mathcal{L}_4 \setminus_c \mathcal{L}_1 \quad = \quad \{\{v_4, v_5\}, \qquad \{v_9, v_{10}, v_{11}\}\}$$
$$\Updownarrow \qquad\qquad\qquad\quad \updownarrow \qquad\qquad\qquad \updownarrow$$
$$\mathcal{L}_3 \quad = \quad \{v_3, \qquad\qquad v_8\}$$

$\mathcal{L}_1$ and $\mathcal{L}_4 \setminus_c \mathcal{L}_1$ are *consistent*.

$$\mathcal{L}_1 \cap_c \mathcal{L}_4 \quad = \quad \{v_1\}$$
$$\Updownarrow \qquad\qquad\qquad\qquad\quad \updownarrow$$
$$\mathcal{L}_2' \quad = \quad \{v_2\}$$
$$\Updownarrow \qquad\qquad\qquad\qquad\quad \updownarrow$$
$$\mathcal{L}_3' \quad = \quad \{v_3\}$$
$$\Updownarrow \qquad\qquad\qquad\qquad\quad \updownarrow$$
$$\mathcal{L}_4' \quad = \quad \{\{v_1, v_4, v_5\}\}$$

$\mathcal{L}_1 \cap_c \mathcal{L}_4$ and L_4' are *consistent*.

$$\mathcal{L}_1 \setminus_c \mathcal{L}_4 \quad = \quad \{v_6\}$$
$$\Updownarrow \qquad\qquad\qquad\qquad\quad \updownarrow$$
$$\mathcal{L}_2' \quad = \quad \{v_7\}$$
$$\Updownarrow \qquad\qquad\qquad\qquad\quad \updownarrow$$
$$\mathcal{L}_3' \quad = \quad \{v_8\}$$
$$\Updownarrow \qquad\qquad\qquad\qquad\quad \updownarrow$$
$$\mathcal{L}_4' \quad = \quad \{\{v_9, v_{10}, v_{11}\}\}$$

$\mathcal{L}_1 \setminus_c \mathcal{L}_4$ and L_4' are *consistent*.

Motifs with quorum $K = 2$.

(a)   (b)

FIGURE 12.17: The solution for the input graph shown in Figure 12.16(a). See text for details.

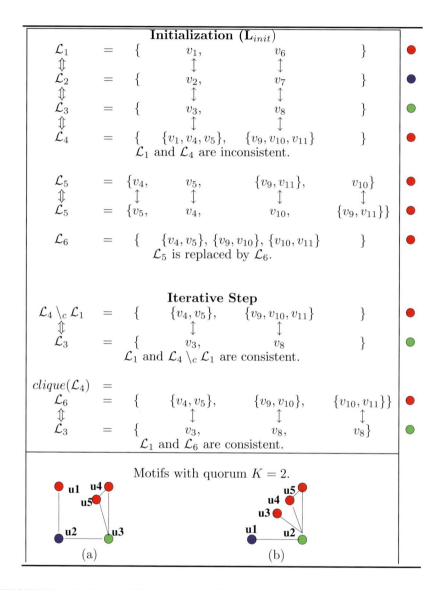

FIGURE 12.18:   The solution for the input graph shown in Figure 12.16(b). Note that singleton location lists are not shown. See text for details.

(b) Let $\mathcal{L}_1$ and $\mathcal{L}_2$ belong to a connected component (of link edges) of the meta-graph.

If $\mathcal{L}_1$ and $\mathcal{L}_2$ are compatible then they are called *consistent* and if they are incompatible, they are called *inconsistent*. If $\mathcal{L}_1$ and $\mathcal{L}_2$ are inconsistent, then

$$(\mathcal{L}_1\mathcal{L}_2) \in \mathbf{E},$$

and the edge type is 'forbidden'. In Figure 12.15(a) this is shown as a bold edge and this is always between vertices (location lists) that have the same attribute.

(c) If

$$\mathcal{L}_2 = \mathcal{L}_1 \setminus_c \mathcal{L}_3,$$

for some $\mathcal{L}_3$, then

$$(\mathcal{L}_1\mathcal{L}_2) \in \mathbf{E},$$

and the edge type is 'subsume'. In Figure 12.15(a) this is shown as a dashed edge and the edge is also always between vertices (location lists) that have the same attribute.

A connected subgraph of $\mathbf{G}(\mathbf{L}, \mathbf{E})$ that has a pair of inconsistent vertices is called an *inconsistent subgraph*. A subgraph that has no inconsistent pair is called a *consistent* subgraph. Why is this property important? [8]

Rationale: If the compact vertices form a cycle (of link edges) on the meta-graph, then these are equivalent to cycles in the motif and then it is important to check that the cycles occur in each occurrence of the motif (by checking the location list).

In the case that the cycle is formed only in some but not all occurrences, then we remove the inconsistency by computing new lists

$$\mathcal{L}_1 \setminus_c \mathcal{L}_2 \quad \text{and} \quad \mathcal{L}_2 \setminus_c \mathcal{L}_1,$$

which excludes these common (offending) vertices. It is easy to see that for an inconsistent pair $\mathcal{L}_1$ and $\mathcal{L}_2$,

1. the two must lie on a cycle (of link edges) in the meta-graph $\mathbf{G}(\mathbf{L}, \mathbf{E})$ and

2. $att(\mathcal{L}_1) = att(\mathcal{L}_2)$.

---

[8]Most patternists (scientists who specialize in pattern discovery in data) with an honest regard for mathematics, shudder at the thought of patterns in graphs because of the sheer promiscuity that the vertices display in terms of how many neighbors (partners) each can sustain. We also get no respite from the consequences of this. So, it is not surprising that both graph-theoretic and set-theoretic tools are used for this (that of *connectedness* in the meta-graph and *complete* intersection of compact lists).

Before we get down to the task of computing the compact motifs, we summarize the important observations as a theorem.

THEOREM 12.2
(Maximal begets maximal theorem) *If $\mathcal{L}_1$ and $\mathcal{L}_2$ are maximal, then the following statements hold.*

1. *Each conjugate list of $\mathcal{L}_1$ is maximal.*

2. *Each list in $Enrich(\mathcal{L}_1)$ is maximal.*

3. *$clique(\mathcal{L}_1)$ is maximal.*

4. *$\mathcal{L}_1 \cap_c \mathcal{L}_2$ is maximal.*

5. *$\mathcal{L}_1 \setminus_c \mathcal{L}_2$ and $\mathcal{L}_2 \setminus_c \mathcal{L}_1$ are maximal, if $\mathcal{L}_1$ and $\mathcal{L}_2$ are inconsistent.*

6. *$\mathcal{L}_i \setminus_c (\mathcal{L}_1 \cap_c \mathcal{L}_2)$ is not necessarily maximal for $i = 1, 2$.*

The following theorem summarizes the observation and is central to extracting the maximal motifs.

THEOREM 12.3
(Compact motif theorem) *Given an input graph, let its meta-graph be given as*

$$\mathbf{G(L, E)}.$$

A subgraph

$$C(V_C, E_C)$$

is a maximal connected consistent subgraph (MCCS) of the meta-graph, if it satisfies the following:

1. *(connected) For any two vertices in V_C there is a path between the two where each edge is of type 'link'.*

2. *(consistent) For no vertices $\mathcal{L}_1, \mathcal{L}_2 \in V_C$ is the edge*

$$(\mathcal{L}_1 \mathcal{L}_2) \in E_C$$

of the type 'forbidden'.

3. *(maximal) No more vertices can be added to V_C satisfying the above two conditions.*

Next,

$$C(V_C, E_C)$$

defines a (maximal) compact motif on input graph $G(V, E)$.

The construction of all the MCCSs in a meta-graph is discussed in Exercise 153. The construction of a maximal motif from each MCSS is discussed below.

MCCS to compact motif. Let the MCCS be given by

$$C(V_C, E_C).$$

We wish to compute

$$M(V_M, E_M)$$

from $C(V_C, E_C)$. See Figure 12.14 for the notation and concrete examples. For a location list $\mathcal{L}_i \in V_C$, let

$$att(\mathcal{L}_i) = a_i \text{ and}$$
$$discSz(\mathcal{L}_i) = d_i.$$

Then the compact vertex U_i corresponding to $\mathcal{L}_i$ is specified as

$$U_i = \{u_{i_1}, u_{i_2}, \ldots, u_{i_{d_i}}\},$$
$$\subset V_M,$$

and for each $1 \leq j \leq d_i$,

$$att\left(u_{i_j} \in V_M\right) = att(U_i).$$

Also if

$$clq(\mathcal{L}_i) = 1,$$

then an edge is introduced in the motif between every pair of vertices, i.e., for each $1 \leq j < k \leq d_i$,

$$(u_{i_j} u_{i_k}) \in E_M.$$

However, if the edge type of $(\mathcal{L}_i \mathcal{L}_j)$ is 'subsume', i.e., one is a complete subset of the other, then without loss of generality let

$$d_i \leq d_j,$$

and d_i vertices in U_i and U_j get the same labels, i.e.,

$$U_i = \{u_{i_1}, u_{i_2}, \ldots, u_{i_{d_i}}\},$$
$$\subset V_M,$$
$$U_j = \{u_{i_1}, u_{i_2}, \ldots, u_{i_{d_i}}, u_{i_{(d_i+1)}}, u_{i_{(d_i+2)}} \ldots, u_{d_j}\}$$
$$\subset V_M.$$

This is demonstrated in Figure 12.15(2b)-(2c). Note that the node marked by 5 ($\mathcal{L}_4 \setminus_c \mathcal{L}_1$) is a complete subset of node marked by 4 ($\mathcal{L}_4$) leading to the labeling of the nodes in Figure 12.15(2c). Notice that for all practical purposes, node 5 may be removed (since it is subsumed by node 4).

Given an input graph $G(V, E)$, we discuss in the next section how to compute the set of all maximal lists **L**.

What's next?

Given an input graph $G(V, E)$ and a quorum K, we have stated the need to discover all the maximal topological motifs (and their occurrences in the graph) that satisfy this quorum constraint. We have then painstakingly convinced the reader that we want these maximal motifs to be in the compact form. The occurrences of these compact motifs on the input graph is described by compact lists.[9]

Now that we know 'what' we want, we must next address the question of 'how'. Thus the next natural step is to explore methods to compute these compact motifs and lists from the input.

12.5   The Discovery Method

Main idea. The method is based on a simple observation that given a graph a topological motif can be represented either as a motif (graph) or as a collection of location lists of the vertices of the motif. By taking a frequentist approach to the problem, the method discovers the multiply occurring motifs by working in the space of location lists. There are two aspects that lend themselves to efficient discovery:

1. The motifs are maximal, so a relatively small number of possibilities are to be explored.

   For instance if a graph has exactly three red vertices with 5, 8 and 10 blue immediate neighbors (along with other edges and vertices with other colors), then any maximal motif with a red vertex can have exactly 5 blue neighbors or exactly 8 blue neighbors, although a subgraph could

---

[9] Note however that the nature of the beast is such that even compact lists may not save the day: see Exercise 154 based on the example of Figure 12.4.

have from 0 to 10 blue neighbors. The compact location list captures this succinctly.

2. A single conservative intersection operation handles all the potential candidates with a high degree of efficiency.

Method overview. The discovery process consists of mainly two steps. In the first step an exhaustive list of potential location lists of vertices of these motifs is computed, which is stored in a compact form, as compact location lists. In the second step this collection of compact location lists computed in the first step is enlarged by including all the nonempty intersections, amongst the location lists computed in the first step. The collection of compact location lists so obtained has the nice property that every compact (maximal) motif has a compact vertex whose location list appear in compact form in this collection. Conversely, every compact location list appearing in this collection is the location list of some compact vertex compact (maximal) motif. The intersection operations at different stages are carried out in an output-sensitive manner: this is possible since we are computing maximal intersections.

Input: The input is a graph $G(V, E)$ where an attribute is associated with each vertex. For the sake of convenience, let B be a two-dimensional array of dimension $|V| \times D$, where D is the total number of distinct attributes, and, which encodes the graph as follows ($1 \leq i \leq |V|, 1 \leq j \leq D$):

$B[i][j]$ is the set of vertices adjacent to v_i having the attribute a_j.

B is called the *adjacency matrix*. Two examples are illustrated in Figure 12.16.
Output: All compact maximal motifs

$$C(V_c, E_c) \text{ along with } \mathcal{L}_U \text{ for each } U \in V_C.$$

12.5.1 The algorithm

The algorithm works by first generating a small collection of maximal lists and their conjugates (which are also maximal),

$$\mathbf{L}_{init},$$

called the *initialization* step. In the iterative step, more maximal lists are carefully added to this initial list through the maximal set generating operations (such as set intersections $\cap_c$, set difference $\backslash_c$, conjugate $conj(\cdot)$, $Enrich(\cdot)$) described in the earlier sections, called the *iterative* step.

12.5.1.1 Initialization: generating $\mathbf{L}_{init}$

At the very first step we compute a collection of compact lists $\mathbf{L}_{init}$. This collection of maximal lists is characterized as follows:

$\mathcal{L} \in \mathbf{L}_{init}$, if and only if there exists no maximal $\mathcal{L}'$ such that $\mathcal{L} \subset_c \mathcal{L}'$.

Recall that we have defined maximality of a location list in terms of the maximal motifs. Then, in the absence of the output motifs, how do we know which list is maximal?

For a pair of attributes, a_i, a_j, two location lists $\mathcal{L}_{a_i}$ and $\mathcal{L}_{a_j}$ are constructed where

$$att(\mathcal{L}_{a_i}) = a_i,$$
$$att(\mathcal{L}_{a_j}) = a_j,$$
$$Conj_{a_i}(\mathcal{L}_{a_j}) = \mathcal{L}_{a_i},$$
$$Conj_{a_j}(\mathcal{L}_{a_i}) = \mathcal{L}_{a_j}.$$

Ensuring the maximality of these lists is utterly simple since it can almost be read off the incidence matrix B. This is best understood by following a simple concrete example. We show two such examples in Figure 12.16. But we must also avoid overcounting, as discussed in the following paragraphs.

Avoiding multiple counting.   We combine maximality with another practical consideration and that is of avoiding multiple reading of the same occurrence. This is achieved by imposing the following constraint on the location lists $\mathcal{L}_1$ and its conjugate $\mathcal{L}_2$:

> If $flat(\mathcal{L}_1) = flat(\mathcal{L}_2)$,
>   then $discSz(\mathcal{L}_1)$ must be different from $discSz(\mathcal{L}_2)$.

To understand this constraint, consider the following scenario. Consider a graph with a single edge

$$G(\{v_1, v_2\}, \{(v_1 v_2)\})$$

and let quorum $K = 2$. Further, let $att(v_1) = att(v_2) = a$, some fixed attribute. A motif

$$M(\{u_1, u_2\}, \{(u_1 u_2)\})$$

with

$$\mathcal{L}_{u_1} = \{v_1, v_2\}$$
$$\Updownarrow \quad \uparrow \quad \uparrow$$
$$\mathcal{L}_{u_2} = \{v_2, v_1\}$$

satisfies the quorum condition since v_1 can be considered either as the start or end vertex of the edge and similarly v_2 can be considered either as the start or end vertex of the edge, giving two apparent occurrences.

Thus it is important to avoid 'over-counting'. But, first we compute the subsets using $Enrich(\cdot)$ and $clique(\cdot)$, along with their conjugates. Then to avoid the over-counting, we replace both $\mathcal{L}_1$ and its conjugate $\mathcal{L}_2$ that have the same flat sets and the same size of discriminant with a single list $\mathcal{L}_3$ that denotes all the edges of the conjugate relationship. It is made maximal, by recognizing the maximal cliques on the vertices of $flat(\mathcal{L}_3)(= flat(\mathcal{L}_1) = flat(\mathcal{L}_2))$. Consider the example in Figure 12.18.

$$
\begin{array}{ccccc}
\mathcal{L}_5 = \{v_4, & v_5, & \{v_9, v_{11}\}, & v_{10}\} & \\
\updownarrow & \uparrow & \uparrow & \uparrow & \uparrow \\
\mathcal{L}_5 = \{v_5, & v_4, & v_{10}, & \{v_9, v_{11}\}\} &
\end{array}
$$

We first compute the maximal subset $\mathcal{L}_5' = Enrich(\mathcal{L}_5)$ and compute its conjugate directly as a subset of $\mathcal{L}_5$ as follows:

$$
\begin{array}{ccc}
\mathcal{L}_5' = & \{\{v_9, v_{11}\}\} & \\
\updownarrow & & \uparrow \\
\mathcal{L}_5'' = & \{v_{10}\} &
\end{array}
$$

Then to avoid multiple counting, $\mathcal{L}_5$ and its conjugate $\mathcal{L}_5$ give

$$\mathcal{L}_6 = \{\{v_4, v_5\}, \{v_9, v_{10}\}, \{v_{10}, v_{11}\}\}$$

with

$$att(\mathcal{L}_6) = \text{red}, \quad discSz(\mathcal{L}_6) = 2, \quad clq(\mathcal{L}_6) = 1.$$

Note that $\mathcal{L}_6$, $\mathcal{L}_5'$ and $\mathcal{L}_5''$ belong to $\mathbf{L}_{init}$.

Back to concrete example. Consider the example G_1 in Figure 12.16(a). For each pair of attributes, the maximal lists are simply 'read off' the adjacency matrix B_1 as follows where the conjugacy relationship of two lists is shown by the '$\Leftrightarrow$' symbol:

	red	blue	green
red	X	$\mathcal{L}_1 \Leftrightarrow \mathcal{L}_2$	$\mathcal{L}_4 \Leftrightarrow \mathcal{L}_3$
blue	-	X	$\mathcal{L}_2 \Leftrightarrow \mathcal{L}_3$
green	-	-	X

Thus

$$\mathbf{L}_{init} = \{\mathcal{L}_1, \mathcal{L}_2, \mathcal{L}_3, \mathcal{L}_4\}$$

and these lists, along with the conjugacy relations, are shown in Figure 12.17.

Next, consider the example G_2 in Figure 12.16(b). Again, for each pair of attributes, the maximal lists are simply 'read off' the adjacency matrix B_2 as follows:

	red	blue	green
red	$\mathcal{L}_5 \Leftrightarrow \mathcal{L}_5$	$\mathcal{L}_1 \Leftrightarrow \mathcal{L}_2$	$\mathcal{L}_4 \Leftrightarrow \mathcal{L}_3$
blue	-	X	$\mathcal{L}_2 \Leftrightarrow \mathcal{L}_3$
green	-	-	X

However, here to avoid over counting, we replace $\mathcal{L}_5 \Leftrightarrow \mathcal{L}_5$, by $\mathcal{L}_6$ as shown, which represents the edges (encoded by $\mathcal{L}_5 \Leftrightarrow \mathcal{L}_5$). Also, notice that it has no conjugate list. But before replacing with $\mathcal{L}_6$, we $Enrich(\mathcal{L}_5)$ to get $\mathcal{L}_5'$ and collect its conjugate $\mathcal{L}_5''$. Thus

$$\mathbf{L}_{init} = \{\mathcal{L}_1, \mathcal{L}_2, \mathcal{L}_3, \mathcal{L}_4, \mathcal{L}_5', \mathcal{L}_5'', \mathcal{L}_6\}$$

and these lists, along with the conjugacy relations, are shown in Figure 12.18.

12.5.1.2 The iterative step

For each new compact (maximal) set $\mathcal{L}$, possible new maximal compact lists (say $\mathcal{L}'$) are generated

1. using the $Enrich(\cdot)$ operations,

2. using the $clique(\cdot)$ operation, and,

3. set difference ($\backslash_c$ operation) if there are inconsistencies in the meta-graph $\mathbf{G}$.

Note that each

$$\mathcal{L}' \subset_c \mathcal{L},$$

and we generate the conjugates of $\mathcal{L}'$ using the conjugates of $\mathcal{L}$.

Also, a collection of p maximal compact lists are used to generate new lists. How do we choose the value of p? And, which collection of p maximal lists do we choose? This is done in the most conservative way using the $Refine(\cdot)$ procedure. It is the process of computing new maximal location lists through compact list intersections. It is formally defined as follows.

Problem 18 *(Refine(C)) The input to the problem is a collection of n compact sets* $\mathbf{C} = \{C_1, C_2, \ldots, C_n\}$. *For a compact set S such that*

$$S = C_{i_1} \cap_c C_{i_2} \cap_c \ldots \cap_c C_{i_p},$$

we denote by

$$I_S = \{i_1, i_2, \ldots, i_p\}.$$

Further, I_S is maximal i.e., there is no I' with

$$I_S \subsetneq I',$$

such that

$$S = C_{j_1} \cap_c C_{j_2} \cap_c \ldots \cap_c C_{j_{p'}}, \text{ where } I' = \{j_1, j_2, \ldots, j_{p'}\}.$$

The output is the set of all pairs (S, I_S).

Given a collection of n compact sets $\mathbf{C} = \{C_1, C_2, \ldots, C_n\}$, we propose a three step solution to this problem.

1. Obtain $\mathbf{C}'$, a set of n flat sets from the given collection

$$\mathbf{C}' = \{flat(C_i) \mid C_i \in \mathbf{C}\}.$$

2. We solve the maximal set intersection problem for $\mathbf{C}'$ (which is defined exactly as $Refine(\cdot)$ except that the sets are flat, thus $\cap$ instead of $\cap_c$ is used).

3. For each solution (a flat set) computed in the last step, we reconstruct a compact set. This is defined by the following problem.

Problem 19 *Given p compact lists, $\mathcal{L}_1, \mathcal{L}_2, \ldots \mathcal{L}_p$, let*

$$\mathcal{L}' = flat(\mathcal{L}_1) \cap flat(\mathcal{L}_2) \cap \ldots \cap flat(\mathcal{L}_p).$$

Using $\mathcal{L}'$, compute $\mathcal{L}_X$ given as

$$\mathcal{L}_X = \mathcal{L}_1 \cap_c \mathcal{L}_2 \cap_c \ldots \cap_c \mathcal{L}_p.$$

We solve this problem by constructing a *neighborhood graph* $G_N(\mathcal{L}', E_N)$. Again, we abuse notation slightly, and each element of $\mathcal{L}'$ is (mapped to) a vertex on this neighborhood graph. Further,

$$E_N = \{(v_1 v_2) \mid \text{for each } 1 \leq i \leq p, \ v_1, v_2 \in L \text{ for some } L \in \mathcal{L}_i\}.$$

Then $\mathcal{L}_X$ is obtained as follows:

$$\mathcal{L}_X = \{L \mid L \subset \mathcal{L}' \text{ is a maximal clique on } G_N(\mathcal{L}', E_N)\}.$$

See Section 12.4.8 for a discussion on maximal cliques.

Putting it all together. These steps are carefully integrated into one coherent procedure outlined as Algorithm (12).

Here *InitMetaGraph(**L**)* is a procedure that generates the meta-graph given the collection of lists, $\mathbf{L}$, and their conjugates. See Section 12.4.11 for details on detecting inconsistencies in a connected component of the meta-graph.

Induct($\mathcal{L}$*, p, $\mathcal{L}_1, \ldots, \mathcal{L}_p$)* is a procedure that introduces the new list $\mathcal{L}$ to the collection. Further, $\mathcal{L} \subset_c \mathcal{L}_1, \ldots, \mathcal{L}_p$ and the conjugates of these p lists are used to compute the conjugates of $\mathcal{L}$.

The remainder of the algorithm is self-explanatory. Figures 12.17 and 12.18 give the solution to the input graphs of Figure 12.16.

Consider Figure 12.17. See also Figure 12.15 for the meta-graph. In the connected component (to avoid clutter, we show only the nodes)

$$(\mathcal{L}_1, \mathcal{L}_2, \mathcal{L}_3, \mathcal{L}_4, \mathcal{L}_4 \setminus_c \mathcal{L}_1),$$

$\mathcal{L}_1$ and $\mathcal{L}_4$ are *inconsistent*. Thus the two maximal subgraphs that are consistent are

1. $(\mathcal{L}_1, \mathcal{L}_2, \mathcal{L}_3, \mathcal{L}_4 \setminus_c \mathcal{L}_1)$ and

2. $(\mathcal{L}_2, \mathcal{L}_3, \mathcal{L}_4, \mathcal{L}_4 \setminus_c \mathcal{L}_1)$.

Note that $\mathcal{L}_4 \setminus_c \mathcal{L}_1$ is a *complete subset* of $\mathcal{L}_4$. Further,

$$(\mathcal{L}_1, \mathcal{L}_2, \mathcal{L}_3)$$

is not maximal. The other two connected components

1. $(\mathcal{L}_1 \cap_c \mathcal{L}_4, \mathcal{L}_2', \mathcal{L}_3', \mathcal{L}_4')$ and

2. $(\mathcal{L}_1 \setminus_c \mathcal{L}_4, \mathcal{L}_2', \mathcal{L}_3', \mathcal{L}_4')$

give motifs with quorum $K < 2$ (the topology of the two connected components of the input graph).

Consider Figure 12.18. In the two connected components

$$(\mathcal{L}_1, \mathcal{L}_2, \mathcal{L}_3, \mathcal{L}_4, \mathcal{L}_4 \setminus_c \mathcal{L}_1, \mathcal{L}_6),$$

$\mathcal{L}_1$ and $\mathcal{L}_4$ are *inconsistent*. The two maximal consistent subgraphs are

1. $(\mathcal{L}_1, \mathcal{L}_2, \mathcal{L}_3, \mathcal{L}_4 \setminus_c \mathcal{L}_1, \mathcal{L}_6)$, and

2. $(\mathcal{L}_2, \mathcal{L}_3, \mathcal{L}_4, \mathcal{L}_4 \setminus_c \mathcal{L}_1, \mathcal{L}_6)$.

Notice that $\mathcal{L}_6$ is a *complete subset* of $\mathcal{L}_4$.

Efficiency in practice. The collection of location lists **L** is partitioned by attribute values $att(\mathcal{L})$ for the $Refine(\cdot)$ operation, since clearly compact lists with distinct attributes have empty intersections.

Also, if $\mathcal{L}$ is such that each $L \in \mathcal{L}$ is a singleton set, then the $Enrich(\mathcal{L})$ and $clique(\mathcal{L})$ do not produce any new lists and these computations can be skipped.

Algorithm 12 *The Compact Motifs Discovery Algorithm*

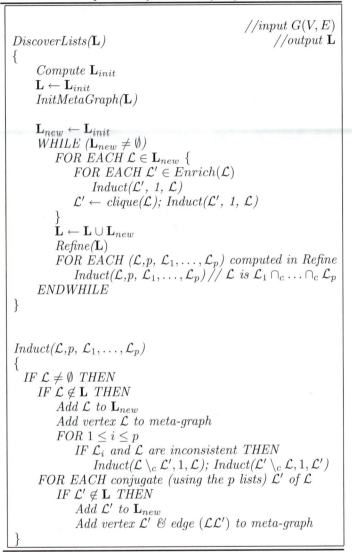

```
                                                    //input G(V, E)
    DiscoverLists(L)                                //output L
    {
        Compute L_init
        L ← L_init
        InitMetaGraph(L)


        L_new ← L_init
        WHILE (L_new ≠ ∅)
            FOR EACH L ∈ L_new {
                FOR EACH L' ∈ Enrich(L)
                    Induct(L', 1, L)
                    L' ← clique(L); Induct(L', 1, L)
            }
            L ← L ∪ L_new
            Refine(L)
            FOR EACH (L,p, L_1,...,L_p) computed in Refine
                Induct(L,p, L_1,...,L_p) // L is L_1 ∩_c ... ∩_c L_p
        ENDWHILE
    }


    Induct(L,p, L_1,...,L_p)
    {
      IF L ≠ ∅ THEN
        IF L ∉ L THEN
            Add L to L_new
            Add vertex L to meta-graph
            FOR 1 ≤ i ≤ p
                IF L_i and L are inconsistent THEN
                    Induct(L \_c L', 1, L); Induct(L' \_c L, 1, L')
            FOR EACH conjugate (using the p lists) L' of L
                IF L' ∉ L THEN
                    Add L' to L_new
                    Add vertex L' & edge (LL') to meta-graph
    }
```

12.6 Related Classical Problems

Before we conclude the chapter, we relate the problem discussed here to other classical problems. To summarize, the automated topological discovery problem is abstracted as follows: *Given an integer $K(>1)$ and a graph*

$G(V, E)$ with labeled vertices and edges, the task is to discover at least K subgraphs that are topologically identical in G. Such subgraphs are termed topological motifs.

It is very closely related to the classical *subgraph isomorphism* problem defined as follows [GJ79]:

Problem 20 *(Subgraph isomorphism) Given graphs $G = (V_1, E_1)$ and $H = (V_2, E_2)$. Does G contain a subgraph isomorphism to H i.e., a subset $V \subseteq V_1$ and a subset $E \subseteq E_1$ such that $|V| = |V_2|$, $|E| = |E_2|$ and there exists a one-to-one function $f : V_2 \to V$ satisfying $\{v_1, v_2\} \in E_2$ if and only if $\{f(v_1), f(v_2)\} \in E$?*

Two closely related problems are as follows [GJ79].

Problem 21 *(Largest common subgraph problem) Given graphs $G = (V_1, E_1)$ and $H = (V_2, E_2)$, positive integer K. Do there exist subsets $E_1' \subseteq E_1$ and $E_2' \subseteq E_2$ with $|E_1'| = |E_2'| \geq K$ such that the two subgraphs $G' = (V_1, E_1')$ and $H' = (V_2, E_2')$ are isomorphic?*

Problem 22 *(Maximum subgraph matching problem) Given directed graphs $G = (V_1, E_1)$ and $H = (V_2, E_2)$, positive integer K. Is there a subset $R \subseteq V_1 \times V_2$ with $|R| \geq K$ such that for all $< u, u' >, < v, v' > \in R, (u, v) \in A$, if and only if $(u', v') \in A_2$?*

All the three problems are NP-complete: each can be transformed from the problem of finding maximal cliques [10] in a graph. The problem addressed in this chapter is similar to the latter two problems. However our interest has been in finding at least K isomorphs and all possible such isomorphs.

12.7   Applications

Understanding large volumes of data is a key problem in a large number of areas in bioinformatics, and also other areas such as the world wide web. Some of the data in these areas cannot be represented as linear strings, which have been studied extensively with a repertoire of sophisticated and efficient algorithms. The inherent structure in these data sets is best represented as graphs. This is particularly important in bioinformatics or chemistry since it

---

[10] The clique problem is a graph-theoretical NP-complete problem. Recall that a clique in a graph is an induced subgraph which is a complete graph. Then, the clique problem is the problem of determining whether a graph contains a clique of at least a given size k. The corresponding optimization problem, the *maximum clique problem*, is to find the largest clique in a graph.

might lead to the understanding of biological systems from indirect evidence in the data. Thus automated discovery of a 'phenomenon' is a promising path to take as is evidenced by the use of motif (substring) discovery in DNA and protein sequences.

Here we give a brief survey of the current use of this discovery problem to answer different biological questions. A protein network is a graph that encodes primarily protein-protein interactions and this is important in understanding the *computations* that happen within a cell [HETC00, SBH+01, MBV05]. A recurring topology or motif in such a setting has been interpreted to act as robust filters in the transcriptional network of *Escherichia coli* [MSOI+02, SOMMA02]. It has been observed that the conservation of proteins in distinct topological motifs correlates with the interconnectedness and function of that motif and also depends on the structure of the topology of all the interactions. This indicates that motifs may represent evolutionary conserved topological units of cellular networks in accordance with the specific biological functions they perform [WOB03, LMF03]. This observation is strikingly similar to the hypothesis in dealing with DNA and protein primary structures.

To study complex relationships involving multiple biological interaction types, an integrated *Saccharomyces cerevisiae* network in which nodes represent genes (or their protein products) and the edges represented different biological interaction types was assembled [ZKW+05]. The authors examined interconnection patterns over three to four nodes and concluded that most of the motifs form classes of higher-order recurring interconnection patterns that encompass multiple occurrences of topological motifs.

Topological motifs are also being studied in the context of structural units in RNA [GPS03] and for structural multiple alignments of proteins [DBNW03]. For yet another application consider a typical chemical data set [CMHK05]: a chemical is modeled as a graph with attributes on the vertices and the edges. A vertex represents an atom and the attribute encodes the atom type; an edge models the bond between the atoms it connects and its attribute encodes the bond type. In such a database, very frequent common topologies could suggest the relationship to the characteristic of the database. For instance, in a toxicology related database, the common topologies may indicate carcinogenicity or any other toxicity.

In *machine learning*, methods have been proposed to search for subgraph patterns which are considered characteristic and appear frequently: this uses an a priori-based algorithm with generalizations from association discovery [IWM03].

In *massive data mining* (where the data is extremely large, of the order of tens of gigabytes) that include the world wide web, internet traffic and telephone call details, the common topologies are used to discover social networks and web communities, among other characteristics [Mur03]. In biological data the size of the database is not as large, yet unsuitable for enumeration schemes. When this scheme was applied researchers had to restrict their motifs to small sizes such as three or four vertices [MSOI$^+$02].

12.8 Conclusion

The potential of an effective automated topological motif discovery is enormous. This chapter presents a systematic way to discover these motifs using compact lists. Most of the proofs of the lemmas and theorems presented in this chapter are straightforward. The only tool they use is 'proof by contradiction', hence they have been left as exercises for the reader.

One of the burning questions is to compute the statistical significance of these motifs. Can compact motifs/lists provide an effective and acceptable method to compute the significance of these complex motifs? We leave the reader with this tantalizing thought.

12.9 Exercises

Exercise 137 (Combinatorics in graphs) *Consider the graph of Figure 12.4 with quorum = 2. Notice that every maximal motif occurs in component labeled 7. Thus the number of distinct collection of components is*

$$2^6 - 1,$$

ignoring the empty set. Since there are four vertices in each component with distinct attributes, the number of location lists of vertices can be estimated as

$$4(2^6 - 1) = 252.$$

However, this number is 212 as shown in Figure 12.8. How is the discrepancy explained?

Hint: Notice that the motifs are (1) connected, (2) maximal and (3) satisfy a given quorum. Thus the numbers are not captured by pure combinatorics, although they are fairly close. The number of distinct motifs can be counted by enumerating i edges, $1 \le i \le 6$, out of 6 possible edges in a motif. Recall that quorum is 2.

$$\binom{6}{5} = 6 \tag{12.5}$$

$$\binom{6}{4} = 15 \tag{12.6}$$

$$\binom{6}{3} = \begin{cases} 16 \\ 4 \end{cases} \tag{12.7}$$

$$\binom{6}{2} - 3 = 12 \tag{12.8}$$

$$\binom{6}{1} = 6 \tag{12.9}$$

$$4\binom{6}{0} = 4 \tag{12.10}$$

The discrepancies are explained as follows:

- (Equation (12.7)): When we choose a motif that has 3 edges, out of a possible 6, then the motif could have 4 vertices or 3 vertices (see row 4 of Figure 12.17), hence the split shown in Equation (12.7) above of 20 as $16 + 4$.

- (Equation (12.8)): Any two edges many not be adjacent, leading to disconnected structures. There are 3 such configurations shown in Figure 12.6.

- (Equation 12.10): Since each connected component of the graph has four distinct colored vertices, Equation (12.10) shows a count of $4 \times 1 = 4$.

Exercise 138 (Graph isomorphisms)

1. *Given graphs G_1, G_2, G_3, show that if G_1 is isomorphic to G_2 and G_2 is isomorphic to G_3, then G_1 is isomorphic to G_3.*

2. *Identify the bijections (f's) that demonstrate that components (1), (4) and (5) are isomorphic in Figure 12.10.*

Exercise 139 (Motif occurrence) *Consider the input graph in (a) below with 10 vertices and two distinct attributes. The attribute of a vertex is denoted by its shape in this figure: 'square' and 'circle'. Let quorum $K = 2$. A motif is shown in (b). In (1)-(10) the occurrences of the motif on the graph are shown in bold.*

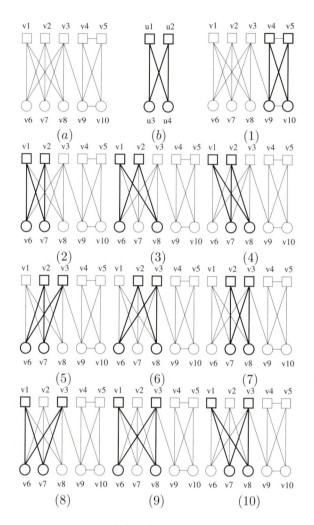

1. For each occurrence i: (a) What is O_i? (b) Define the map F_i (of Definition (12.1)).

2. Is this motif maximal? Why?

3. Obtain the compact notation of the motif and the location lists of the compact vertices.

Exercise 140 (Motif occurrence) *Consider the input graph of Problem (139) and let quorum $K = 2$. (a) below shows a motif with six vertices. The occurrences (in bold) of the motif are shown in (1)-(5). The dashed-bold indicates multiple occurrences shown in the same figure: the occurrence of the motif is* *to be interpreted as all the solid vertices and edges and one of the dashed edges and the connected dashed/solid vertex.*

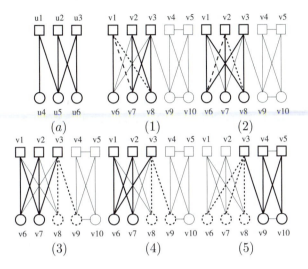

1. *How many distinct occurrences does the motif have?*

2. *Is this motif maximal? Why?*

3. *Obtain the compact notation of this motif and the location lists of the compact vertices.*

Hint: Note that the motif edge $(u_1 u_5)$ must also be represented in the compact motif notation giving at least two compact vertices $U_1 = \{u_5\}$ and $U_2 = \{u_5, u_6\}$ although $U_1 \subset U_2$.

Exercise 141 (Directed graph) *Let $G(V, E)$ be an undirected graph with attributes on both vertices and edges. Devise a scheme to construct an undirected graph*

$$G'(V', E')$$

with attributes only on the vertices so that a maximal motif occurring in G can be constructed from a maximal motif occurring in $G'(V', E')$.

1. *What is the size of V' in terms of V?*

2. *What is the size of E' in terms of E?*

Hint: Fill in the details for a scheme outlined below.

1. (Annotate G): Introduce suffixes to common vertex and edge attributes.

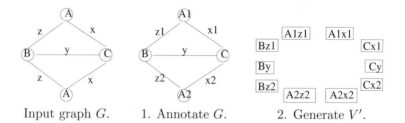

| | Input graph G. | 1. Annotate G. | 2. Generate V'. |

2. (Generate V'): For each vertex with attribute A_j and incident edge with attribute x_i, create a node with attribute $A_j x_i$ in G'.

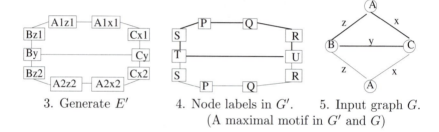

3. Generate E' 4. Node labels in G'.   5. Input graph G.
(A maximal motif in G' and G)

3. (Generate E'): For each pair of nodes with labels $.y_k$ and $.y_k$, introduce an undirected edge. Similarly for each pair of nodes with labels $A_i.$ and $A_i.$, introduce an undirected edge.

4. (Node labels in G'): Two node labels of the form $A_{j_1} y_{k_1}$ and $A_{j_2} y_{k_2}$ are deemed to have the same node attribute in G'.

Exercise 142 (Directed, labeled graph) *Let $G(V, E)$ be a directed graph with attributes on both vertices and edges. Devise a scheme to construct an undirected graph*

$$G'(V', E')$$

with attributes only on the vertices so that a maximal motif occurring in G can be constructed from a maximal motif occurring in $G'(V', E')$.

1. *What is the size of V' in terms of V?*

2. *What is the size of E' in terms of E?*

Hint: Fill in the details for a scheme outlined below.

1. (Annotate G): Introduce suffixes to common vertex and edge attributes.

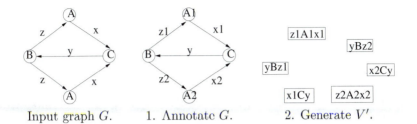

Input graph G.　　1. Annotate G.　　2. Generate V'.

2. (Generate V'): For each incoming edge with attribute x_i, vertex with attribute A_j and outgoing edge with attribute y_k, create a node with attribute $x_i A_j y_k$ in G'.

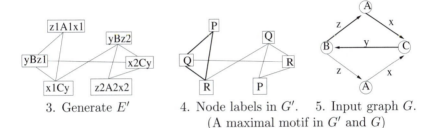

3. Generate E'　　4. Node labels in G'.　　5. Input graph G.
(A maximal motif in G' and G)

3. (Generate E'): For each pair of nodes with labels $\cdots y_k$ and $y_k \cdots$, introduce an undirected edge.

4. (Node labels in G'): Two node labels of the form $x_{i_1} A_{j_1} y_{k_1}$ and $x_{i_2} A_{j_2} y_{k_2}$ are deemed to have the same node attribute in G'.

Exercise 143 *Consider the initialization step discussed in Section 12.5.1.1 to compute $\mathbf{L}_{init}$ and recall property 2 as:*

> 2. *If $M(V_M, E_M)$ is a any maximal motif on the input graph and $U \subset V_M$ is a compact vertex then*
>
> $$\mathcal{L}_U \subset_c \mathcal{L}, \text{ for some } L \in \mathbf{L}_{init}.$$

Show that the same property holds even when motif $M(V_M, E_M)$ is not maximal.

Exercise 144 *Using the definition of the intersection of two compact lists, define the intersection of p compact lists.*

Exercise 145 *Using the definitions of the set operations on compact lists, show that given compact lists $\mathcal{L}_1$, $\mathcal{L}_2$ and $\mathcal{L}_3$, if*

$$\mathcal{L}_3 \subset_c \mathcal{L}_1, \mathcal{L}_2,$$

then

$$\mathcal{L}_3 \subset_c (\mathcal{L}_1 \cap_c \mathcal{L}_2).$$

Exercise 146 *Given a compact list $\mathcal{L}$, do the following statements hold?*

1. $\mathcal{L} \cup_c \mathcal{L} =_c \mathcal{L}$.

2. $\mathcal{L} \cap_c \mathcal{L} =_c \mathcal{L}$.

3. $\mathcal{L} \setminus \mathcal{L} =_c \emptyset$.

Exercise 147 *Given two compact lists $\mathcal{L}_1$ and $\mathcal{L}_2$, prove the following statements.*

1. $flat(\mathcal{L}_1 \cap_c \mathcal{L}_2) = flat(\mathcal{L}_1) \cap flat(\mathcal{L}_2)$.

2. $flat(\mathcal{L}_1 \cup_c \mathcal{L}_2) = flat(\mathcal{L}_1) \cap flat(\mathcal{L}_2)$.

3. $flat(\mathcal{L}_1 \setminus_c \mathcal{L}_2) = flat(\mathcal{L}_1) \setminus flat(\mathcal{L}_2)$.

Hint: Use the definitions of the set operations.

Exercise 148   *(a) Prove Theorem (12.2).*

(b) Can you relax the definition of intersection $\cap_c$, to $\cap_{new}$ in such a manner that an intersection is not necessarily maximal, i.e., for maximal sets $\mathcal{L}_1$ and $\mathcal{L}_2$, $\mathcal{L}_1 \cap_{new} \mathcal{L}_2$ may not be maximal.

Hint: (a) (1-4) Use proof by contradiction. (5) Construct an example.
(b) What aspect of the definition lends maximality to the intersection set?

Exercise 149 *Let $\mathcal{L}_1'$ be a conjugate of $\mathcal{L}_1$ and and $\mathcal{L}_2'$ be a conjugate of $\mathcal{L}_2$. Then prove that*

$$if \ \mathcal{L}_2 \subset_c \mathcal{L}_1, \ then \ \mathcal{L}_2' \subset_c \mathcal{L}_1'.$$

Exercise 150 *Construct an example to show that the* enrich *operation (Section 12.4.8) is essential to obtain all the maximal topological motifs.*

Hint: Consider the following input graph with three connected components.

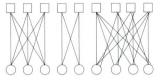

The attribute of a vertex is represented by the shape of the vertex (square or circle). Let quorum $K = 2$. How many maximal motifs occur on this graph?

Exercise 151 *Let $\mathcal{L}_0 \in \mathbf{L}_{init}$ with $att(\mathcal{L}_0) = att(\mathcal{L})$ and $discSz(\mathcal{L}_0) > 1$. Then, show that*

$$clique(\mathcal{L}) = \mathcal{L} \cap_c \mathcal{L}_0.$$

Hint: Given a graph $G(V, E)$ and an attribute a, let

$$V_a = \{v \in V \mid att(v) = a\}.$$

Further, let the input graph be such that the induced subgraph on V_a has k cliques with the following sizes of the cliques:

$$1 < d_1 \leq d_2 \leq \ldots \leq d_k.$$

Does there exist $\mathcal{L}_0 \in \mathbf{L}_{init}$ with $discSz(\mathcal{L}_0) > 1$? Why? If yes, then determine the following characteristics of $\mathcal{L}_0$: $flat(\mathcal{L}_0)$, $discSz(\mathcal{L}_0)$, $att(\mathcal{L}_0)$ and $clq(\mathcal{L}_0)$.

Exercise 152 *We give a new definition of topological motif, by modifying Definition (12.1) as follows: the mapping H_i is not mandated to be bijective but just a total function or total mapping. Recall that bijective mapping implies that every vertex u in the motif is mapped to a unique vertex v in the occurrence O. A total mapping implies that multiple vertices in the motif may be mapped to the same vertex in O.*

What are the implications of this new motif definition? How does the discovery algorithm change?

Hint: See Figure 12.13. Is the structure in (d) a topological motif in the input graph in (a)? Are the motifs in (b) and (c) maximal by the new definition?

Exercise 153 (MCCS) *Given a meta-graph, devise a method to extract all the MCCSs.*

1. What is the time complexity of the method?

2. Comment on the following approach to the problem:

   (a) Obtain the largest subgraph G' with no forbidden edges.

   (b) Check for connectivity in G'.

Hint: Given a graph, an *independent set* is a subset of its vertices that are pairwise not adjacent. In other words, the subgraph induced by these vertices has no edges, only isolated vertices. Then, the independent set problem is as follows: Given a graph G and an integer k, does G have an independent set of size at least k ? The corresponding optimization problem is the *maximum independent set problem*, which attempts to find the largest independent set in a graph. The independent set problem is known to be NP-complete.

The connectedness of a graph, on the other hand, can be computed in linear time using a BFS or a DFS traversal.

Exercise 154 (Flat list) *A compact list $\mathcal{L}$ where each $L \in \mathcal{L}$ is a singleton set is termed a* flat *list.*

1. Show that if $\mathcal{L}' \subset_c \mathcal{L}$ and $\mathcal{L}$ is a flat list, then so is $\mathcal{L}'$.

2. Consider the graph with seven connected components in Figure 12.4. We follow a convenient notation to denote a vertex of this given graph G as follows. Since each connected component has exactly one vertex of a fixed color, a vertex denoted as c_X uniquely identifies a vertex with color X in component numbered c. We follow the convention:

$$X = \begin{cases} r & \text{denotes red,} \\ b & \text{denotes blue,} \\ g & \text{denotes green, and} \\ d & \text{denotes black (dark).} \end{cases}$$

   Thus, $v = 4_r$ denotes the red vertex in the component numbered 4. Hence, although slightly unusual, we adopt this notation for this input graph.

   In the following example, identify $\mathbf{L}_{init}$. What can be said about each $\mathcal{L} \in \mathbf{L}_{init}$? Retrace the steps in the discovery algorithm and also re-generate the maximal motifs from the exhaustive list of compact vertices shown here.

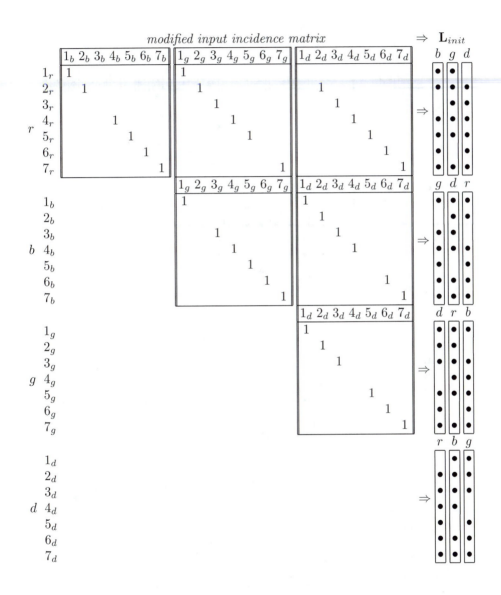

modified input incidence matrix $\Rightarrow$ $\mathbf{L}_{init}$

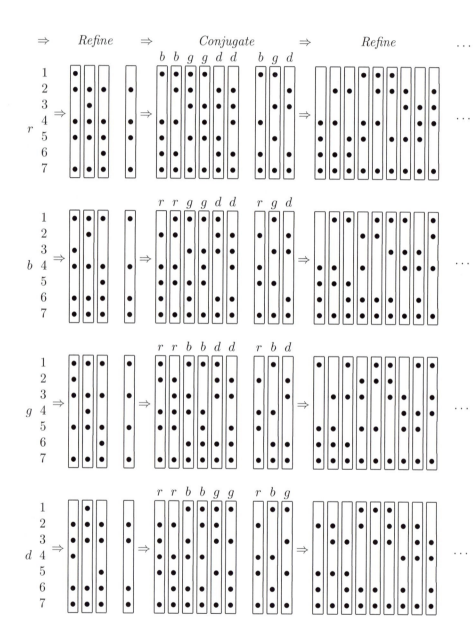

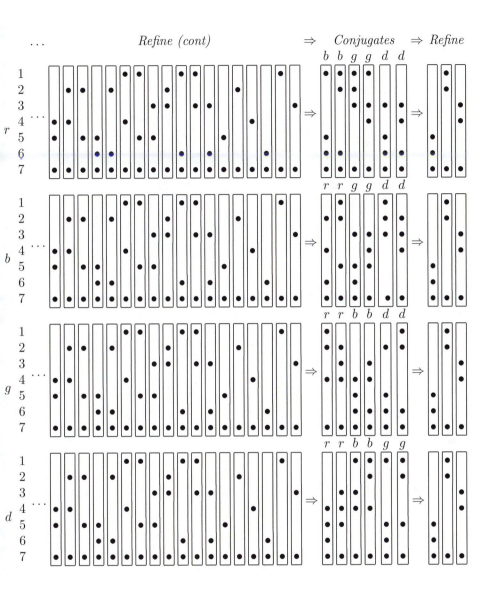

Exercise 155 *Consider the following two examples where the vertices in the graphs have only one attribute. Assume quorum $K = 2$. Compute all the maximal motifs in their compact form.*

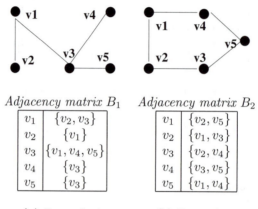

Adjacency matrix B_1	
v_1	$\{v_2, v_3\}$
v_2	$\{v_1\}$
v_3	$\{v_1, v_4, v_5\}$
v_4	$\{v_3\}$
v_5	$\{v_3\}$

Adjacency matrix B_2	
v_1	$\{v_2, v_5\}$
v_2	$\{v_1, v_3\}$
v_3	$\{v_2, v_4\}$
v_4	$\{v_3, v_5\}$
v_5	$\{v_1, v_4\}$

(a) Example 1.   *(b) Example 2.*

Hint: $\mathbf{L}_{init}$ for Example 1.

$$\mathcal{L}_{11} = \{\{v_1\}, \quad \{v_3\}, \quad \{v_2, v_3\}, \{v_1, v_4, v_5\}\}$$
$$\Updownarrow \qquad \updownarrow \qquad \updownarrow \qquad \updownarrow$$
$$\mathcal{L}_{11} = \{\{v_2, v_3\}, \{v_1, v_4, v_5\}, \quad \{v_1\}, \quad \{v_3\}\}$$
$$\mathcal{L}_{12} = \{\{v_1, v_2\}, \quad \{v_1, v_3\}, \quad \{v_3, v_5\}, \{v_3, v_4\}\}$$

$\mathbf{L}_{init}$ for Example 2.

$$\mathcal{L}_{21} = \{\{v_2, v_4\}, \{v_1, v_3\}, \{v_2, v_5\}, \{v_1, v_5\}, \{v_3, v_4\}\}$$
$$\Updownarrow \qquad \updownarrow \qquad \updownarrow \qquad \updownarrow \qquad \updownarrow$$
$$\mathcal{L}_{22} = \{\{v_1\}, \quad \{v_2\}, \quad \{v_3\}, \quad \{v_4\}, \quad \{v_5\}\}$$

Do the following (motifs) satisfy the quorum constraint? Are they maximal?

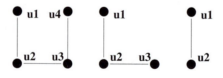

Exercise 156 ** *Let G be a graph and let $\mathcal{M}$ be the collection of all topological motifs on G that satisfy a quorum K. Design an algorithm to update $\mathcal{M}$ when a new vertex v and all its incident edges are added to G.*

Hint: This is also known as an *incremental* discovery algorithm.

Comments

The reader might find this the most challenging chapter in the book. But he/she can seek solace in the fact that even experts have had difficulty with this material. However, the ideas presented in this chapter are simple, though perhaps not obvious.

It is amazing to what lengths scientists are willing to inconvenience themselves to gain only a sliver of understanding of the nature of biology–brute-force enumeration of topological motifs have been used, albeit for ones with very few vertices.

The natural question that might arise in a curious reader's mind: *Why not use multiple BFS (breadth first search) traversals to detect the recurring motifs?* In fact, a survey of literature will reveals that such approaches have been embraced. However, the problem of explosion due to occurrence-isomorphism will cripple such a system in data sets with rampant common attributes. Of course, it is quite possible to generate instances of input that would be debilitating even to the approach presented here.

Chapter 13

Set-Theoretic Algorithmic Tools

The devil is in the details,
and so is science.

- anonymous

13.1   Introduction

Time and again, one comes across the need for an efficient solution to a task that leaves one with an uncomfortable sense of déjà vu. To alleviate such discomfort, to a certain extent, most compilers provide libraries of routines that perform oft-used tasks. Most readers may be familiar with string or input-output libraries. Taking this idea further, packages such as Maxima, a symbolic mathematics system, R, a language and environment for statistical computing,[1] and other such tools provide invaluable support and means for solving difficult problems.

In the same spirit what are the nontrivial tasks, requiring particular attention, in the broad area of pattern discovery that one encounters over and over again?

This chapter discusses sets, their interesting structures (such as orders, partial orders) and efficient algorithms for simple, although nontrivial, tasks (such as intersections, unions). This book treats lists as sets. Thus a location list, which is usually a sorted list of integers (or tuples), is treated as a set for practical purposes.

---

[1]More about Maxima: http://maxima.sourceforge.net/
More about R: http://www.r-project.org/.

13.2 Some Basic Properties of Finite Sets

Let the sets be defined on some finite alphabet

$$\Sigma = \{\sigma_1, \sigma_2, \ldots, \sigma_L\},$$

where

$$L = |\Sigma|.$$

Let S_1 and S_2 be two nonempty sets. Then only one of the following three holds:

1. S_1 and S_2 are *disjoint* if and only if

$$S_1 \cap S_2 = \emptyset.$$

   For example,

$$S_1 = \{a, b, c\},$$
$$S_2 = \{d, e\},$$

   are disjoint since

$$\{a, b, c\} \cap \{d, e\} = \emptyset.$$

2. Without loss of generality, S_1 is *contained* or *nested* in S_2 if and only if

$$S_1 \subseteq S_2.$$

   For example,
$$S_1 = \{a, b\}$$

   is contained in
$$S_2 = \{a, b, d, e\},$$

   since
$$\{a, b\} \subset \{a, b, d, e\}.$$

3. S_1 and S_2 *straddle* if and only if the two set differences are nonempty, i.e.,

$$S_1 \setminus S_2 \neq \emptyset \text{ and}$$
$$S_2 \setminus S_1 \neq \emptyset.$$

   For example,

$$S_1 = \{a, b, c, e\}, \text{ and}$$
$$S_2 = \{a, b, d\}$$

straddle since

$$\{a, b, c, e\} \setminus \{a, b, d\} = \{c, e\}, \text{ and}$$
$$\{a, b, d\} \setminus \{a, b, c, e\} = \{d\}.$$

In other words, for some $x, y \in \Sigma$,

$$x \in S_1 \setminus S_2 \text{ and}$$
$$y \in S_2 \setminus S_1.$$

Two sets S_1 and S_2 are said to *overlap* if

1. without loss of generality, S_1 is contained in S_2, or

2. S_1 and S_2 straddle.

Given a collection of sets $\mathcal{S}$, the following collection of two-tuples are termed the *containment information* or C:

$$C(\mathcal{S}) = \{(S_1, S_2) \mid S_1 \subset S_2 \text{ and } S_1, S_2 \in \mathcal{S}\}. \tag{13.1}$$

13.3 Partial Order Graph G($\mathcal{S}$, E) of Sets

For a collection of sets $\mathcal{S}$, define a directed graph, $\mathbf{G}(\mathcal{S}, \mathbf{E})$, where with a slight abuse of notation, each vertex in the graph is an element of $\mathcal{S}$. The edge set E is defined as follows. If

$$S_1 \subsetneq S_2$$

holds, then there is a directed edge from S_2 to S_1, written as

$$S_2 S_1 \in E.$$

The graph $G(\mathcal{S}, E)$ is called $\mathcal{S}$'s *partial order* (graph). Let

$$S_2 S_1 \in E,$$

Then the following terminology is used.

1. S_1 is called the *child* of S_2.

2. S_2 is called the *parent* of S_1.

3. If two nodes S_1 and S_3 have a common parent S_2, i.e.,

$$S_2 S_1, S_2 S_3 \in E,$$

then S_1 and S_3 are called *siblings*.

4. If there is a directed path from S_2 to S_1 then

    (a) S_1 is a *descendent* of S_2 and

    (b) S_2 is an *ascendant* of S_1.

LEMMA 13.1
(The descendent lemma) *If S_1 is a descendent of S_2, then*

$$S_1 \subsetneq S_2.$$

Let (see Equation (13.1))

$$C(\mathcal{S}) = C(G(\mathcal{S}, E)).$$

Next, we ask the question:

> *Can some edges be removed without losing the subset information on any pair of sets?*

In other words, is there some

$$E' \subset E,$$

such that

$$C(G(\mathcal{S}, E)) = C(G(\mathcal{S}, E')).$$

13.3.1   Reduced partial order graph

Consider the partial order graph

$$G(\mathcal{S}, E).$$

Define a set of edges $E_r(\subseteq E)$ as follows:

$$S_2 S_1 \in E_r$$

$$S_2 S_1 \in E_r \Leftrightarrow \text{there is no } S' \in \mathcal{S} \text{ such that } \begin{cases} S' \text{ is a descendant of } S_2, \text{ and} \\ S_1 \text{ is a descendant of } S'. \end{cases}$$

The *transitive reduction*[2] of $G(\mathcal{S}, E)$ is written as

$$G(\mathcal{S}, E_r).$$

For convenience, in the rest of the chapter this is also called the *reduced* partial order (graph) of $\mathcal{S}$. Figure 13.1 gives a concrete example.

---

[2] A binary relation R is *transitive* if $(A, B), (B, C) \in R \implies (A, C) \in R$.

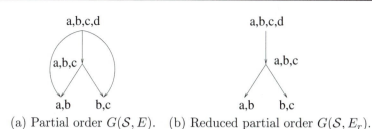

(a) Partial order $G(\mathcal{S}, E)$.   (b) Reduced partial order $G(\mathcal{S}, E_r)$.

FIGURE 13.1: Here $\mathcal{S} = \{\{a, b, c, d\}, \{a, b, c\}, \{a, b\}, \{b, c\}\}$. Note that some edges are missing in (b), yet $C(G(\mathcal{S}, E)) = C(G(\mathcal{S}, E_r))$.

Note that the reduced partial order encodes exactly the same subset information as the partial order but with possibly fewer edges, i.e.,

$$E \subset E_r.$$

In fact, a stronger result holds.

LEMMA 13.2
(Unique transitive reduction lemma) E_r, *the smallest set of edges satisfying*

$$C(G(\mathcal{S}, E)) = C(G(\mathcal{S}, E_r)),$$

is unique.

The proof is left as Exercise 157 for the reader.

13.3.2   Straddle graph

We next address the problem of partitioning $\mathcal{S}$ such that any pair in each partition straddle.

The sets (nodes) in $\mathcal{S}$ that straddle, can be partitioned using the reduced partial order graph

$$G(\mathcal{S}, E_r).$$

Each partition is a connected graph, induced by any set (node), say S, in the partition. This is best defined as an iterative process as follows which constructs one connected graph corresponding to one of the partitions of $\mathcal{S}$.

Algorithm 13 *Straddle Graph Construction*

Initialize $V_S = \{S\}$ and $E_S = \emptyset$.
REPEAT
    IF $S' \in \mathcal{S} \setminus V_S$ and $S \in V_S$ have a common child, THEN
        Add S' to V_S
        Add edge SS' to E_S
UNTIL no new node can be added to V_S

We call this graph the *straddle graph* induced by S, written as,

$$G_{straddle}(V_S, E_S).$$

Thus the straddle graph is defined on some subset of $\mathcal{S}$ that straddle. Figure 13.2 shows a concrete example. Let

$$\begin{aligned}
\mathcal{S} = \{&\{a\}, \{b\}, \{c\}, \{d\}, \{e\}, \\
&\{a,b\}, \{b,c\}, \{c,d\}, \{c,e\}, \\
&\{a,b,c\}, \\
&\{a,b,c,d\}, \{a,b,c,e\}, \\
&\{a,b,c,d,e\}\}.
\end{aligned}$$

The reduced partial order graph, $G(\mathcal{S}, E_r)$ is shown in (a); (b) shows the edges that connect nodes with common children (these edges are shown as dashed edges), and (c) shows the two nonsingleton connected straddle graphs. The singleton straddle graphs (graphs whose node set V has only one element) are:

$$G(\{\{a,b,c,d,e\}\}, \emptyset),$$
$$G(\{\{a,b,c\}\}, \emptyset),$$
$$G(\{\{a\}\}, \emptyset), \ G(\{\{b\}\}, \emptyset), \ G(\{\{c\}\}, \emptyset), \ G(\{\{d\}\}, \emptyset), \ G(\{\{e\}\}, \emptyset).$$

LEMMA 13.3
(Unique straddle graph lemma) *If*

$$S, S' \in V_S,$$

then

$$G_{straddle}(V_S, E_S) = G_{straddle}(V_{S'}, E_{S'}).$$

This can be verified and we leave that as Exercise 160 for the reader.

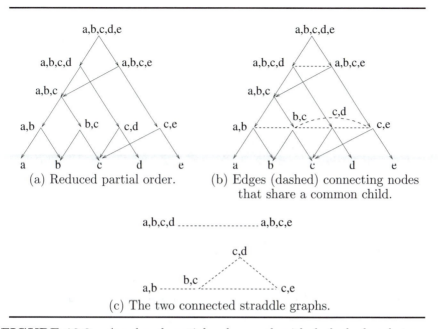

(a) Reduced partial order.

(b) Edges (dashed) connecting nodes that share a common child.

(c) The two connected straddle graphs.

FIGURE 13.2: A reduced partial order graph with dashed edges between nodes that have common children.

13.4 Boolean Closure of Sets

Let $\mathcal{S}$ be a collection of sets. Then the boolean closure of sets, $B(\mathcal{S})$, is defined in terms of its *intersection* and *union* closures.

13.4.1 Intersection closure

The intersection closure $B_\cap(\mathcal{S})$ is defined as follows:

$$B_\cap(\mathcal{S}) = \left\{ S_1 \cap S_2 \cap \ldots \cap S_l \mid S_i \in \mathcal{S}, 1 \leq i \leq l, \text{ for some } l \geq 1 \right\}.$$

In other words, this is the collection of all possible intersection of the sets. For example, let

$$\mathcal{S} = \{\{a, b, c, d\}, \{a, b, c, e\}, \{b, c, f\}, \{e, f\}\}.$$

Then,

$$
\begin{array}{ll}
B_\cap(\mathcal{S}) = \{\ \{e\}, & (\{a,b,c,e\} \cap \{e,f\}) \\
\{f\}, & (\{b,c,f\} \cap \{e,f\}) \\
\{b,c\}, & (\{a,b,c,d\} \cap \{a,b,c,e\} \cap \{b,c,f\}, \text{ or} \\
 & \{a,b,c,d\} \cap \{b,c,f\}, \text{ or} \\
 & \{a,b,c,e\} \cap \{b,c,f\}) \\
\{a,b,c\}, & (\{a,b,c,d\} \cap \{a,b,c,e\}) \\
\{e,f\}, & (\text{in } \mathcal{S}) \\
\{b,c,f\}, & (\text{in } \mathcal{S}) \\
\{a,b,c,d\}, & (\text{in } \mathcal{S}) \\
\{a,b,c,e\}\ \}. & (\text{in } \mathcal{S})
\end{array}
$$

13.4.2 Union closure

The union closure $B_\cup(\mathcal{S})$ is defined as follows:

$$
B_\cup(\mathcal{S}) = \left\{ S_1 \cup S_2 \cup \ldots \cup S_l \;\middle|\; \begin{array}{c} \text{for some } l \geq 1 \text{ and for each } S_j \\ \text{there exists some } S_k \text{ such that} \\ S_j \cap S_k \neq \emptyset \end{array} \right\},
$$

Note that we define the union only over a collection of overlapping sets. For example, when

$$\mathcal{S} = \{\{a,b,c,d\}, \{a,b,c,e\}, \{b,c,f\}, \{e,f\}, \{g,h\}\},$$

then,

$$
\begin{array}{ll}
B_\cup(\mathcal{S}) = \{\ \{e,f\}, & (\text{in } \mathcal{S}) \\
\{g,h\}, & (\text{in } \mathcal{S}) \\
\{b,c,f\}, & (\text{in } \mathcal{S}) \\
\{a,b,c,d\}, & (\text{in } \mathcal{S}) \\
\{a,b,c,e\}, & (\text{in } \mathcal{S}) \\
\{b,c,e,f\}, & (\{b,c,f\} \cup \{e,f\}) \\
\{a,b,c,d,e\}, & (\{a,b,c,d\} \cup \{a,b,c,e\}) \\
\{a,b,c,d,f\}, & (\{a,b,c,d\} \cup \{b,c,f\}) \\
\{a,b,c,e,f\}, & (\{a,b,c,e\} \cup \{b,c,f\} \cup \{e,f\}) \\
\{a,b,c,d,e,f\}\ \}. & (\{a,b,c,d\} \cup \{a,b,c,e\} \cup \{b,c,f\})
\end{array}
$$

Can the following be a member of $B_\cup(\mathcal{S})$,

$$\{a,b,c,d\} \cup \{g,h\}\ ?$$

The answer is no, since the two sets have no overlap. Note that the set $\{g,h\}$ does not overlap with any of the other set in $\mathcal{S}$.

Then, how about the following:

$$\{a,b,c,d,e,f\} = (\{a,b,c,d\} \cup \{e,f\}. \tag{13.2}$$

But
$$\{a, b, c, d\} \cap \{e, f\} = \emptyset,$$
i.e., the two have no overlap, so this union also cannot be considered for membership in $B_\cup(\mathcal{S})$. However, consider
$$\{a, b, c, d\}, \{a, b, c, e\}, \{b, c, f\}.$$
Here
$$\{a, b, c, d\} \cap \{a, b, c, e\} \neq \emptyset \text{ and}$$
$$\{a, b, c, e\} \cap \{b, c, f\} \neq \emptyset.$$
Thus the union of the three sets can be considered as an element of the union closure:
$$\{a, b, c, d, e, f\} = \{a, b, c, d\} \cup \{a, b, c, e\} \cup \{b, c, f\}. \tag{13.3}$$
Note that Equations (13.2) and (13.3) give rise to the same set, but only the second union is acceptable by the definition of our union closure.

Back to boolean closure. The boolean closure is the union of the intersection closure, $B_\cap(\mathcal{S})$, and the union closure, $B_\cup(\mathcal{S})$,
$$B(\mathcal{S}) = B_\cap(\mathcal{S}) \cup B_\cup(\mathcal{S}),$$
Thus, boolean closure is all the possible intersection and union of the overlapping sets.

LEMMA 13.4
(Straddle graph properties lemma) *Let $\mathcal{S}$ be a collection of sets defined on alphabet Σ such that no two sets in $\mathcal{S}$ straddle.*

1. *Then the boolean closure, $B(\mathcal{S})$, is the same as $\mathcal{S}$, i.e.,*
$$\mathcal{S} = B(\mathcal{S}) = B_\cap(\mathcal{S}) = B_\cup(\mathcal{S}).$$

2. *Then the reduced partial order graph*
$$G(\mathcal{S}, E_r) = G(B(\mathcal{S}), E_r)$$
$$= G(B_\cap(\mathcal{S}), E_r)$$
$$= G(B_\cup(\mathcal{S}), E_r)$$
   is acyclic.

3. *Then the elements of Σ can be consecutively arranged as a string say s, such that for each $S \in \mathcal{S}$, its members appear consecutive[3] in s.*

The proof is left as Exercise 161 for the reader.

---

[3]We later introduce the notation '$s \in \mathcal{F}(\mathcal{S})$', to articulate the same condition.

13.5   Consecutive (Linear) Arrangement of Set Members

One of the statements in Lemma (13.4) is about the consecutive arrangement of the alphabet Σ. We study the necessary and sufficient condition for a consecutive arrangement by considering a a classical problem studied in combinatorics.

Problem 23 *(The general consecutive arrangement (GCA) problem)*
Given a finite set Σ and a collection $\mathcal{S}$ of subsets of Σ, find

$$\mathcal{F}(\mathcal{S})$$

the collection of permutations s of Σ in which the members of each subset $S \in \mathcal{S}$ appear as a consecutive substring of s.

For example, consider

$$\mathcal{S}_1 = \{\{a, b, c\}, \{a, c, d\}, \{a, c\}\}.$$

Here

$$\Sigma = \{a, b, c, d\}.$$

It can be verified that

$$\mathcal{F}(\mathcal{S}_1) = \{bacd, \quad dcab, \quad bcad, \quad dacb\}.$$

Note that each set $S \in \mathcal{S}$ is such that its members appear consecutively in each $s \in \mathcal{F}(\mathcal{S}_1)$. Next, consider

$$\mathcal{S}_2 = \{\{a, b, c\}, \{a, c, d\}, \{b, d\}\}.$$

Note that

$$\mathcal{F}(\mathcal{S}_2) = \emptyset,$$

i.e., it is not possible to arrange the members of Σ in such a way that each set $S \in \mathcal{S}$ appears consecutively.

Can a systematic approach be developed to solve the GCA problem. PQ tree is a data structure introduced by Booth and Leukar [BL76] to solve the GCA problem in linear time.

13.5.1   PQ trees

A PQ tree, T, is a directed acyclic graph with the following properties.

1. T has one root (no incoming edges).

2. The leaves (no outgoing edges) of T are labeled bijectively by Σ.

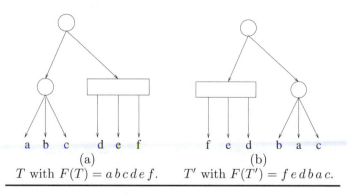

(a)

(b)

T with $F(T) = a\,b\,c\,d\,e\,f.$ T' with $F(T') = f\,e\,d\,b\,a\,c.$

FIGURE 13.3:   Two equivalent PQ trees, $T' \equiv T$ and their frontiers.

3. It has two types of internal (not leaf) nodes:

> P-node: The children of a P-node occur in no particular order. This node is denoted by a circle.
>
> Q-node: The children of a Q-node appear in a left to right or right to left order. This node is denoted by a rectangle.

An example PQ tree is shown in Figure 13.3(a) where

$$\Sigma = \{a, b, c, d, e, f\}.$$

The root node is a P-node and it has six ($|\Sigma|$) leaves mapped bijectively to the elements of Σ. The *frontier* of a tree T denoted by

$$F(T),$$

is the permutation of Σ obtained by reading the labels of the leaves from left to right. Note that this definition is valid for *any tree*, even when the tree is not a PQ tree.

The frontiers of the PQ trees are shown in Figure 13.3. Two PQ trees T and T' are *equivalent*, denoted

$$T \equiv T',$$

if one can be obtained from the other by applying a sequence of the following transformation rules:

1. Arbitrarily permute the children of a P-node, and

2. Reverse the children of a Q-node.

Any frontier obtainable from a tree equivalent with T is called *consistent* with T, and $\mathcal{F}(T)$ is defined as follows:

$$\mathcal{F}(T) = \{F(T')|T' \equiv T\}.$$

In Figure 13.3,

$$\mathcal{F}(T) = \mathcal{F}(T')$$
$$= \{abcde, \quad abced, \quad cbade, \quad cbaed,$$
$$deabc, \quad decba, \quad edabc, \quad edcba\}.$$

In the remainder of the discussion, we omit the directions of the edges in the PQ tree, since the direction is obvious from the PQ tree diagram.

Back to consecutive arrangement.   Recall from last section:

> If no two sets in $\mathcal{S}$ straddle, then the reduced partial graph, $G(\mathcal{S}, E_r)$, is acyclic.

The following is straightforward to verify and we leave this as Exercise 163 for the reader.

LEMMA 13.5
(Frontier lemma) *For a tree T, whose leaves are labeled bijectively with the elements of Σ and each internal node represents the collection of the labels of the leaf nodes reachable from this node, say*

$$S \in \mathcal{S},$$

a consecutive arrangement of Σ that respects S is the frontier $F(T)$.

Since $G(\mathcal{S}, E_r)$ is some tree, then a consecutive arrangement is the frontier

$$F(G(\mathcal{S}, E_r)).$$

Next, consider a pair of straddling sets as follows:

$$\mathcal{S} = \{\{a, b, c, d, e\}, \{d, e, f, g, h\}\}.$$

As we have seen before, $s_1, s_2 \in \mathcal{F}(\mathcal{S})$, where

$$s_1 = \boxed{\text{a b c d e}}\,\text{f g h}$$
$$s_2 = \text{a b c}\,\boxed{\text{d e f g h}}$$

This shows that some straddling sets do allow a consecutive arrangement. What is that condition(s) on straddling sets?

13.5.2 Straddling sets

We give an exposition on the solution to the GCA problem using *boolean closure* of $\mathcal{S}$,

$$B(\mathcal{S}),$$

and its *reduced partial order graph*,

$$G(B(\mathcal{S}), E_r).$$

Note that this is *not* an algorithm to solve the GCA problem. Actually, it is possible to translate the exposition into an algorithm, but is not the most efficient.

If every node in $G(B(\mathcal{S}), E_r)$ has at most one parent, then it is a strict hierarchy or a tree and the frontier

$$F(G(B(\mathcal{S}), E_r))$$

is the consecutive arrangement. Can some nodes have multiple parents and yet a consecutive arrangement of the elements possible?

A graph $G(V, E)$ is a *chain* or a *total order* if all its vertices can be arranged along a path. In other words, there are exactly two end vertices, with degree one, and every other vertex has degree two.

The following theorem gives the conditions under which

$$\mathcal{F}(\mathcal{S}) \neq \emptyset,$$

i.e., a linear consecutive arrangement of elements is possible.

THEOREM 13.1
(Set linear arrangement theorem) *Given $\mathcal{S}$, a collection of sets on alphabet Σ,*

$$\mathcal{F}(\mathcal{S}) \neq \emptyset,$$

(i.e., there exists a consecutive arrangement of the n members of Σ) if and only if every straddle graph,

$$G_{straddle}(V_S, E_S),$$

for each $S \in \mathcal{S}$, in the reduced partial order graph

$$G(B(\mathcal{S}), E_r)$$

is a chain.

PROOF   For $S_i \in \mathcal{S}$, let the straddle graph be

$$G_{straddle}(V_{S_i}, E_{S_i}).$$

We make the two following observations.

1. For each S_i, since
$$G_{straddle}(V_{S_i}, E_{S_i})$$
is a chain, without loss of generality, the chain is given as:
$$S_{i_1} S_{i_2} \dots S_{i_l},$$
for some l. This gives a consecutive arrangement of the elements of the sets in V_{S_i}, written as,
$$s_{S_i} = (S'_{i_0})\text{-}(S'_{i_1})\text{-}(S'_{i_2})\text{-}(S'_{i_3})\text{-}\dots\text{-}(S'_{i_{l-2}})\text{-}(S'_{i_{l-1}})\text{-}(S'_{i_l})\text{-}(S'_{i_{l+1}}) \qquad (13.4)$$
where
$$\begin{aligned} S'_{i_0} \cup S'_{i_1} &= S_{i_1}, \\ S'_{i_1} \cup S'_{i_2} \cup S'_{i_3} &= S_{i_2}, \\ S'_{i_2} \cup S'_{i_3} \cup S'_{i_4} &= S_{i_3}, \\ S'_{i_3} \cup S'_{i_4} \cup S'_{i_5} &= S_{i_4}, \end{aligned}$$
$$\dots$$
$$\begin{aligned} S'_{i_{l-2}} \cup S'_{i_{l-1}} \cup S'_{i_l} &= S_{i_{l-1}}, \\ S'_{i_l} \cup S'_{i_{l+1}} &= S_{i_l}. \end{aligned}$$

and (S'_{i_j}) is simply a consecutive arrangement of the members of S'_{i_j} in any order. However, each (S'_{i_j}) must follow the left to right order as given by the index ordering, i.e., if $j < k$ then (S'_{i_j}) is to the left of (S'_{i_k}).

The proof that such a consecutive arrangement s_{S_i} is possible is left as Exercise 165 for the reader.

2. We claim that there exists some arrangement of all the n members of Σ, without loss of generality, as
$$\sigma_1 \sigma_2 \dots \sigma_n.$$
We call this the *consensus arrangement*, and this arrangement satisfies the following property. For each
$$1 \le p < q < r \le n,$$
if for some i,
$$\sigma_p \in S_{i_j}, \sigma_q \in S_{i_k}, \text{ and } \sigma_r \in S_{i_\ell},$$
then the following holds:
$$j < k < \ell \text{ or } \ell < k < j$$
must hold. In other words, the ordering in the elements in the consensus arrangement does not violate the ordering suggested by each s_{S_i}.

Again, the proof that such a consensus arrangement exists is left as Exercise 165 for the reader.

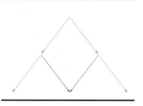

FIGURE 13.4:   Since the collection of sets is closed under union of straddling sets, if a node has two parents then it *must* be part of a pyramid structure
as shown.

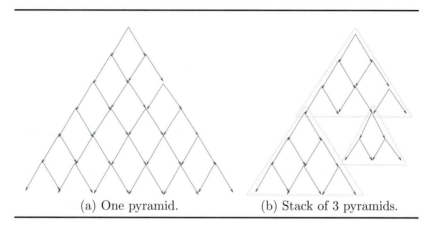

(a) One pyramid.   (b) Stack of 3 pyramids.

FIGURE 13.5:   Reduced partial order when the sets can be linearly arranged.

This concludes the proof. □

Figures 13.4 and 13.5(a) show examples of a single pyramid and Figure 13.5(b)
shows possible stacking of three pyramids in a reduced partial order graph.

So what does a Q-node encapsulate?   Consider the following collection
of sets:

$$S =$$
$$\{\,\{a,b\}, \{b,c\}, \{c,d\}, \{d,e\},$$
$$\{a,b,c\}, \{b,c,d\}, \{c,d,e\},$$
$$\{a,b,c,d\}, \{b,c,d,e\},$$
$$\{a,b,c,d,e\}\,\}.$$

Does there exist a sequence s of $\{a, b, c, d, e\}$ in which the members of each
subset $S \in \mathcal{S}$ appear as a consecutive substring of s? It is easy to see that

there are exactly two such sequences $\mathcal{F}(\mathcal{S}) = \{s_1, s_2\}$ where

$$s_1 = abcde, \text{ and}$$
$$s_2 = edcba.$$

The reduced partial order graph is shown in Figures 13.6(a) and (b). The exact same sets are denoted by a single Q-node as shown in (c) of the figure. If the number of leaf nodes in the reduced partial order graph is k, then the number of nonleaf nodes nodes is given as:

$$\mathcal{O}(k^2).$$

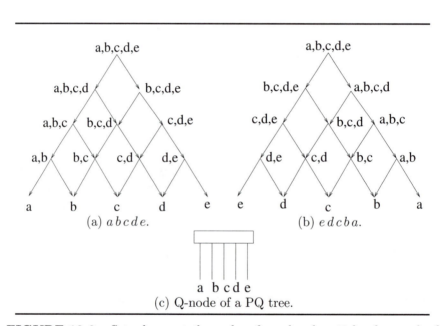

(a) $abcde$.

(b) $edcba$.

(c) Q-node of a PQ tree.

FIGURE 13.6: Sets shown at the nodes of a *reduced partial order* graph of a sequence (a) and its reversal (b). This entire information can be represented by a single Q-node of a PQ tree as shown in (c).

Back to the PQ tree. To summarize:

If the reduced partial order graph of the boolean closure of $\mathcal{I}$,

$$G(B(\mathcal{I}), E_r),$$

is such that graph can be partitioned into

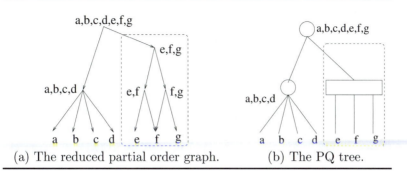

(a) The reduced partial order graph.   (b) The PQ tree.

FIGURE 13.7: If the elements (a, b, c, d, e, f, g) can be organized as a sequence, then the reduced partial order graph can be encoded as a PQ tree as shown in the above example.

- *'pyramid' structures and*
- *'tree' structures,*

then the elements of Σ *can be linearly arranged as some string* s.

By partition we mean that the edge set

$$E_r = E_1 \cup E_2 \ldots \cup E_l,$$

where

$$E_i \cap E_j = \emptyset \text{ with } 1 \le i < j \le l,$$

and each partition is defined on E_i. By 'tree structure' we mean that no node in the (sub)graph has more than one parent.

Thus we conclude:

Given $\mathcal{I}$*, if the elements of* X*, can be organized as a linear string, then the reduced partial order graph can be encoded as a PQ Tree.*

As the final example, consider

$$\mathcal{S} = \{ \ \{a, b, c, d\}, \tag{13.5}$$
$$\{e, f\}, \{f, g\}, \{e, f, g\}, \tag{13.6}$$
$$\{a, b, c, d, e, f, g\}\}. \tag{13.7}$$

The reduced partial order graph of this collection is shown in Figure 13.7(a). It can be partitioned into one 'pyramid structure' (shown enclosed by a dashed rectangle) and a 'tree structure'.

The PQ encoding is as follows: The tree has one Q-node, two P-node and seven leaf nodes. Set shown as (13.5) is encoded as a single P-node, sets shown as (13.6) are encoded as a single Q-node.

13.6   Maximal Set Intersection Problem (maxSIP)

Consider the following scenario. Let

$$\Sigma = \{a, b, c, d, e, f, g\},$$

be a finite alphabet with a collection of sets, $\mathbf{C}$, defined on it as follows: $\mathbf{C} = \{C_1, C_2, C_3, C_4, C_5\}$, where

$$
\begin{aligned}
C_1 &= \{g, b, c, d, e, a\},\\
C_2 &= \{a, f, c, d\},\\
C_3 &= \{a, b, d, e, c\},\\
C_4 &= \{b, d, e\},\\
C_5 &= \{f, a, b, c, d, e\}.
\end{aligned}
$$

We seek all 'maximal' set intersections, denoted as S, of at least K (called *quorum*) elements of $\mathbf{C}$, denoted as I_S. To gain an informal understanding of *maximality* of the pair (S, I_S), consider the following two cases. Let quorum $K = 3$.

1. Let $S' = \{b, d\}$,

   $$\text{then index set } I_{S'} = \{1, 3, 4, 5\},$$

   with $|I_{S'}| \geq 3$. However, $(S', I_{S'})$ is not maximal since there exists

   $$S_1 \supset S'$$

   with

   $$S_1 = \{b, d, e\}, \text{ and } I_{S_1} = I_{S'} = \{1, 3, 4, 5\}.$$

   The pair (S_1, I_{S_1}) is maximal.

2. Let an index set be $I' = \{1, 2, 5\}$, then the corresponding set S_2

   $$S_2 = \{a, d, c\}.$$

   Clearly, the pair (S_2, I') is not maximal, since

   $$S \subset C_3 \text{ and the index set must contain } 3.$$

   If $I_{S_2} = \{1, 2, 3, 5\}$, then the pair (S_2, I_{S_2}) is maximal.

This demonstrates that both a set S and an index set I must be 'expanded' to ensure maximality.

Further, the only other maximal pair for this example is

$$S_3 = \{a, b, c, d, e\} \text{ with } I_{S_3} = \{1, 3, 5\}.$$

The problem is formally stated as follows.

Problem 24 *(Maximal Set Intersection Problem (maxSIP(**C**,K)))* *The input to the problem is a collection of n sets* $\mathbf{C} = \{C_1, C_2, \ldots, C_n\}$, *where each* $C_i \subset \Sigma$, $|\Sigma| = m$, *and a quorum* $K > 0$. *For a set S such that*

$$S = C_{i_1} \cap C_{i_2} \cap \ldots \cap C_{i_p},$$

we denote by I_S *the set of indices*

$$I_S = \{i_1, i_2, \ldots, i_p\}.$$

Further, I_S *is maximal i.e., there is no* I' *with*

$$I_S \subsetneq I',$$

such that

$$S = C_{j_1} \cap C_{j_2} \cap \ldots \cap C_{j_{p'}}, \text{ where } I' = \{j_1, j_2, \ldots, j_{p'}\}.$$

The output is the set of all maximal pairs (S, I_S) *such that* $|S| \geq K$.

Further, we say S' is nonmaximal with respect to S if

$$S' \subsetneq S, \text{ and } I_{S'} = I_S.$$

13.6.1 Ordered enumeration trie

Given an input collection of sets, how do we detect *all* such maximal pairs? We propose a simple scheme to enumerate the maximal pairs (S, I_S). To avoid multiple enumerations, we follow some arbitrary (but fixed) order on the alphabet. Without loss of generality, let

$$\sigma_1 < \sigma_2 < \ldots < \sigma_m.$$

Given a set

$$S = \{\sigma_{i_1} < \sigma_{i_2} < \ldots < \sigma_{i_l}\},$$

define a sequence, $seq(S)$, as

$$seq(S) = \sigma_{i_1} \sigma_{i_2} \ldots \sigma_{i_l}.$$

Given an input **C** and a quorum K, consider the following collection of sequences:

$$\mathcal{S}_{\mathbf{C},K} = \left\{ S \,\middle|\, \begin{array}{l} (S, I_S) \text{ is a maximal pair} \\ \text{for input } \mathbf{C} \text{ and quorum } K \end{array} \right\}. \tag{13.8}$$

Also,

$$Sq_{\mathbf{C},K} = \{seq(S) \mid S \in \mathcal{S}_{\mathbf{C},K}\}. \tag{13.9}$$

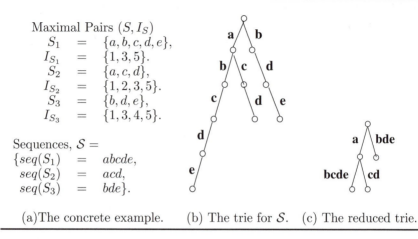

Maximal Pairs (S, I_S)

S_1 = $\{a, b, c, d, e\}$,
I_{S_1} = $\{1, 3, 5\}$.
S_2 = $\{a, c, d\}$,
I_{S_2} = $\{1, 2, 3, 5\}$.
S_3 = $\{b, d, e\}$,
I_{S_3} = $\{1, 3, 4, 5\}$.

Sequences, $\mathcal{S}$ =
$\{seq(S_1)$ = $abcde$,
$seq(S_2)$ = acd,
$seq(S_3)$ = $bde\}$.

(a)The concrete example.   (b) The trie for $\mathcal{S}$.   (c) The reduced trie.

FIGURE 13.8: Continuing the concrete example of Section 13.6: The set of sequences $\mathcal{S}$, following the ordering $a < b < c < d < e < f < g$ and the corresponding trie in (b). The reduced trie where each internal node has at least two children is shown in (c).

The *trie*, $\mathbf{T}_{\mathbf{C},K}$, of the elements of $Sq_{\mathbf{C},K}$ is termed the *ordered enumeration trie* or simply the *trie* of the input $\mathbf{C}$ with quorum K. Figure 13.8 displays the trie for a simple example. What is the size of this trie?

LEMMA 13.6

(Enumeration-trie size lemma) *The number of nodes in trie* $\mathbf{T}_{\mathbf{C},K}$ *is no more than*

$$\sum_{s \in Sq_{\mathbf{C},K}} |s| = \sum_{S \in \mathcal{S}_{\mathbf{C},K}} |S|,$$

where $\mathcal{S}_{\mathbf{C},K}, Sq_{\mathbf{C},K}$ *are as defined in Equations (13.8) and (13.9).*

13.6.2   Depth first traversal of the trie

Given $\mathbf{C}$ and K, we propose an algorithm, which traverses the ordered enumeration trie, $\mathbf{T}_{\mathbf{C},K}$, in a depth first order. The pseudocode in Algorithm (14) implicitly generates this tree and produces the maximal pairs. Before we delve into the details of the algorithm, we note two of its important characteristics.

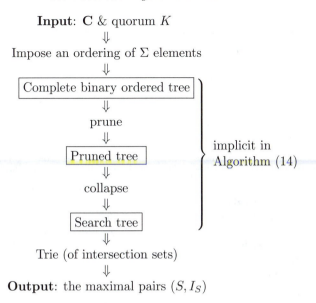

Input: **C** & quorum K

⇓

Impose an ordering of Σ elements

⇓

| Complete binary ordered tree |

⇓

prune

⇓

| Pruned tree |

⇓

collapse

⇓

| Search tree |

} implicit in Algorithm (14)

⇓

Trie (of intersection sets)

⇓

Output: the maximal pairs (S, I_S)

FIGURE 13.9: Overall scheme of the enumeration algorithm.

1. The running time complexity of the algorithm is *output-sensitive* i.e., the amount of work done is (almost) linear with the size of the output.

2. It is possible to output the maximal sets as the algorithm is executed, without ever having a need to backtrack. In other words, the enumeration is done in such a way that once a set $(S$, with its $I_S)$ is ascertained to be maximal, it remains so till the end of the execution of the algorithm.

How does the algorithm work? This is best understood as a scheme described in Figure 13.9.

Note that the algorithm does not follow the stages–for efficiency reasons it directly constructs the trie after ordering the elements of Σ. The 'boxed' stages aid in proving the correctness of the algorithm and in the analysis of running time complexity of the algorithm.

Algorithm 14 *(Maximal Set Intersection Problem)*

```
maxSIP(C, K, S, I_S, j)
{
IF j > 0 AND |I_S| ≥ K
    I_{S_new} ← {i | i ∈ I_S AND σ_j ∈ (C_i ∈ C)}        //takes O(n) time
    IF (|I_{S_new}| ≥ K) {
        S_new ← S ∪ {σ_j}
        Terminate ← FALSE
        IF (I_{S_old} = Exists(T, I_{S_new}))              //takes O(log n) time
            IF |I_{S_new}| = |I_S|                         //immediate parent;
                                                           //(S_new, I_S) is possibly maximal
                Replace(T, S_old, S_new)                   //takes O(log n) time
            ELSE Terminate ← TRUE //(S_new, I_{S_new}) is nonmaximal,
                                                           //hence terminate this branch
        ELSE Add(T, S_new)                                 //takes O(log n) time
        IF NOT Terminate
            maxSIP(C, K, S_new, I_{S_new}, j-1)            //left-child call
    }

    OUTPUT(S, I_S)                                         //(S, I_S) is certainly maximal

    IF |I_{S_new}| ≠ |I_S| OR j = 1                        //right & left subtrees the same
                                                           //OR very last right-child call
        maxSIP(C, K, S, I_S, j-1)                          //right-child call
}
```

Input parameters. The routine in Algorithm (14) takes the following four parameters:

1. A collection of n sets

$$\mathbf{C} = \{C_i, C_2, \ldots, C_n\},$$

   where each C_i is defined on some alphabet $\Sigma = \{\sigma_1, \sigma_2, \ldots, \sigma_m\}$, say and $1 \leq i \leq n$.

2. The quorum K that restricts every output set S to have at least K elements.

3. The pair S and I_S computed until this point. The algorithm is initiated with the following settings:

   (a) $S \leftarrow \emptyset$ and $I_S \leftarrow U = \{1, 2, 3, \ldots, n\}$.

   (b) $j = m$ (recall $|\Sigma| = m$).

4. The symbol $\sigma_j \in \Sigma$ which is given as simply j in the call.

The Complete Binary Ordered Tree $\mathbf{B_C}$. Recall that, in general

$$\sigma_1 < \sigma_2 < \ldots < \sigma_m.$$

A small example is discussed in Figure 13.10. The input sets are shown in (a) where the alphabet is

$$\Sigma = \{a < c < d\}.$$

A complete binary ordered tree, $\mathbf{B}_{C,K}$, for this input is shown in (b). The root of this tree is labeled with the pair (S, I_S) where

$$S = \emptyset \text{ and } I_S = U = \{1, 2, \ldots, n\},$$

where $n = |\mathbf{C}|$. Every internal node (including the root node) has exactly two children: the left child is labeled with some $\sigma_j \in \Sigma$, $1 \le j \le m$, and the right-child is unlabeled and is shown as a dashed edge in the figure.

For a node v at depth of j from the root, define $path_v$, the path from the root node down to node v on this complete binary ordered tree, by the labels on the edges written as

$$path_v = X_1 X_2 \ldots X_j,$$

where, for each $1 \le k \le j$,

$$X_k = \begin{cases} \{\sigma_k\} & \text{if the edge label } X_k \text{ is a left child with label } \sigma_j, \\ \emptyset & \text{if the edge label } X_k \text{ is a right child.} \end{cases}$$

The node is labeled by the pair (S_v, I_{S_v}) which is defined as:

$$S_v = \Sigma_v \cap \left(C_{i_1} \cap C_{i_2} \cap \ldots \cap C_{i_p} \right), \text{ and}$$
$$I_{S_v} = \{i_1, i_2, \ldots, i_p\}.$$

where

$$\Sigma_v = X_1 \cup X_2 \cup \ldots \cup X_j.$$

In other words,

1. the left child corresponds to 'presence of σ_j', and

2. the right child corresponds to 'ignoring σ_j' (note that it is not 'absence of σ_j').

LEMMA 13.7

Each dashed edge is such that the label, (S, I_S), on both the nodes of this edge are the same.

$$C_1 = \{a, c\},$$
$$C_2 = \{a, d\},$$
$$C_3 = \{a, c, d\}.$$

(a) The input $\mathbf{C} = \{C_1, C_2, C_3\}$, with quorum $K = 2$.

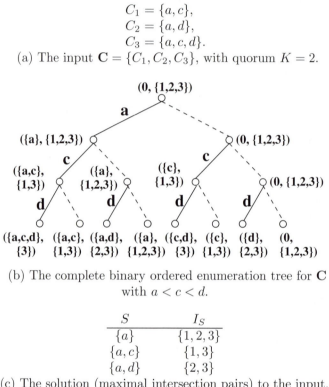

(b) The complete binary ordered enumeration tree for $\mathbf{C}$
with $a < c < d$.

S	I_S
$\{a\}$	$\{1, 2, 3\}$
$\{a, c\}$	$\{1, 3\}$
$\{a, d\}$	$\{2, 3\}$

(c) The solution (maximal intersection pairs) to the input.
(note that the pair $(\{c\}, \{1, 3\})$ is not maximal)

FIGURE 13.10: (a) An input, (b) the corresponding complete binary ordered enumeration tree, and (c) the solution.

See Figure 13.10(b) for a concrete example. It can be verified that the ordered enumeration trie, $\mathbf{T}_{C,K}$, is 'contained' in this complete binary ordered tree $\mathbf{B_C}$.

We now relate $\mathbf{B_C}$ to the pseudocode in Algorithm (14). This tree is implicitly generated by the algorithm. Note that this is the complete tree, but the code will prune the tree (as discussed later) for efficiency purposes. In detail, the algorithm can be understood as follows. The routine makes at most two recursive calls

1. marked as *left-child call*, and

2. marked as the *right-child call*.

Thus each left-child edge of $\mathbf{B_C}$ is marked with $\{\sigma_j\}$ while the right-child edge is labeled (implicitly) with an empty set $\emptyset$ (or unlabeled in Figure 13.10(b)). For easy retrieval the pair (S_v, I_{S_v}) at node v is stored in a balanced tree data structure $\mathcal{T}$ (see below).

Pruning $\mathbf{B_C}$. Next, we explore how this complete binary ordered tree is pruned. This is done by identifying a set of *traversal terminating conditions*. Every recursive routine must have a mandatory termination condition for obvious reasons.[4] For efficiency purposes, the routine has four additional terminating conditions. The terminations of a recursive call corresponds to the pruning of the complete binary search tree. All the terminating conditions are discussed below.

1. The 'mandatory' terminating condition when all the alphabet has been explored (the $j > 0$ condition).

2. We discuss two terminating conditions here, both arising when the computed $I_{S_{new}}$ is found to be the same as some existing set ($I_{S_{old}}$ or I_S). This condition ensures that *a nonmaximal set S is detected no more than once*, giving an asymptotic improvement in the overall efficiency (leading to an output-sensitive time complexity) of the algorithm.

   (a) Case $I_{S_{new}} = I_{S_{old}}$, i.e., the freshly computed $I_{S_{new}}$ already exists in the data structure $\mathcal{T}$.
   Further if

   $$S_{old} \supset S_{new},$$

   then clearly S_{new} is nonmaximal. In this case not only S_{new} can be discarded but also, the traversal can be terminated here, since it is guaranteed that all the subsequent sets detected on this subtree (of the complete binary ordered tree) will be nonmaximal.
   However, if

   $$S_{old} \not\supset S_{new},$$

---

[4]Otherwise, the run-time stack will overflow eventually crashing the system.

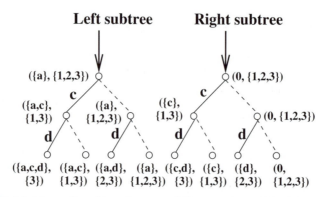

FIGURE 13.11:   Consider the tree of Figure 13.10. The left and the right subtrees of the root node when the set of indies I is the same on both nodes. Notice that the index sets I are identical on the nodes in both the subtrees but the sets S are nonmaximal in the right subtree, i.e., the set S^R on the right subtree is a subset of the corresponding S^L on the left subtree ($S^R \subset S^L$).

then S_{new} is *possibly* [5] maximal. What can we say about S_{old}? The answer lies in the following question: Is it possible that

$$S_{old} \not\supseteq S_{new} \text{ and } S_{new} \not\supseteq S_{old} \text{ but } I_{S_{new}} = I_{S_{old}}?$$

If this is the case then there must exist

$$S' \supseteq S_{old} \cup S_{new},$$

with

$$I_{S'} = I_{S_{new}} = I_{S_{old}}.$$

Hence it must be that

$$S_{old} \subset S_{new}$$

and S_{new} replaces S_{old} in the data structure $\mathcal{T}$.

We use the variable *Terminate*, in the pseudocode to check for this condition.

(b) Case $I_{S_{new}} = I_S$ (at the 'right-child' call).

In fact, it is adequate to simply check if the set sizes are the same, i.e.,

$$|I_S| = |I_{S_{new}}|,$$

which can be done in $\mathcal{O}(1)$ time.

Why can we safely terminate the search of this branch ('right-child') on the complete binary ordered tree? Since $I_S = I_{S_{new}}$, the

---

[5] In other words S_{new} is *maximal until this point in the execution*. The possibility remains that sometime later in the execution, it may turn out to be indeed *nonmaximal*.

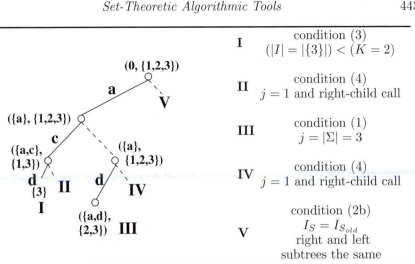

I	condition (3) $(	I	=	\{3\}	) < (K = 2)$
II	condition (4) $j = 1$ and right-child call
III	condition (1) $j =	\Sigma	= 3$
IV	condition (4) $j = 1$ and right-child call
V	condition (2b) $I_S = I_{S_{old}}$ right and left subtrees the same

FIGURE 13.12: The pruned search tree of Figure 13.10. The order of symbols is $a < c < d$ and quorum $K = 2$. The different terminating or pruning conditions are explained above. See text for more details.

two subtrees corresponding to the left and the right child rooted at this node are identical. Further, all the sets (S) associated with the nodes of the right subtree are nonmaximal. See Figure 13.11 for a concrete example of nonmaximal sets in the right subtree.

3. If the set size, $|I_{S_{new}}|$, falls below quorum, K, the search on the complete binary ordered tree can be terminated. This gives rise to efficiency in practice but no asymptotic improvements can be claimed due to this condition.

4. The very last (when $j = 1$) 'right-child' call need not be made, since it does not give rise to any new sets S. Again, this gives rise to efficiency in practice but no asymptotic improvements can be claimed due to this condition.

For a complete example see Figure 13.12: it shows the pruned version of the complete binary tree shown in Figure 13.10, along with the various terminating conditions.

Collapsing the pruned enumeration tree. The dashed edges of the pruned tree can be collapsed, so that each node in this collapsed tree has a unique label given as (S, I_S) and the collapsed tree is shown in Figure 13.13(b).

We call this the *search tree.* We study the characteristics of this tree

1. We identify three kinds of vertices in the search tree:

    (a) Solid circles: these correspond to the maximal sets.

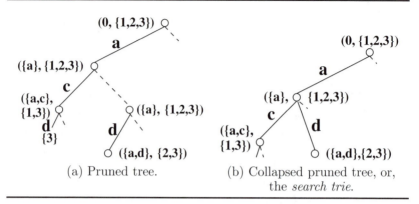

(a) Pruned tree.

(b) Collapsed pruned tree, or,
the *search trie*.

FIGURE 13.13:   The pruned tree of Figure 13.12 can be collapsed to the search trie shown on the right. The dashed edge has identical labels $(\{a\}, \{1, 2, 3\})$ at both its end-points (see Lemma (13.7)) and represents a nonexistent call. This dashed edge is removed to produce a collapsed tree with unique labels at each node. It retains the *stub* edges to keep track of the (number of) terminated calls.

    (b) Hollow circles: these correspond to the nonmaximal sets.

    (c) Little squares: these are the 'nonexistent' nodes, since the recursive call is terminated due to different terminating conditions.

2. We identify two kinds of edges in the reduced tree:

    (a) *Stubs*: These are edges that are incident on the *little square* nodes.

    (b) Regular: These are the ones that are not stubs.

The search tree without the stubs is indeed the trie that we discussed earlier in the section.

The edges in the search tree (both regular and *stub*) correspond to the number of recursive calls in Algorithm (14). Using Lemma (13.6), we obtain the following:

LEMMA 13.8

The number of regular and stub *edges in the search tree is no more than*

$$|\Sigma| \sum_{S \in \mathcal{S}_{\mathbf{C}, K}} |S|.$$

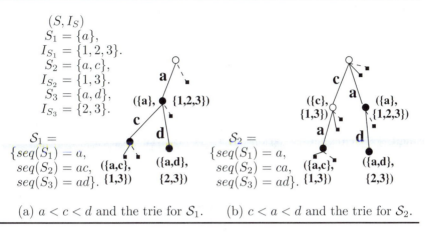

(a) $a < c < d$ and the trie for $\mathcal{S}_1$.   (b) $c < a < d$ and the trie for $\mathcal{S}_2$.

FIGURE 13.14:   The maximal sets shown as solid circles and nonmaximal as hollow circles in the trie (or the search tree with the stubs). Different orderings leading to different tries, but the same maximal pairs.

Different ordering of elements of Σ.   Figure 13.14 shows the search tree for two different orderings of the alphabet Σ. Clearly, the tries are topologically different (i.e., not isomorphic to each other), but the resulting maximal sets are the same. This is straightforward to verify and we leave that as an exercise for the reader.

Order of maximal sets.   The boxed statement OUTPUT(S, I_S) outputs all the maximal pairs (S, I_S). It turns out that these sets correspond exactly to the sets stored in $\mathcal{T}$. In other words, the routine can simply output the maximal set S (at the boxed statement) with never a need to undo or augment the set later.

Indeed, this is an optional statement in the routine. This just indicates that the maximal sets can be 'output' as they are generated.

Operations of the data structure $\mathcal{T}$.   For efficiency purposes, the algorithm uses a (balanced) tree data structure $\mathcal{T}$ to store the sets represented by (S, I_S), sorted by I_S. The different routines on $\mathcal{T}$ are as follows. Each routine takes $\mathcal{O}(\log n)$ time (recall that $|\mathbf{C}| = n$).

1. $I_S' = \text{Exists}(\mathcal{T}, I_S)$ checks if I_S exists in the data structure and if it does, returns it as I_S'.

2. Replace$(\mathcal{T}, S, S')$, replaces S in the data structure $\mathcal{T}$ with S'.

3. Add$(\mathcal{T}, (S, I_S))$ adds (S, I_S) as a new node in the data structure $\mathcal{T}$ using I_S as the key.

Algorithm time complexity.   Let the size of the input N_I and the size of the output N_O be given as

$$N_I = \sum_{C \in \mathbf{C}} |C|,$$

$$N_O = \sum_{S \in \mathcal{S}_{\mathbf{C},K}} (|S| + |I_S|),$$

using $\mathcal{S}_{\mathbf{C},K}$ defined in Equation (13.8). For convenience, let

$$N_{O_1} = \sum_{S \in \mathcal{S}_{\mathbf{C},K}} |S|,$$

$$N_{O_2} = \sum_{S \in \mathcal{S}_{\mathbf{C},K}} |I_S|.$$

Thus

$$N_O = N_{O_1} + N_{O_2}.$$

The number of calls is bounded by the total number of edges in the search tree given in Lemma (13.6) as $\mathcal{O}(N_{O_1}|\Sigma|)$. Each routine call, which corresponds to a node in this tree, takes

1. $\mathcal{O}(N_I)$ time to read the input,

2. $\mathcal{O}(n)$ time to compute $I_{S_{new}}$, and

3. each operation on the (balanced) tree $\mathcal{T}$ takes $\mathcal{O}(\log |\mathbf{C}|) = \mathcal{O}(\log n)$ time.

Thus the amount of work done on all the nodes of the search tree is

$$\mathcal{O}\left(N_{O_1}(N_I + n + \log n)|\Sigma|\right).$$

However, a more careful counting significantly tightens this bound.

To improve the bound, we use N_{O_2} in a more meaningful manner. We ask the question: How many times is an index set, I_S, encountered (or 'read') in the course of the execution of the algorithm? It turns out that this number is $\mathcal{O}(N_S)$ where N_S is the the number of nonmaximal sets S' of S encountered in the search tree. For any S, the number of nonmaximal S' with respect to S is $\mathcal{O}(|\Sigma|)$, as a rough estimate. Also, notice in the pseudocode of Algorithm (14), that the input is also read along with the index set. Thus the overall time taken for reading, *all* the index sets along with the input is

$$\mathcal{O}((N_{O_2} + N_I)|\Sigma|).$$

Combining, this with the time taken for the remainder of the routine, the overall time complexity of Algorithm (14) is given as:

$$\mathcal{O}((N_{O_2} + N_I)|\Sigma| + N_{O_1} \log n |\Sigma|)$$
$$= \mathcal{O}((N_{O_2} + N_I + N_{O_1} \log n)|\Sigma|)$$
$$= \mathcal{O}(N_I + N_O \log n)$$

This time complexity is considered *output-sensitive* since the amount of work done is (almost) linear with the size of the output N_O. We say 'almost linear' since strictly speaking, there is also a logarithmic term, $\log n$, in the formula.

13.7 Minimal Set Intersection Problem (minSIP)

Problem 25 *(Minimal Set Intersection Problem (minSIP(**C**,K)))* *The input to the problem is a collection of n sets* $\mathbf{C} = \{C_1, C_2, \ldots, C_n\}$, *and a quorum* $K > 0$. *For a set S such that*

$$S = C_{i_1} \cap C_{i_2} \cap \ldots \cap C_{i_p},$$

we denote by I_S the set of indices

$$I_S = \{i_1, i_2, \ldots, i_p\}.$$

Further, I_S is minimal i.e., there is no I' with

$$I' \subsetneq I_S,$$

such that

$$S = C_{j_1} \cap C_{j_2} \cap \ldots \cap C_{j_{p'}}, \ \text{where } I' = \{j_1, j_2, \ldots, j_{p'}\}.$$

The output is the set of all pairs (S, I_S) such that $|S| \geq K$.

Notice here that it is possible to have distinct minimal sets $S_1, S_2, \ldots, S_p$, such that

$$I_{S_1} = I_{S_2} = \ldots, = I_{S_p}.$$

In other words, multiple sets may be associated with a single index set I.

13.7.1 Algorithm

We design an algorithm along the lines of Algorithm (14) as follows here. The algorithm (Algorithm (15)) is almost the same except for two differences:

1. There is only one terminating condition in the routine here which is when the size of the set $I_{S_{new}}$ falls below the quorum K. In Algorithm (14), there is yet another terminating condition which is when the set S_{new} is nonmaximal.

2. Since multiple sets can be associated with one index set I, note that $\mathcal{S}_{old}$ is a collection of sets. Further we use a new routine on the balanced binary tree $\mathcal{T}$ called Append($\mathcal{T}, \mathcal{S}_{old}, S_{new}$). This routine first removes any $S \in \mathcal{S}_{old}$ from the collection $\mathcal{S}_{old}$ satisfying

$$S \supset S_{new}.$$

Then S_{new} is added to the collection $\mathcal{S}_{old}$ if there is no $S \in \mathcal{S}_{old}$ such that

$$S \subset S_{new}.$$

Clearly, the computation of the minimal sets is more expensive because of the Append($\cdot$) routine and the lack of quick termination due to nonmaximality or such similar conditions. We leave the time complexity analysis of this algorithm as an exercise for the reader. Instead, we describe a scheme where the minimal sets are computed from the maximal sets.

Algorithm 15 *(Minimal Set Intersection Problem)*

```
minSIP(C, K, S, I_S, j)
{
    IF (j ≤ 0) EXIT
    I_{S_new} ← {i | i ∈ I_S AND σ_j ∈ (C_i ∈ C)}
    IF (|I_{S_new}| ≥ K) {
        S_new ← S ∪ {σ_j}
        IF (I_{S_old} = Exists(T, I_{S_new}))
            IF |I_{S_new}| = |I_S| //immediate parent, so dont update T
            ELSE Append(T, S_old, S_new) //multiple minimal sets
                        //if subset of existing set S', remove S' from T
        ELSE Add(T, S_new)
        minSIP(C, K, S_new, I_{S_new}, j-1)
    }
    minSIP(C, K, S, I_S, j-1)
}
```

13.7.2 Minimal from maximal sets

In this section we discuss how to extract the minimal sets while computing the maximal sets. This is best explained through a concrete example. See Figure 13.15 for a portion of a search trie. This tree is not in reduced form, i.e., it retains the internal nodes that have only one child. In other words, every node of this tree is labeled by a single character.

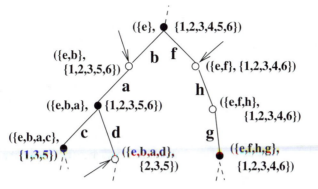

FIGURE 13.15: A search trie with singleton labels on the edges. The solid nodes represent maximal set pairs. The hollow nodes with a pointing arrow represent minimal set pairs.

Let the label of a node A be the pair

$$(S_A, I_{S_A}).$$

The following observation follows directly from the construction of the trie.

LEMMA 13.9

(Trie partial-order lemma) *Node B is a descendent of node A in the trie, if and only if the following two statements hold:*

$$S_B \supset S_A \ \text{and}$$
$$I_{S_B} \subseteq I_{S_A}.$$

Then the following is straightforward to see and we leave the proof as an exercise for the reader (Exercise 171).

1. Node A denotes a *maximal* intersection pair (S_A, I_{S_A}) if and only if

   A has more than one child.

2. Node A denotes a *minimal* intersection pair (S_A, I_{S_A}) if and only if

   the parent of A has multiple children and

   A has a single child.

13.8   Multi-Sets

We next consider multi-sets, where the multiplicity of the elements (sometimes also called *copy number*) is taken into account. For example a multi-set is given as:

$$S = \{\sigma(k) \mid \sigma \in \Sigma, k \geq 1\}.$$

In other words, each element σ also has a copy number stating the number of times σ appears in the set. The sets considered in Section 13.6 were such that $k = 1$ for each σ. For example, if

$$S = \{a(2), b(4)\},$$

then multi-set S has two copies of a and four copies of b. Let

$$\Sigma' = \{\sigma(c) \mid \sigma \in \Sigma \text{ and } c \text{ is a copy number in the data}\}.$$

Set operations of multi-sets. Given multi-set S, let the closure of S be defined as follows:

$$cls(S) = S \cup \{\sigma(0) \mid \sigma \in \Sigma \text{ and } \sigma(\text{-}) \notin S\}.$$

1. Given two multi-sets S_1 and S_2,

$$S_1 \subset S_2 \Leftrightarrow \text{ for each } \sigma(k) \in cls(S_1), \text{ there exists } \sigma(k' \geq k) \in S_2.$$

2. Given multi-sets $S_1, S_2, \ldots, S_p$,

$$S_1 \cap S_2 \cap \ldots \cap S_p = \{\sigma(k_{\min}) \mid k_{\min} = \min_{1 \leq i \leq p} (k_i), \text{ where } \sigma(k_i) \in S_i\}.$$

The following are some illustrative examples. Here $\Sigma = \{a, b, c\}$.

$$\{a(2), b(4)\} = \{b(4), a(2)\},$$
$$\{a(2), b(2)\} \subset \{b(2), a(3)\},$$
$$\{a(2), b(2)\} \subset \{b(2), a(3), c(1)\},$$
$$\{a(2), b(2)\} \not\subset \{b(2), c(1)\}, \text{ and}$$
$$\{b(2), c(1)\} \not\subset \{a(2), b(2)\}.$$

The maximal multi-set intersection problem is exactly along the lines of Problem (24) and is stated below.

Problem 26 *(Maximal Multi-Set Intersection Problem (maxMIP(**S**,K)))* *The input to the problem is a collection of n multi-sets* $\mathbf{S} = \{C_1, C_2, \ldots, C_n\}$, *and a quorum $K > 0$. For a multi-set S such that*

$$S = C_{i_1} \cap C_{i_2} \cap \ldots \cap C_{i_p},$$

we denote by I_S the set of S indices

$$I_S = \{i_1, i_2, \ldots, i_p\}.$$

Further, I_S is maximal i.e., there is no I' with

$$I_S \subsetneq I',$$

such that

$$S = C_{j_1} \cap C_{j_2} \cap \ldots \cap C_{j_{p'}}, \text{ where } I' = \{j_1, j_2, \ldots, j_{p'}\}.$$

The output is the set of all pairs (S, I_S) such that $|S| \geq K$.

For example, consider three multi-sets with $\Sigma = \{a, b, d\}$:

$$S_1 = \{a(2), b(6)\},$$
$$S_2 = \{a(1), b(6), d(2)\}, \text{ and}$$
$$S_3 = \{a(3), b(2), d(2)\}.$$

Let quorum $K = 2$. What are the maximal multi-sets? Using the problem specification, the maximal intersection multi-sets are given below.

$$S_1 = \{a(1), b(2)\} \qquad \text{with } I_{S_1} = \{1, 2, 3\},$$
$$S_2 = \{a(1), b(6)\} \qquad \text{with } I_{S_2} = \{1, 2\},$$
$$S_3 = \{a(2), b(2)\} \qquad \text{with } I_{S_3} = \{1, 3\}, \text{ and}$$
$$S_4 = \{a(1), b(2), d(2)\} \quad \text{with } I_{S_4} = \{2, 3\}.$$

We use this as the running example in the remainder of the discussion.

13.8.1 Ordered enumeration trie of multi-sets

Renaming scheme. Can we treat each symbol σ with its copy number c, say $\sigma(c)$ as a new symbol, say σ_c? For a concrete example let a symbol a have two different copy numbers say 3 and 7. Consider the following simple example:

$$C_1 = \{a(3)\} \text{ and } C_2 = \{a(7)\}.$$

Now if we replace a with two new symbols a_3 and a_7, then the new sets are

$$C_1' = \{a_3\} \text{ and } C_2' = \{a_7\},$$

with

$$C_1' \cap C_2' = \emptyset \text{ but } C_1 \cap C_2 = \{a(3)\}.$$

Thus, the original input set C_1 and C_2 has one intersection set given as:

$$S = \{a(3)\} \text{ with } I_S = \{1, 2\},$$

Clearly this renaming scheme does not work. We must take the 'interactions' of $a(3)$ and $a(7)$ into account.

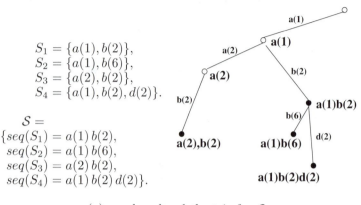

$S_1 = \{a(1), b(2)\},$
$S_2 = \{a(1), b(6)\},$
$S_3 = \{a(2), b(2)\},$
$S_4 = \{a(1), b(2), d(2)\}.$

$S =$
$\{seq(S_1) = a(1)\,b(2),$
$seq(S_2) = a(1)\,b(6),$
$seq(S_3) = a(2)\,b(2),$
$seq(S_4) = a(1)\,b(2)\,d(2)\}.$

(a) $a < b < d$ and the trie for $\mathcal{S}$.

FIGURE 13.16:   Given four multi-sets S_1, S_2, S_3 and S_4. The corresponding collection of sequences $\mathcal{S}$ and the trie for $\mathcal{S}$.

Trie of multi-sets.   As for the other sets, we first define an ordering on the elements of Σ. Let

$$\Sigma = \{\sigma_1 < \sigma_2 < \ldots < \sigma_m\}.$$

Given a set

$$S = \{\sigma_{i_1}(c_{i_1}), \sigma_{i_2}(c_{i_2}), \ldots, \sigma_{i_l}(c_{i_l})\},$$

with

$$\sigma_{i_1} < \sigma_{i_2} < \ldots \sigma_{i_l},$$

define a sequence, $seq(S)$, as

$$seq(S) = \sigma_{i_1}(c_{i_1})\,\sigma_{i_2}(c_{i_2}) \,\ldots\, \sigma_{i_l}(c_{i_l}).$$

In the sequence, each element σ_i appears exactly once and is annotated with the copy number c_i. A trie of the sequences is a tree satisfying the following properties.

1. Each sequence must be represented by a unique path from the root to the leaf node. In the trie for the multi-set, if a symbol σ appears multiple times with different copy numbers

$$c_{\min} = c_1 < c_2 < \ldots < c_l = c_{\max},$$

then we interpret that as a single symbol $\sigma(c_{\max})$.

2. No two siblings in the trie have a label of the form $\sigma(\text{-})$, i.e, the same symbol σ.

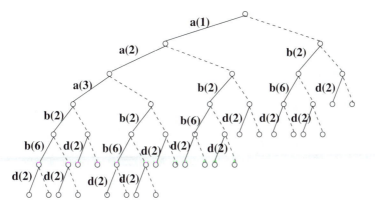

FIGURE 13.17: Let $\Sigma = \{a < b < d\}$ where the data shows these copy numbers: $1, 2, 3$ for a; $2, 6$ for b; 2 for d. The complete binary ordered enumeration tree for such an input is shown above. To avoid clutter, the labels on the nodes are omitted.

The running example is discussed in Figure 13.16. This shows how the sequence is 'represented' by the trie (encoded at the nodes). Each edge is labeled by the element σ and its copy number. What is the size of the trie? In fact, it is the same as before and Lemma (13.6) holds even for the multi-sets.

13.8.2 Enumeration algorithm

We again propose a simple ordered enumeration scheme along the lines of Algorithm (14). See Figure 13.9 for the overall scheme.

The complete binary ordered tree. We define the *Complete Binary Ordered Tree* as before with a few additional properties due to the multiplicities. For each $\sigma_j \in \Sigma$, let the j_n copy numbers of σ_j be as

$$c_{j_1} < c_{j_2} < \ldots < c_{j_n}.$$

1. As before, every internal node (including the root node) has exactly two children. The edges are labeled as follows. The right-child (edge) is unlabeled and is shown as a dashed edge in the figure. The left-child (edge) is labeled as follows.

   (a) The left child of the root node is labeled with $\sigma_1(c_{1_1})$.

   (b) Let an internal node v have an incoming edge labeled as $\sigma_j(c_{j_l})$, then its left child has the following label:

$$\begin{cases} \sigma_j(c_{j_{l+1}}), & \text{if } (l+1) \leq j_n, \\ \sigma_{j+1}(c_{j_1}), & \text{otherwise.} \end{cases}$$

2. Every node is labeled by the pair (S, I_S) as before. S is determined exactly as in the trie for multi-sets.

See Figure 13.17 for the running example. The tree is pruned using exactly the same terminating conditions as before and the pruned tree for the running example is shown in Figure 13.18(a) and the corresponding collapsed tree, or the search tree, is shown in Figure 13.18(b). Figure 13.19 shows the search tree for a different ordering of the elements of Σ.

Reorganizing the input. For efficiency purposes, we reorganize the input sets. For each $\sigma_j \in \Sigma$, a 'dictionary' list, Dic_{σ_j}, is built. Let the j_n copy numbers of σ_j be as

$$c_{j_1} < c_{j_2} < \ldots < c_{j_n}.$$

This is best explained through an example. Continuing the running example, the dictionary lists are as follows:

$$
\begin{array}{ccc}
1 & 2 & 3 \\
\downarrow & \downarrow & \downarrow \\
Dic_a \rightarrow \boxed{2} \rightarrow \boxed{1} \rightarrow \boxed{3} \dashv
\end{array}
\qquad
\begin{array}{cc}
2 & 6 \\
\downarrow & \downarrow \\
Dic_b \rightarrow \boxed{3} \rightarrow \boxed{1} \rightarrow \boxed{2} \dashv
\end{array}
\qquad
\begin{array}{c}
2 \\
\downarrow \\
Dic_d \rightarrow \boxed{2} \rightarrow \boxed{3} \dashv
\end{array}
$$

The sublist corresponding to c_{j_r} is denoted as $Dic_{\sigma_j(c_{j_r})}$. Note that each sublist includes *all* elements to the right, thus with an abuse of notation if $Dic_{\sigma(c)}$ denotes the set of the elements in the list, then

$$
\begin{aligned}
Dic_{a(1)} &= \{2, 1, 3\}, \\
Dic_{a(2)} &= \{1, 3\}, \\
Dic_{a(3)} &= \{3\},
\end{aligned}
$$

with $Dic_{a(1)} \supsetneq Dic_{a(2)} \supsetneq Dic_{a(3)}$.

For convenience, we maintain this as a single list and the algorithm uses the three sublists in three consecutive recursive calls (unless the calls are terminated by other conditions). The pseudocode for this for complete ordered enumeration is given in Algorithm (16). This follows exactly along the lines on Algorithm (14). The major difference in the algorithms is in the use of copy numbers along with the symbols. These statements are shown boxed in the code. This code is geared towards using the sorted list Dic described above.

Algorithm 16 *(Maximal Multi-Set Intersection Problem (MIP))*

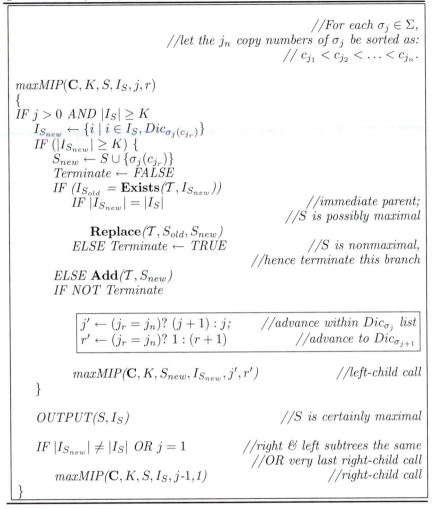

```
                                                    //For each σ_j ∈ Σ,
                                      //let the j_n copy numbers of σ_j be sorted as:
                                                  // c_{j_1} < c_{j_2} < ... < c_{j_n}.

maxMIP(C, K, S, I_S, j, r)
{
IF j > 0 AND |I_S| ≥ K
    I_{S_new} ← {i | i ∈ I_S, Dic_{σ_j(c_{j_r})}}
    IF (|I_{S_new}| ≥ K) {
        S_new ← S ∪ {σ_j(c_{j_r})}
        Terminate ← FALSE
        IF (I_{S_old} = Exists(T, I_{S_new}))
            IF |I_{S_new}| = |I_S|                        //immediate parent;
                                                          //S is possibly maximal

                Replace(T, S_old, S_new)
            ELSE Terminate ← TRUE                         //S is nonmaximal,
                                                          //hence terminate this branch

        ELSE Add(T, S_new)
        IF NOT Terminate

            ┌────────────────────────────────────────────────────────────────┐
            │ j' ← (j_r = j_n)? (j + 1) : j;    //advance within Dic_{σ_j} list │
            │ r' ← (j_r = j_n)? 1 : (r + 1)     //advance to Dic_{σ_{j+1}}      │
            └────────────────────────────────────────────────────────────────┘

            maxMIP(C, K, S_new, I_{S_new}, j', r')        //left-child call
    }

    OUTPUT(S, I_S)                                        //S is certainly maximal

    IF |I_{S_new}| ≠ |I_S| OR j = 1           //right & left subtrees the same
                                              //OR very last right-child call
        maxMIP(C, K, S, I_S, j-1, 1)          //right-child call

}
```

13.9   Adapting the Enumeration Scheme

The enumeration scheme presented here is a depth first traversal of the trie. It is the power of such a scheme that a code of less than fifteen lines can accurately encode the solution to a problem as complex as Problem (24).

Recall that the routine in Algorithm (14) is recursive, enabling the depth first traversal of the implicit trie. It invokes many instances of itself (no

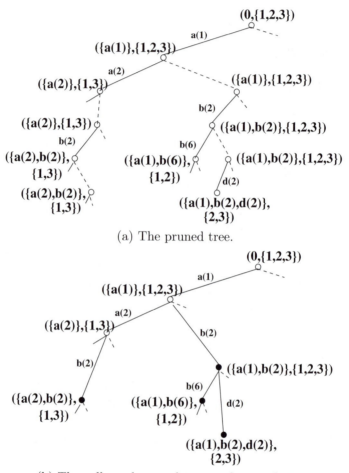

(a) The pruned tree.

(b) The collapsed pruned tree or the search tree.

FIGURE 13.18:   The complete binary tree of Figure 13.17 has been pruned by using the different terminating conditions in the routine. This tree has been further collapsed to obtain the search tree. The maximal sets are shown as solid circles.

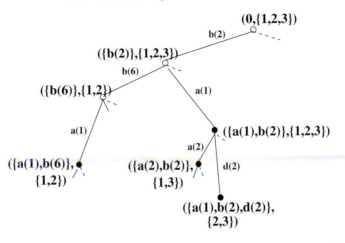

FIGURE 13.19: The ordered enumeration trie with $b < a < d$. The nodes with maximal pairs as shown as solid circles.

more than two at each call). At some time point, these instances, due to the recursive nature of the process, are partially executed and are waiting for other instances to complete their execution before they can complete their own. For very large data sets, the number of such partially executed routines may pile up, eventually running out of space in the computer memory.

Is the depth-first order of traversal really crucial? There are at least two implications of this order of traversal.

1. Run time efficiency: The order ensures that a maximal set has been seen before its nonmaximal versions. This results in an effective pruning of the trie due to termination conditions (2) of Section 13.6. The only exception is when the intersection set S is being built, one element σ at a time.

2. OUTPUT statement in the algorithm description: This order of traversal guarantees that the routine can spit out the maximal sets, without ever having to backtrack.

It is always possible to rewrite the depth first traversal code as a nonrecursive procedure. This code maintains the trie explicitly. It is also possible to carry out a breadth first traversal, at the cost of a less time efficient algorithm. We leave these as exercises for the reader (Exercise 169).

These are two systematic ways of dealing with run-time space issues for very large data sets. Sometimes in practice, two sets, S_1 and S_2, are deemed equal ($S_1 \approx S_2$), if they have a large number of common elements. Quantitatively,

$$(S_1 \approx S_2) \Leftrightarrow \frac{|S_1 \cap S_2|}{|S_1 \cup S_2|} \geq \delta$$

for some fixed $0 < \delta \leq 1$. There is no guarantee that such an assumption will reduce the number of solutions, but may be tractable in some problem domain. Of course, the question whether the algorithm reports *all* solutions that fit the definition must be addressed.

13.10   Exercises

Exercise 157 (Unique transitive reduction lemma)

1. Show that E_r (of Section 13.3) is the smallest set of edges such that

$$C(G(\mathcal{S}, E)) = C(G(\mathcal{S}, E_r)).$$

2. If E' is another set of edges with

$$|E_r| = |E'|, \text{ and}$$
$$C(G(\mathcal{S}, E_r)) = C(G(\mathcal{S}, E')),$$

then show that

$$E' = E_r.$$

Hint: Use proof by contradiction.

Exercise 158 (Size of reduced partial order) *Consider a reduced partial order graph $G(\mathcal{S}, E_r)$. Let $|\mathcal{S}| = n$. In the figures below, two new terminating nodes (shown as solid circles), a 'start' node and and an 'end' node are used for convenience.*

1. *The partial order is called a* total order *if for any two nodes $S_1, S_2 \in \mathcal{S}$ there is*

   (a) *a path from S_1 to S_2 or*
   (b) *a path from S_2 to S_1.*

   In this case,

   $$|E_r| = n + 1,$$

   as shown below. However, without the terminating nodes, the number of edges is $n - 1$.

2. *The partial order is called an* empty order *if for any two nodes* $S_1, S_2 \in$ *$\mathcal{S}$, there is*

    (a) neither a path from S_1 to S_2

    (b) nor a path from S_2 to S_1.

In this case,

$$|E_r| = 2n,$$

as shown below. However, without the terminating nodes, the number of edges is zero.

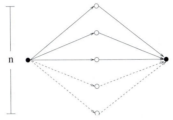

In the worst case, how many edges can a reduced partial order with n nodes have?

Hint: Consider the following example. Is it a reduced partial order? How many edges does the graph have?

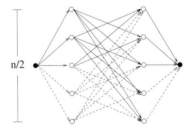

Let

$$\Sigma = \{\sigma_1, \sigma_2, \ldots, \sigma_{n/2}\}.$$

Let a node in the first column be $S_{1,i}$ and in the second column be $S_{2,i}$, then for $i = 1, 2, \ldots, n/2$,

$$S_{1,i} = \{\sigma_i\},$$
$$S_{2,i} = \Sigma \setminus \{\sigma_i\}.$$

Exercise 159 (Incomparable nodes) *Two nodes* $S_1, S_2 \in \mathcal{S}$ *in the reduced partial order* $G(\mathcal{S}, E_r)$ *are* incomparable, *if both of the following conditions hold:*

> *there is no path from* S_1 *to* S_2, *and*

> *there is no path from* S_2 *to* S_1.

1. *In Figure 13.1(a), are the nodes* $\{a, b\}$ *and* $\{b, c\}$ *incomparable? Why?*

2. *Show that any pair of siblings in a reduced partial order graph are incomparable.*

Hint: (2) Does the statement hold when the partial order is not reduced?

Exercise 160 (Straddle graph) *Given a reduced partial order* $G(\mathcal{S}, E_r)$, *let*

$$\mathcal{S}_m \subset \mathcal{S}$$

be the set of nodes that have at least one child with multiple parents. Then show that $\mathcal{S}_m$ *can be partitioned as*

$$\mathcal{S}_m = \mathcal{S}_1 \cup \mathcal{S}_2 \cup \ldots \cup \mathcal{S}_h,$$

where each partition $\mathcal{S}_i$, $1 \le i \le h$ *induces a connected straddle graph.*

Hint: If $S, S' \in V_S$, then show that

$$V_S = V_{S'} \text{ and } E_S = E_{S'}.$$

Exercise 161 (Boolean closure) *Let* $\mathcal{S}$ *be a collection of sets defined on alphabet* Σ *such that no two sets in* $\mathcal{S}$ *straddle.*

1. *Then show that the boolean closure,* $B(\mathcal{S})$, *is the same as* $\mathcal{S}$, *i.e.,*

$$\mathcal{S} = B(\mathcal{S}) = B_\cap(\mathcal{S}) = B_\cup(\mathcal{S}).$$

2. *Then show that the reduced partial order graph*

$$G(\mathcal{S}, E_r) = G(B(\mathcal{S}), E_r) = G(B_\cap(\mathcal{S}), E_r) = G(B_\cup(\mathcal{S}), E_r)$$

is acyclic.

3. *Show that the elements of Σ can be consecutively arranged as a string say s, such that for each $S \in \mathcal{S}$, its members appear consecutive in s.*

Hint: 1. What is $S_1 \cup S_2$ or $S_1 \cap S_2$ for any $S_1, S_2 \in \mathcal{S}$?
2. Does any node have multiple parents? Why?
3. Use the fact that the reduced partial order graph is a tree.

Exercise 162 (Frontiers) *Enumerate $\mathcal{F}(T)$ for the PQ tree, T, shown below.*

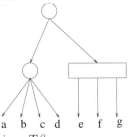

a b c d   e f g

What is $|\mathcal{F}(T)|$, for the given T?

Exercise 163 *For a tree T,*

1. *whose leaves are labeled bijectively with the elements of Σ and*

2. *each internal node represents a unique set, $S \in \mathcal{S}$, which is the collection of the labels of the leaf nodes reachable from this node,*

show that a consecutive arrangement of Σ that respects $\mathcal{S}$ is the frontier $F(T)$.

Hint: Show that either the sets are disjoint or one is contained in the other. Thus this nested containment has a linear representation.

Exercise 164 (Linear arrangement) *Consider the following reduced partial order of a collections of sets. Can the sets be arranged along a line? Why?*

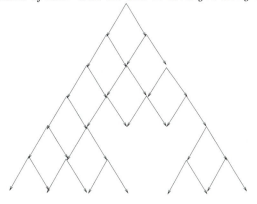

Exercise 165 (Set linear arrangement theorem) *Consider Theorem (13.1) and its proof presented in Section 13.1.*

1. *Prove that a consecutive arrangement s_S of Equation (13.4) is always possible.*

2. *Prove that a consensus arrangement exists when each straddle graph is a chain.*

Hint: 1. Recall that the graph is a reduced partial order of the boolean closure of $\mathcal{S}$. Finally, use proof by contradiction.

2. If there is only one nonsingleton straddle graph, then we are done. Assume there are at least two distinct nonsingleton straddle graphs, say $G(V_1, E_1)$ and $G(V_2, E_2)$. Let

$$\sigma_p \in S_{i_j}, \ \sigma_q \in S_{i_k} \text{ and } \sigma_r \in S_{i_\ell},$$

where $S_{i_j}, S_{i_k}, S_{i_\ell} \in V_1$ and $j < k < \ell$. Next let

$$\sigma_p \in S_{i_{j'}}, \ \sigma_q \in S_{i_{k'}} \text{ and } \sigma_r \in S_{i_{\ell'}},$$

where $S_{i_{j'}}, S_{i_{k'}}, S_{i_{\ell'}} \in V_2$ and

$$\text{neither } j' < k' < \ell' \text{ nor } \ell' < k' < j' \text{ holds.} \tag{13.10}$$

If $j = k$ or $k = \ell$, we are done. Also, if $j' = k'$ or $k' = \ell'$, we are done. Show that

$$(S_{i_t} \in V_1) \subset (S_{i_{t'}} \in V_2), \text{ for } t = j, k, \ell.$$

and the assumption (13.10) must be wrong.

Exercise 166 (Local graph structures) *Some properties of a graph can be determined by studying its local structures. Consider two such structures shown below for nodes with multiple parents in a reduced partial order of the boolean closure of $\mathcal{S}$:*

$$G(B(\mathcal{S}), E_r).$$

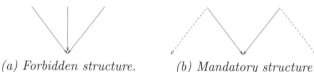

(a) Forbidden structure.   *(b) Mandatory structure (siblings, indicated by dashed edges, possibly empty).*

1. *Show that if $G(B(\mathcal{S}), E_r)$ has a forbidden structure, then*

$$\mathcal{F}(\mathcal{S}) = \emptyset,$$

i.e., the sets cannot be consecutively arranged.

2. *If every node with two parents has the mandatory structure, then is*

$$\mathcal{F}(\mathcal{S}) \neq \emptyset,$$

i.e., can the sets be consecutively arranged?

3. *Is it possible to conclude if*

$$\mathcal{F}(\mathcal{S}) = \emptyset,$$

by studying local properties of

$$G(B(\mathcal{S}), E_r).$$

Hint: (1) The forbidden structure shows nodes with more than two parents. Note that the graph is a reduced partial order of the boolean closure. (2) Consider the example shown below. Does each node with multiple parents, respect the *mandatory structure*? What is $\mathcal{F}(\mathcal{S})$?

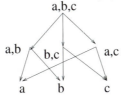

Exercise 167 (Minimal set intersection algorithm)

1. *What is the running time complexity of Algorithm (15), the Minimal Set Intersection Algorithm?*

2. *How does this compare with the running time complexity of the Maximal Set Intersection Algorithm (14)?*

Hint: Is it better to compute the minimal from the maximal sets?

Exercise 168 (Maximal intersection trie enumeration) *Consider Problem (24), the maximal set intersection problem.*

1. *Can the search tree be made any more conservative? In other words, is there room to improve the time complexity of Algorithm (14) by introducing more terminating conditions? Why?*

2. *If*

$$m \gg n,$$

i.e., the alphabet is much larger than the number of sets, how can Algorithm (14) be modified to retain an efficient run time complexity?

3. If

$$|\Sigma| = \mathcal{O}(N_I),$$

how can the enumeration be changed to exploit this fact?

Exercise 169 (Maximal intersection trie enumeration) *Consider Algorithm (14).*

1. *How are the results affected if the order of the recursive calls marked 'left-child call' and 'right-child call' are switched?*

2. *Show that by changing the order of alphabet Algorithm (14) gives the same maximal sets but possibly different search tree (or tries).*

3. *Re-write Algorithm (14) as a nonrecursive procedure.*

4. *Re-write Algorithm (14) as a nonrecursive procedure that traverses the complete binary ordered tree $\mathbf{B_C}$ in a breadth first order.*

Hint: 3. Build the trie $\mathbf{T_{C,K}}$ as an explicit structure and traverse it in a depth first order.

4. Build $\mathbf{B_C}$ as an explicit structure and traverse it in a breadth first order. Also, identify the tree pruning conditions to make the procedure efficient.

Exercise 170 (Maximal multi-set intersection trie enumeration)

1. *Let $\sigma \in \Sigma$ and consider sets, C_i, $1 \le i \le 8$, with*

$$\sigma \notin C_1, \quad \sigma(c_2) \in C_2, \quad \sigma(c_1) \in C_3, \quad \sigma(c_2) \in C_4,$$
$$\sigma(c_2) \in C_5, \quad \sigma(c_3) \in C_6, \quad \sigma(c_1) \in C_7, \quad \sigma(c_3) \in C_8.$$

The list is sorted as follows for Algorithm (16).

   (a) *What does the list assume about the ordering of the elements*

$$c_1, c_2, c_3 \quad ?$$

   (b) *What is important about this ordering? In other words, how is Algorithm (16) affected if this ordering is not maintained in the list?*

2. *Consider the pruned tree in Figure 13.18. Identify all the terminating conditions in Algorithm (16) that bring about this pruning.*

Exercise 171 (Maximal to minimal) *Consider a search trie T where the label of each node is a single character. Node A on this tree has the label (S_A, I_{S_A}).*

1. *Show that the following two conditions on node A are equivalent. Condition (1):*

    (a) *the parent of A has multiple children and*

    (b) *A has a single child.*

    Condition (2):

    (a) *Node B has multiple children and*

    (b) *A is the closest ascendant of B such that*

        i. *A has a single child and*

        ii. *the parent (immediate ascendant) of A has multiple children.*

2. *Node A denotes a minimal intersection pair*

$$(S_A, I_{S_A})$$

   if and only if any one of conditions (1) and (2) hold.

Hint: For a fixed index set I', how many nodes in the tree have label (S, I_S) with $I' = I_S$? How are these nodes arranged on the tree? Use Lemma (13.9).

Exercise 172 (Multi-set intersection problem)

1. *Assume the input is a collection of multi-sets. The ordered enumeration trie for two different orders*

$$a < b < d,$$
$$b < a < d,$$

   shown in Figures 13.18(b) and (13.19) respectively, are isomorphic to each other.
   Does this always hold? Why?

2. *Identify all the terminating conditions, marked by Roman numerals I, II, ... VIII, in the figure below for the example of Section 13.8.*

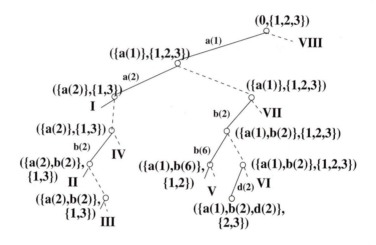

Exercise 173 (Maximal string intersection) *Consider an input of m sequences s_i, $1 \leq i \leq m$ on some alphabet Σ. Then the pair $(p, \mathcal{L}_p)$, $p \subset \Sigma$ and $\mathcal{L}_p \subset \{1, 2, \ldots n\}$, is a maximal intersection when*

$$p = \{\sigma \mid i \in \mathcal{L}_p \text{ and there is some } k \text{ such that } s_i[k] = \sigma\},$$

and p is maximal when there exists no distinct $p' \supset p$ with

$$\mathcal{L}_{p'} = \mathcal{L}_p.$$

Algorithm 17 *Maximal String Intersection Algorithm (maxStIP)*

$$Dic_{\sigma_j} = \{i \mid s_i[k] = \sigma_j \text{ for some } k\} \qquad //dictionary$$
$$\mathcal{L}_p \leftarrow \{1, 2, \ldots, m\}$$
$$p \leftarrow \emptyset$$
$$Mine\pi Pat(K, \sigma_1, \mathcal{L}, p) \qquad //main\ call$$

$$maxStIP(K, \sigma_j, \mathcal{L}_p, p)$$
$$\quad IF\ (j \leq |\Sigma|)\ AND\ (|\mathcal{L}_p| \geq K)$$
$$\qquad \mathcal{L}_{p_{sav}} \leftarrow \mathcal{L}_p,\ p_{sav} \leftarrow p$$
$$\qquad FOR\ EACH\ i \in \mathcal{L}_{p_{sav}}$$
$$\qquad\quad IF\ i \notin Dic_{\sigma_j}\ \mathcal{L}_p \leftarrow \mathcal{L}_p \setminus \{i\}$$
$$\qquad p \leftarrow p \cup \{\sigma_j\}$$
$$\qquad Quit \leftarrow ((|\mathcal{L}_p| < K)\ OR\ (p \subset \textbf{ExistPat}(\mathcal{L}_p)))$$
$$\qquad IF\ NOT\ Quit$$
$$\qquad\quad \textbf{StorePat}(\mathcal{L}_p, p) \qquad //new\ or\ updated\ motif$$
$$\qquad\quad maxStIP(K, \sigma_{j+1}, \mathcal{L}_p, p) \qquad //with\ \sigma_j$$
$$\qquad IF\ |\mathcal{L}_p| < |\mathcal{L}_{p_{sav}}| \qquad //only\ if\ the\ two\ are\ distinct$$
$$\qquad\quad maxStIP(K, \sigma_{j+1}, \mathcal{L}_{p_{sav}}, p_{sav}) \qquad //ignoring\ \sigma_j$$

1. *For each input string s_i $(1 \leq i \leq m)$, construct a set as follows*

$$C_i = \{\sigma \mid s_i[k] = \sigma \text{ for some } k\}$$

Further let

$$\mathbf{C} = \{C_1, C_2, \ldots, C_m\}.$$

Does this algorithm produce the same results on the strings as Algorithm (14) on $\mathbf{C}$?

2. *Identify the 'left' and the 'right' child calls.*

3. *Identify the recursive call terminating conditions.*

4. *Outline the steps in **StorePat** and **ExistPat**.*

5. *Compare this algorithm with Algorithm (14). Which is more efficient? Why? Is there an asymptotic improvement in one algorithm over the other?*

Exercise 174 (Tree data structure) *For an input $\mathbf{C}$ and quorum K, what is the relationship between the trie*

$$\mathbf{T}_{\mathbf{C}, K}$$

and the data structure $\mathcal{T}$ to store the pairs (S, I_S)?

Hint: Consider the reduced trie where each internal node as at least two children. Can the trie be balanced?

Comments

The material in this chapter is fairly straightforward. At first blush, it even seems like it does not deserve the dignity of a dedicated chapter. However, I have seen the very same problem pop up in so many hues and shapes that perhaps an unobstructed treatment of the material, within its very own chapter, will do more good than harm.

Chapter 14

Expression & Partial Order Motifs

It is better to understand some of the questions,
than to know all of the answers.
- adapted from James Thurber

14.1 Introduction

Consider the task of capturing the commonality across data sequences in a variety of scenarios. Depending on the data and the domain, the questions change.

1. Total order: Segments appear exactly at each occurrence and these are called the *string patterns*.

   Certain wild cards may be allowed or even flexible extension of the gap regions (called *extensible motifs*).

2. No order (but proximity): If groups of elements appear together, even if they respect no order, these clusters may be of interest. These are called *permutation patterns*.

   Again, they may show some substructures of proximity within them (as PQ structures).

3. Partial order: Is it possible that key players are only partially ordered?

   The key players themselves could be as simple as motifs or clusters or as complex as a boolean expression.

   Further, if the input is organized as a sequence and this order must be important, these can be modeled as the mathematical structure *partial order*.

   Of course, a more general order is defined as a graph and *topological motifs* can be discovered from this organization of data possibly providing some insight into the process that produced this data.

As the landscape changes, the questions change and so do the answers. There is an interesting interplay of different ideas such as permutations of motifs or extensible motifs of permutations or partial orders of expressions and so on.

469

This chapter focuses on boolean expressions (on possibly string motifs); partial orders of motifs and finally, partial orders of expressions on motifs. We motivate the reader with a brief example below and details of the main ideas are discussed in the subsequent sections.

14.1.1   Motivation

In the following, *mini-motifs* refer to string motifs. Consider the results obtained by mining patterns in binding site layouts in four yeast species as studied in Kellis et al. [KPE⁺03]: *S. cerevisiae, S. paradoxus, S. mikatae,* and *S. bayanus.*

Out of 45,760 mini-motifs, some 2419 significantly conserved mini-motifs are grouped into 72 consensus motifs. [1] For the small fraction of sequences where some motifs occur more than once, only the position closest to the TATA box is utilized. Many of these motifs correspond to known transcription factor binding sites [ZZ99] whereas others are new and putative, supported indirectly by co-expression or functional category enrichment.

We use the number id's for the motifs and show an example below. Is there more structure than just the cluster?

$$37 \wedge 66 \wedge 5$$

The symbol $\wedge$ denotes 'and' and $\vee$ denotes 'or'. Notice that motif 37 always precedes motif 66, but motif 5 could be in any relative position, in each of the clusters. This is captured in the partial order shown to the right below. Symbols S and E are meta symbols that denote the left and right ends respectively.

```
Spar (YDR034C-A): 37 66 5
Spar(YJL008C):     5 37 66
Spar (YJL007C):    5 37 66
Spar (YMR083W):   37 66 5
Spar (YOR377W):   37 5 66
```
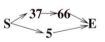

   (1) Input sequence data.   (2) Partial order motif.

Another example with the cluster

$$48 \wedge 55 \wedge 37 \wedge 5 \wedge 24$$

is shown below. Here motif 48 always precede motif 55 and motif 37 always precedes motif 24. The common ordering of the elements is captured in the

---

[1] See [KPE⁺03] for more details.

partial order shown on the right.

```
Spar (YDR034C-A):   48 37 24 55 5
Spar (YJL008C):     5 37 48 55 24
Spar (YJL007C):     5 37 48 55 24
Spar (YMR083W):     24 55 37 48 5
Spar (YOR377W):     48 55 37 5 24
```

(1) Input sequence data.

(2) Partial order motif.

Thus, informally speaking, a *partial order motif* is a decorated cluster.

Partial order of boolean expressions. As a further generalization, each node in the partial order is a boolean expression of multiple elements. The final example here illustrates the use of disjunctions in the motif expressions. The common expressions are as follows:

$$24 \wedge 68 \wedge 19 \wedge (50 \vee 55 \vee 39)$$

$$24 \wedge 68 \wedge 19 \wedge (40 \vee 54)$$

The partial ordering of these expressions are shown below (S, E, S1, E1 are meta symbols). Note the use of 'OR' in the boolean expressions which are shown as '|' in the nodes of the partial order.

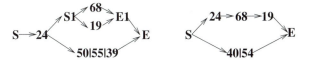

The genes and motifs in this example are enriched for multiple stress response pathways (e.g., HSPs, Ras) and sensing of extracellular amino acids.

14.2   Extracting (Monotone CNF) Boolean Expressions

Expression mining exhibits traits of conceptual clustering, constructive induction, and logical formula discovery [Fis87, Mic80]. Inducing understandable definitions of classes using a set of features [VPPP00] is a goal common to all descriptive classification applications.

The given input of n sequences (S) on the set of m motifs (F) is modeled as an $(n \times m)$ boolean incidence matrix I, without order information, as:

$$I[i, j] = \begin{cases} 1 & \text{if motif } m_j \text{ occurs in sequence } s_i \in S, \\ 0 & \text{otherwise.} \end{cases}$$

In this section, a motif is treated as a boolean variable. Thus a boolean expression of the form

$$e = m_1 \wedge m_2 \vee m_3$$

implies that either both m_1 and m_2 occur or m_3 occurs in a sequence for which this expression e holds. Alternatively, the same expression can be written as

$$e = m_1 m_2 + m_3.$$

The negation of a variable, m, is written as

$$\overline{m}.$$

DEFINITION 14.1 *(expression e, features $\Pi(e)$, objects $O(e)$) e is a boolean expression on a set of motifs*

$$V \subseteq F.$$

Given e, we denote the set of motifs involved in e by

$$\Pi(e)$$

and the set of sequences it represents by $O(e)$.

We also use the following intuitive convention:

$$\begin{aligned}
O(e) &= O(e = m_1 \wedge m_2 \vee m_3) \\
&= O(m_1) \cap O(m_2) \cup O(m_3) \\
&= O(m_1 m_2 + m_3).
\end{aligned}$$

For simplicity of notation,

$$\begin{aligned}
e &= m_1 \wedge m_2 \vee m_3 \\
&= m_1 \cap m_2 \cup m_3 \\
&= m_1 m_2 + m_3.
\end{aligned}$$

The set difference is written as

$$O(m_1) \setminus O(m_2)$$

or simply

$$m_1 \setminus m_2.$$

Two expressions e_1 and e_2 defined over V_1 and V_2 respectively are distinct (denoted as $e_1 \neq e_2$), if one of the following holds:

(i) $V_1 \neq V_2$, or

(ii) there exists some input I for which $O(e_1) \neq O(e_2)$.

$$I = \begin{bmatrix} m_1 & m_2 \\ 0 & 0 \\ 0 & 1 \\ 1 & 0 \\ 1 & 1 \end{bmatrix}.$$

e	$\Pi(e)$	$O(e)$	
0	$\emptyset$	$\emptyset$	$\checkmark$
$\overline{m}_1\overline{m}_2$	$\{m_1, m_2\}$	$\{1\}$	
$\overline{m}_1 m_2$	$\{m_1, m_2\}$	$\{2\}$	
$m_1\overline{m}_2$	$\{m_1, m_2\}$	$\{3\}$	
$m_1 m_2$	$\{m_1, m_2\}$	$\{4\}$	$\checkmark$
$\overline{m}_1$	$\{m_1\}$	$\{1, 2\}$	
$\overline{m}_2$	$\{m_2\}$	$\{1, 3\}$	
$\overline{m}_2\overline{m}_2 + m_1 m_2$	$\{m_1, m_2\}$	$\{1, 4\}$	
$\overline{m}_1 m_2 + m_1\overline{m}_2$	$\{m_1, m_2\}$	$\{2, 3\}$	
m_2	$\{m_2\}$	$\{2, 4\}$	$\checkmark$
m_1	$\{m_1\}$	$\{3, 4\}$	$\checkmark$
$\left.\begin{matrix} \overline{m}_1 + m_1\overline{m}_2 \\ \overline{m}_2 + \overline{m}_1 m_2 \end{matrix}\right\}$	$\{m_1, m_2\}$	$\{1, 2, 3\}$	
$\left.\begin{matrix} \overline{m}_1 + m_1 m_2 \\ m_2 + \overline{m}_1\overline{m}_2 \end{matrix}\right\}$	$\{m_1, m_2\}$	$\{1, 2, 4\}$	
$\left.\begin{matrix} m_1 + \overline{m}_1\overline{m}_2 \\ \overline{m}_2 + m_1 m_2 \end{matrix}\right\}$	$\{m_1, m_2\}$	$\{1, 3, 4\}$	
$\left.\begin{matrix} m_2 + m_1\overline{m}_2 \\ m_1 + \overline{m}_1 m_2 \end{matrix}\right\}$	$\{m_1, m_2\}$	$\{2, 3, 4\}$	
1	$\{m_1, m_2\}$	$\{1, 2, 3, 4\}$	$\checkmark$

FIGURE 14.1: An incidence matrix I and all possible expressions, e, defined on the variables m_1 and m_2. The monotone expressions are marked with $\checkmark$.

Notice that this condition rules out tautologies. For example, using the set notation, the expressions

$$m_1 \cap m_4$$

and

$$m_1 \setminus (m_1 \setminus m_4)$$

are not distinct.

An expression e is in *conjunctive normal form* (CNF) if it is a conjunction of clauses, where a clause is a disjunction of literals. For example,

$$e = (m_1 + m_2)(m_3 + m_4)\overline{m}_5 m_6 (\overline{m}_1 + m_4)$$

is in CNF form.

An expression e is in *disjunctive normal form* (DNF) if it is a disjunction of clauses, where a clause is a conjunction of literals. For example,

$$e = m_1 m_2 + m_3 m_4 + \overline{m}_5 + m_6 + \overline{m}_1 m_4$$

is in DNF form.

It is straightforward to prove that any expression e can be written either in a CNF or a DNF form and we leave this as an exercise for the reader (Exercise 176).

A boolean expression is very powerful since it has the capability of expressing very complex interrelationships between the variables (motifs in this case). See Figure 14.1 for an example. However, it is this very same power that renders it ineffective: it can be shown that there always exists a boolean expression that precisely represents any collection of rows in any incidence matrix I. See and Exercise 175 and Figure 14.1 for an example.

So we focus on a particular subclass of boolean expression called *monotone expression* [Bsh95]. This is a subclass of boolean expressions that uses no negation. In other words, it uses only conjunctions and disjunctions. See the marked expressions in the example of Figure 14.1.

However, an expression is called monotone because it displays *monotonic* behavior (see Exercise 177), thus care needs to be taken to determine if an expression e is monotone. For example consider

$$e = m_1 m_2 + \overline{m}_1 m_2,$$

that appears not to be monotone due to the presence of $\overline{m}_1$. However

$$\begin{aligned}
e &= m_1 m_2 + \overline{m}_1 m_2 \\
&= (m_1 + \overline{m}_1) m_2 \\
&= 1\, m_2 \\
&= m_2.
\end{aligned}$$

Thus expression $e = m_1 m_2 + \overline{m}_1 m_2$ is indeed monotone.

Now we are ready to define the central task:

Given an incidence matrix I, and a quorum K, the task is to find all monotone expressions e in the CNF form such that

$$|O(e)| \geq K.$$

Note that since the expressions are restricted to be monotone, this specification is *nontrivial*. We say it is nontrivial since it is possible that there exists a collection of rows, V, of I such there exist no expression e with

$$O(e) = V.$$

In other words, the solution for the problem is not simply all K'-sized subsets of the rows where

$$K' \geq K.$$

The algorithm to detect monotone expressions uses the intermediate notion of *biclusters*.

14.2.1 Extracting biclusters

Given I, a bicluster is a nonempty collection of columns V and a nonempty collection of rows O satisfying the following. For each $j \in V$, let a constant c_j be such that

$$O = \{i \mid I[i,j] = c_j, \text{ for each } j \in V\}.$$

Thus O can be determined uniquely by defining V. Note the similarity of this form with a pattern p defined by the column values c_j for each $j \in V$. Then the location list $\mathcal{L}_p$ is simply O and thus a quorum constraint could be imposed on the size of O. Again in practice, usually the interest is in biclusters with $|V| \geq 2$.

The bicluster is *maximal* (also sometimes called a *closed itemset* [Zak04]) if this pattern cannot be expanded with more columns without reducing the number of rows in O. In other words, there does not exist

$$j' \notin V \text{ with } I[i,j'] = c_{j'},$$

for each $i \in O$ and some fixed $c_{j'}$. These conditions define the 'constant columns' type of biclusters. See [MO04] for different flavors of biclusters used in the bioinformatics community.

The bicluster is *minimal* if for each $j \in V$, the collection of rows O and the collection of columns

$$V \setminus \{j\}$$

is no longer a bicluster. In other words, the collection of rows can be expanded for this smaller collection of columns.

$$I = \begin{bmatrix} m_1 & m_2 & m_3 & m_4 & m_5 & m_6 & m_7 \\ 0 & 1 & 1 & 1 & 1 & 1 & 1 \\ 1 & 0 & 0 & 1 & 0 & 0 & 0 \\ 0 & 1 & 1 & 1 & 0 & 1 & 1 \\ 1 & 0 & 0 & 1 & 0 & 1 & 1 \\ 0 & 1 & 1 & 1 & 1 & 1 & 0 \\ 0 & 1 & 0 & 0 & 0 & 0 & 1 \\ 0 & 1 & 0 & 1 & 1 & 1 & 0 \\ 1 & 0 & 0 & 0 & 0 & 0 & 1 \end{bmatrix}.$$

m_1	m_2	m_3	m_4	m_5	m_6	m_7	i
0	1	1	1	1	1	1	← 1
	0		1		0		2
0	1	1	1	0	1	1	← 3
	0		1		1		4.
0	1	1	1	1	1	0	← 5
	1		0		0		6
0	1	0	1	1	1	0	← 7
	0		0		0		8

$I' = $ (the above)

A maximal bicluster:
$$V_1 = \{m_2, m_4, m_6\},$$
$$O_1 = \{1, 3, 5, 7\}.$$

A minimal bicluster:
$$V_2 = \{m_2, m_4\},$$
$$O_2 = \{1, 3, 5, 7\}.$$

Corresponding conjunctive form in I:
$$e_1 = m_2 m_4 m_6,$$
$$O(e_1) = O_1 = \{1, 3, 5, 7\}.$$

Corresponding disjunctive form in $\overline{I}$:
$$e_2 = \overline{m}_2 + \overline{m}_4,$$
$$O(e_2) = \overline{O}_2 = \{2, 4, 6, 8\}.$$

FIGURE 14.2: I is an incidence matrix. Let quorum $K = 3$. I' shows the rows and the columns corresponding to the bicluster V_1, O_1.

Relationship between biclusters and expressions.   Let V, O be a bicluster and let expression e be the conjunction of the literals corresponding to the columns in V. Then O is the same as the support

$$O(e) = \{i \mid \text{row } i \text{ satisfies } e\}.$$

Thus conjunction of literals (or columns in I) correspond naturally to biclusters.

In the spirit of *irredundancy* of Chapter 4, it can be argued that maximal biclusters when the corresponding expression is a conjunction of literals, and minimal biclusters when the expression is a disjunction, can be considered irredundant. Note that all the other *redundant* expressions (any subset of the columns of V for conjunctions and any superset of V for disjunctions) can be trivially obtained from these in both the cases.

Thus it is meaningful to have maximal biclusters for conjunctions of literals but minimal biclusters for disjunctions of literals. But, to compute disjunctions from the incidence matrix I, we must resort to careful negations as summarized in the following lemma.

LEMMA 14.1
(Flip lemma)   *Given an $(n \times m)$ incidence matrix I, for some $1 \leq l \leq m$,*

$$e = f_1 \vee f_2 \vee \ldots \vee f_l,$$

is a minimal disjunction with support $O(e)$ if and only if

$$\overline{e} = \overline{f}_1 \wedge \overline{f}_2 \wedge \ldots \wedge \overline{f}_l$$

is a miniimal conjunction with $O(\overline{e}) = \{i \mid 1 \leq i \leq m \ AND \ i \notin O(e)\}$.

Figure 14.2 shows an example of maximal and minimal biclusters and the corresponding expressions. Note that $\overline{I}$ is defined as

$$\overline{I}_{ij} = \begin{cases} 1 & \text{if } I_{ij} = 0, \\ 0 & \text{if } I_{ij} = 1. \end{cases}$$

The reader is directed to Chapter 13 for algorithms on finding maximal and minimal biclusters, which is mapped to the problem of finding maximal and minimal set intersections respectively. We leave the mapping of this construction as Exercise 180 for the reader. Using Lemma (14.1), the mining of monotone CNF expressions is staged as follows.

1. Find all minimal monotone disjunctions in I, by performing the following two substeps:

   (a) Find all minimal conjunctions on $\overline{I}$.

   (b) Extract all minimal monotone disjunctions by negating each of these computed minimal conjunctions stored in $\mathcal{T}$ (see Lemma 14.1). For example, if the minimal conjunction is $e = \overline{f}_1 \wedge \overline{f}_2$ (since $\overline{I}$ is used) then the minimal disjunction is $e' = f_1 \vee f_2$. Let the number of minimal disjunctions computed be d.

2. Copy matrix I to I'. Augment this new matrix I' with the results of the last step as follows. For each minimal disjunction form e', introduce a new column c in I' with

$$I'[i,c] = \begin{cases} 1 & \text{if } i \in O(e'), \\ 0 & \text{otherwise.} \end{cases}$$

The augmented matrix, I', is then of size $n \times (m + d)$. Next, find all monotone conjunctions as maximal biclusters in I'.

A concrete example is shown below. To avoid clutter, $\overline{I}$ is not shown. However $e = \overline{m}_1 \overline{m}_2$ is detected as a minimal (conjunction) bicluster in $\overline{I}$ with support $\{3,4\}$. Thus I' has the new column $m_1 + m_2$ with support $\{1,2,5\}$. Some solutions on I' are shown below.

	m_1	m_2	m_3
1	1	0	1
2	0	1	1
3	0	0	1
4	0	0	0
5	1	0	1
		I	

$\Rightarrow$

	m_1	m_2	m_3	m_1+m_2
1	1	0	1	1
2	0	1	1	1
3	0	0	1	0
4	0	0	0	0
5	1	0	1	1
			I'	

$\Rightarrow$

Some solutions:

$e_1 = m_1 m_3$,
$O(e_1) = \{1,5\}$.

$e_2 = m_3(m_1+m_2)$,
$O(e_2) = \{1,2,5\}$.

14.2.2   Extracting patterns in microarrays

The similarity of extracting patterns or biclusters from a real matrix M (unlike the binary matrix I) with the task discussed in the last section is startling, so we briefly digress here to compare and contrast the two.

Microarrays.   A *DNA microarray*, also known as *gene chip* or a *DNA chip*, is a matrix of microscopic DNA spots attached to a solid surface, such as glass or silicon, for the purpose of monitoring expression levels of a large number of genes simultaneously. This is called *expression profiling*.   The affixed DNA oligomers are known as *probes*.

Measuring gene expression using microarrays is relevant to many areas of biology. For instance, microarrays can potentially be used to identify genes causing diseases by comparing the expression profiles of disease affected and normal cells.

For our purposes, the real matrix M is a direct conversion of the extent of gene expression in position (i, j) which is the expression of gene numbered j in the cell numbered i. However, it is important to keep in mind that the raw data from the microarray must undergo some form of *normalization* before subjecting it to any form of analysis. Literature abounds with proposed normalization methods and a healthy debate continues over what an appropriate model is.

Notwithstanding the unresolved debate, it is important to note that the microarray technology provides an astounding increase in the dimensionality of the data over a traditional experiment such as a clinical study. Such a study may gather hundreds of data items for each patient. However, even a medium-sized microarray has the capability and can obtain thousands of data items (like gene expression) per sample. The need to extract 'knowledge' from this data set is undoubtedly a challenging task.

In the discussion here we focus on the normalized real matrix M and the task is to extract patterns seen in the arrays. A pattern is a bicluster V, O where V is a collection of genes and O is a collection of samples. In other words,

> *A bicluster (V, O) denotes the set of genes represented by V that show similar expression in the samples represented by O.*

Back to eliciting biclusters.   Consider column j of the real $n \times m$ matrix M. Assume that for this column (gene numbered j), two expression values, v_1 and v_2 are considered to be equal if and only if, for some fixed $\delta_j \geq 0$,

$$|v_1 - v_2| < \delta_j.$$

Then using the technique, described in Section (6.6.2), we convert this to a column defined on some alphabet F_j. Let the size of this alphabet be n_j, then clearly,

$$n_j = |F_j| \le n.$$

Let

$$F_j = \{\sigma_{j_1}, \sigma_{j_2}, \ldots, \sigma_{j_{n_j}}\}.$$

Thus using some m fixed values

$$\delta_1, \delta_2, \ldots, \delta_m,$$

the real matrix M is transformed to a matrix Q with discrete values, i.e.,

$$Q[i, j] \in F_j \text{ for each } i \text{ and } j.$$

Next, we stage the bicluster detection problem (or pattern extraction from microarrays) as a maximal set intersection problem in the following steps.

1. For each column j and for each symbol σ_{jk} where $1 \le k \le j_n$, compute the following sets of rows:

$$S_{j\sigma_{jk}} = \{i \mid Q[i, j] = \sigma_{jk}\}.$$

This gives mn_j nonempty sets.

2. Invoke an instance of the maximal set intersection problem (of Section (13.6), with the mn_j nonempty sets and quorum K.

3. The solution of Step 2 is mapped back to the solution of the original (bicluster) problem. This is a straightforward process and we leave the details as an exercise for the reader (Exercise 180).

A simple concrete example is shown below. Let $\delta_j = 0.5$, for $1 \le j \le 3$. Only nonsingleton sets are shown in Step 1. The two bicluster patterns are shown in the input array at the bottom.

	g_1	g_2	g_3			g_1	g_2	g_3				
1	1.0	3.1	2.85		1	a	d	a		$S_{1a} = \{1, 3, 4\}$,		$S_{3a} = \{1, 3\}$,
2	2.0	2.5	3.4	$\Rightarrow$	2	b	c	b, c	$\Rightarrow$	$S_{1b} = \{2, 4\}$.	$\cap$	$S_{3b} = \{2, 3\}$,
3	1.25	1.9	3.1		3	a	b	a, b				$S_{3c} = \{2, 4\}$.
4	1.5	0.7	3.7		4	a, b	a	c				

$$M \qquad\qquad Q \qquad\qquad \text{Maximal Set Intersection Problem}$$

	g_1	g_2	g_3				g_1	g_2	g_3	
1	**1.0**	**3.1**	**2.85**	$\leftarrow$		1	1.0	3.1	2.85	
2	2.0	2.5	3.4			2	**2.0**	**2.5**	**3.4**	$\leftarrow$
3	**1.25**	**1.9**	**3.1**	$\leftarrow$		3	1.25	1.9	3.1	
4	1.5	0.7	3.7			4	**1.5**	**0.7**	**3.7**	$\leftarrow$

$\Rightarrow$

$$\uparrow \qquad \uparrow \qquad\qquad\qquad \uparrow \qquad \uparrow$$
$$S_{1a} \cap S_{3b} = \{1, 3\} \qquad\qquad S_{1b} \cap S_{3c} = \{2, 4\}$$

This method of detecting bicluster patterns has even been applied to protein folding data in an attempt to understand the folding process at a higher level [PZ05, ZPKM07].

14.3   Extracting Partial Orders

In this step, we restore the order information among motifs, for each mined expression. We group disjunctions into a 'meta-motif' (as done in the example in Section (14.1.1)) so that we can view all gene sequences as sequences of the same length over the alphabet of meta-motifs. Before we detail the specifics, we briefly review partial orders and related terminology.

14.3.1   Partial orders

Let F be a finite alphabet F,

$$F = \{m_1, m_2, \ldots, m_L\},$$

where

$$L = |F|.$$

A binary relation B,

$$B \subset F \times F,$$

(a subset of the Cartesian product of F) is a *partial order* if it is

1. reflexive,

2. antisymmetric, and

3. transitive.

For any pair $m_1, m_2 \in F$,

$$m_1 \preceq m_2 \text{ if and only if } (m_1, m_2) \in B.$$

In other words. $(m_1, m_2) \notin B$ if and only if $m_1 \npreceq m_2$.

A string q is *compatible* with B, if for no pair m_2 preceding m_1 in q, $m_1 \preceq m_2$ holds in B. In other words, the order of the elements in q does not violate the precedence order encoded by B. A compatible q is also called an *extension* of B. q is a *complete extension* of B, if q contains all the elements of the alphabet F. Such a q is also called a *permutation* on F. Also,

$$Prm(F) = \{q \mid q \text{ is permutation on } F\},$$
$$Cex(B) = \{q \mid q \text{ is a complete extension of } B\}.$$

$Prm(F)$ is the set of all possible permutations on F. [2] Thus

$$Cex(B) \subseteq Prm(B).$$

---

[2]If $|F| = n$, $Prm(F)$ can be related to S_n in combinatorics, which is the group of permutations of $\{1, 2, \ldots, n\}$.

B can also be represented by a directed graph $G(F, E')$ where edge

$$(m_1 m_2) \in E', \text{ if and only if } m_1 \preceq m_2.$$

Since B is antisymmetric, it is easy to see that $G(F, E')$ is a directed acyclic graph (DAG).

If E is the smallest set of edges such that when $m_1 \preceq m_2$, there is a path from m_1 to m_2 in E, then $G(F, E)$ is called the *transitive reduction* of B.

The following property of the edge set E of a transitive reduction of a partial order is easily verified (Exercise 182).

LEMMA 14.2
(Reduced-graph lemma) *If $G(F, E)$ is the transitive reduction of a partial order, then for any pair, $m_1, m_2 \in F$, if there is a directed path of length larger than 1 from m_1 to m_2 in $G(F, E)$, then edge $(m_1 m_2) \notin E$.*

In the following discussion, the transitive reduction of a partial order B will be denoted by its DAG representation $G(F, E)$. If the directed edge $(m_1, m_2) \in E$, then m_1 is called the *parent* of m_2. Similarly, m_2 is called the *child* of m_1. If there is a directed path from m_1 to m_2 in E, then m_2 is called a *descendant* of m_1. Further, we say sequence $q \in G(F, E)$ if q is compatible with the partial order represented by $G(F, E)$.

14.3.2 Partial order construction problem

Problem 27 *Given m permutations q_i, $1 \leq i \leq m$, each defined on F, $1 \leq i \leq m$, the task is to construct the transitive reduction of a partial order*

$$G(F, E),$$

satisfying the following. For each pair $m_1, m_2 \in F$, m_2 is a descendant of m_1 if and only if m_1 precedes m_2 in all the given m permutations.

Does such a DAG always exist given some (nonempty) input permutations? In fact, the solution to Problem 27 always exists and this transitive reduction DAG

$$G(F, E)$$

is unique. Our interest is in constructing this transitive reduction of the partial order DAG. The following property is crucial and is also central to the algorithm design.

THEOREM 14.1
(Pair invariant theorem) *Let*

$$G(F, E)$$

be the solution to Problem 27. $(m_1 m_2) \in E$, if and only if for each q_i,

1. m_1 precedes m_2 and

2. the following set is empty:

$$L(m_1 \preceq m_2) = \left\{ m \in F \;\middle|\; \begin{array}{l} m_1 \text{ precedes } m \text{ and} \\ m \text{ precedes } m_2, \text{ in each } q_i \end{array} \right\}.$$

The proof is left as an exercise for the reader (Exercise 184). Note the equivalence of the following sets:

$$L(m_1 \preceq m_2) = \left\{ m \in F \;\middle|\; \begin{array}{l} m_1 \text{ precedes } m \text{ and} \\ m \text{ precedes } m_2, \text{ in each } q_i \end{array} \right\}$$

$$= \left\{ m \in F \;\middle|\; \begin{array}{l} m \text{ is a descendent of } m_1 \text{ and} \\ m_2 \text{ is a descendent of } m, \text{ in } G(F, E) \end{array} \right\}.$$

Partial order construction algorithm.   Theorem (14.1) is used to design an incremental algorithm. Each q_i is padded with a start character

$$S \notin F$$

and an end character

$$E \notin F.$$

This ensures the two following properties:

1. The resulting DAG has exactly one connected component.

2. The only vertex with no incoming edges is S and the only vertex with no outgoing edges is E.

The algorithm is described using a concrete example in Figure 14.3. The DAG is initialized as a 'chain' representing q_1. The generic step for q_i, $1 < i \leq m$ is defined as follows. At each iteration

$$G^{i+1}(F, E^{i+1})$$

is constructed from

$$G^i(F, E^i)$$

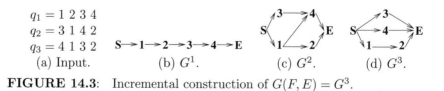

$q_1 = 1\ 2\ 3\ 4$
$q_2 = 3\ 1\ 4\ 2$
$q_3 = 4\ 1\ 3\ 2$ S→1→2→3→4→E

(a) Input.   (b) G^1.   (c) G^2.   (d) G^3.

FIGURE 14.3: Incremental construction of $G(F, E) = G^3$.

of the previous step. Finally,

$$G^m(F, E^m) = G(F, E)$$

is the required solution to the problem. At step $i+1$,

$$E^{i+1} = E' \cup E'',$$

where E' and E'' are defined (constructed) as follows.

1. E' is the set of edges that 'survive' the new permutation q_{i+1} and is defined (constructed) as follows:

$$E' = \{(m_1 m_2) \in E^i \mid m_1 \text{ precedes } m_2 \text{ in } q_{i+1}\}.$$

2. E'' is the set of new edges that must be added to the DAG due to the q_{i+1} and is defined (constructed) as follows. A pair of characters m_1 and m_2 are *imm-compatible* if all of the following hold:

   (a) m_1 precedes m_2 in q_{i+1} with

   $$L' = \{m \mid m_1 \text{ precedes } m \text{ and } m \text{ precedes } m_2 \text{ in } q_{i+1}\}.$$

   (b) m_1 is an ancestor of m_2 in $G^i(F, E^i)$ with

   $$L'' = L(m_1 \preceq m_2).$$

   (c) $L' \cap L'' = \emptyset$.

   Then

   $$E'' = \{(m_1 m_2) \notin E^i \mid m_1 \text{ and } m_2 \text{ are } imm\text{-}compatible\}.$$

The proof of correctness of the algorithm is left as an exercise for the reader (Exercise 185).

What is the size of a reduced partial order graph, in the worst case? See Exercise 158 of Chapter 13.

14.3.3 Excess in partial orders

Consider the concrete example discussed in the last section and reproduced here in Figure 14.4. The reduced partial order graph has the smallest number of edges that respect the order information in all the input permutations, yet

$$q_4, q_5, q_6 \in G(F, E),$$

which are not part of the input collection of permutations. Thus it is clear that the partial order captures some but not all of the subtle ordering information in the data. Let I be the collection of input permutations then

$$Cex(B) \supseteq \mathcal{I}.$$

The gap,

$$Cex(B) \setminus I,$$

is termed *excess*.

Handling excess with PQ structures. A heuristic using PQ trees to reduce excess is discussed here.

1. The alphabet F is augmented with a new set of characters F' to obtain

$$F \cup F'.$$

Also

$$|F'| = \mathcal{O}(|F|).$$

2. A complete extension q' on $(F \cup F')$ is converted to q on F by simply removing all the

$$m' \in F'.$$

This is best explained through an example. Let I be given as follows.

$$
\begin{aligned}
q_1 &= \text{a b c d e g f,} \\
q_2 &= \text{b a c e d f g,} \\
q_3 &= \text{a c b d e f g,} \\
q_4 &= \text{e d g f a b c,} \\
q_5 &= \text{d e g f b a c}
\end{aligned}
$$

with

$$F = \{\text{a, b, c, d, e, f, g.}\}.$$

However, notice that certain blocks appear together as follows:

$$
\begin{aligned}
q_1 &= \boxed{\text{a b c}}\ \boxed{\text{d e}}\ \boxed{\text{g f}}\,, \\
q_2 &= \boxed{\text{b a c}}\ \boxed{\text{e d}}\ \boxed{\text{f g}}\,, \\
q_3 &= \boxed{\text{a c b}}\ \boxed{\text{d e}}\ \boxed{\text{f g}}\,, \\
q_4 &= \boxed{\text{e d}}\ \boxed{\text{g f}}\ \boxed{\text{a b c}}\,, \\
q_5 &= \boxed{\text{d e}}\ \boxed{\text{g f}}\ \boxed{\text{b a c}}\,.
\end{aligned}
$$

The reduced partial order graph for this input is shown in Figure 14.5(a). In G', the alphabet is augmented with

$$F' = \{S1, E1, S2, E2, S3, E3\},$$

to obtain a 'tighter' description of I.

$q_1 = 1\ 2\ 3\ 4$
$q_2 = 3\ 1\ 4\ 2$
$q_3 = 4\ 1\ 3\ 2$
(a) Input I.

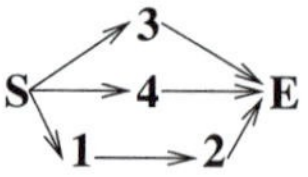

(b) The reduced partial order DAG G.

$q_4 = 4\ 1\ 2\ 3$
$q_5 = 1\ 3\ 4\ 2$
$q_6 = 1\ 4\ 2\ 3$
(c) Compatible q's.

FIGURE 14.4: Incremental construction of $G(F, E) = G^3$.

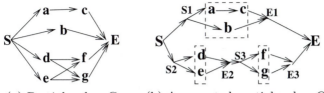

(a) Partial order, G (b) Augmented partial order, G'

FIGURE 14.5: G' is a specialization of G: If $q \in G'$, then $q \in G$, but not *vice-versa*.

The boxed elements in Figure 14.5(b) are clusters that *always* appear together, i.e., are uninterrupted in I. For example, if

$$q = a\ b\ d\ e\ f\ c\ g,$$

then clearly

$$q \in G,$$

but since the cluster $\{f, g\}$ is interrupted by c and also the cluster $\{a, b, c\}$ is interrupted,

$$q \notin G'.$$

These clusters are flanked by meta symbols, Si on the left and Ei on the right, in the augmented partial order. Thus this scheme forces the elements of the cluster to appear together, thus reducing excess.

We leave the details of this scheme as Exercise 188 for the reader.

14.4   Statistics of Partial Orders

We need compute the number of permutations that are compatible with a given partial order B.

Figure 14.6 shows examples of some partial orders, B along with $Prm(B)$ and $Cex(B)$.

An *empty partial order* is the one whose DAG has only two vertices, representing the start and the end symbol. The *inverse* partial order $\overline{B}$, is obtained by

1. switching the start and end symbols and

2. reversing the direction of every edge in the partial order DAG.

Figure 14.7 shows an example of a partial order and its inverse. If a partial order B is such that

$$Cex(B) = Cex(\overline{B}),$$

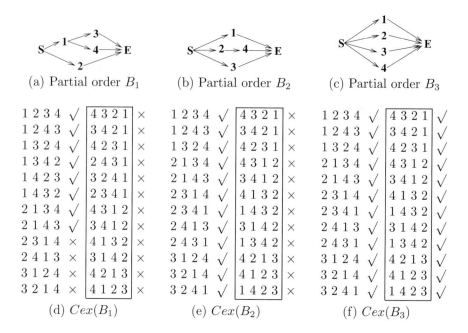

1 2 3 4 ✓	4 3 2 1 ×	1 2 3 4 ✓	4 3 2 1 ×	1 2 3 4 ✓	4 3 2 1 ✓
1 2 4 3 ✓	3 4 2 1 ×	1 2 4 3 ✓	3 4 2 1 ×	1 2 4 3 ✓	3 4 2 1 ✓
1 3 2 4 ✓	4 2 3 1 ×	1 3 2 4 ✓	4 2 3 1 ×	1 3 2 4 ✓	4 2 3 1 ✓
1 3 4 2 ✓	2 4 3 1 ×	2 1 3 4 ✓	4 3 1 2 ×	2 1 3 4 ✓	4 3 1 2 ✓
1 4 2 3 ✓	3 2 4 1 ×	2 1 4 3 ✓	3 4 1 2 ×	2 1 4 3 ✓	3 4 1 2 ✓
1 4 3 2 ✓	2 3 4 1 ×	2 3 1 4 ✓	4 1 3 2 ×	2 3 1 4 ✓	4 1 3 2 ✓
2 1 3 4 ✓	4 3 1 2 ×	2 3 4 1 ✓	1 4 3 2 ×	2 3 4 1 ✓	1 4 3 2 ✓
2 1 4 3 ✓	3 4 1 2 ×	2 4 1 3 ✓	3 1 4 2 ×	2 4 1 3 ✓	3 1 4 2 ✓
2 3 1 4 ×	4 1 3 2 ×	2 4 3 1 ✓	1 3 4 2 ×	2 4 3 1 ✓	1 3 4 2 ✓
2 4 1 3 ×	3 1 4 2 ×	3 1 2 4 ✓	4 2 1 3 ×	3 1 2 4 ✓	4 2 1 3 ✓
3 1 2 4 ×	4 2 1 3 ×	3 2 1 4 ✓	4 1 2 3 ×	3 2 1 4 ✓	4 1 2 3 ✓
3 2 1 4 ×	4 1 2 3 ×	3 2 4 1 ✓	1 4 2 3 ×	3 2 4 1 ✓	1 4 2 3 ✓

(d) $Cex(B_1)$	(e) $Cex(B_2)$	(f) $Cex(B_3)$

FIGURE 14.6: Partial orders, each on elements $F = \{1, 2, 3, 4\}$ shown in the top row. The bottom row shows all 24 permutations of F. The permutations compatible with the partial order are marked with $\sqrt{}$ and the rest are marked with $\times$. Also, each permutation in the boxed array is the inverse of the one to its left in the unboxed array.

then B is a *degenerate* partial order. The partial order B_3 in Figure 14.6 is degenerate. Note that

$$Cex(B_3) = Cex(\overline{B}_3) = Prm(B_3).$$

The proof, that this is the only nonempty degenerate partial order, is straightforward and we leave it as an exercise for the reader (Exercise 189). In other words, we can focus on just nondegenerate partial orders.

For a nondegenerate partial order B, what is the relationship between

$$Cex(B) \text{ and } Cex(\overline{B})?$$

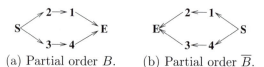

(a) Partial order B.	(b) Partial order $\overline{B}$.

FIGURE 14.7: A partial order B and its inverse partial order $\overline{B}$.

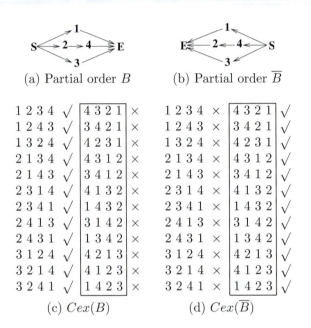

(a) Partial order B (b) Partial order $\overline{B}$

1 2 3 4	✓	4 3 2 1	×	1 2 3 4	×	4 3 2 1	✓

(c) $Cex(B)$ (d) $Cex(\overline{B})$

FIGURE 14.8: A partial order and its inverse is shown in the top row. The next row shows all 24 permutations in $Prm(\cdot)$ for each. The elements of $Cex(\cdot)$ are marked with ✓ and the rest are marked with ×. Also, each permutation in the boxed array is the inverse of the one to its left in the unboxed array.

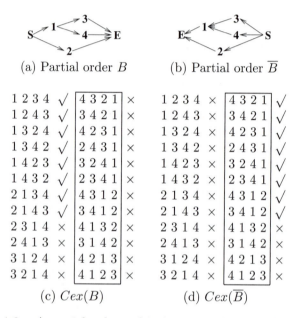

(a) Partial order B (b) Partial order $\overline{B}$

1 2 3 4	√	4 3 2 1	×	
1 2 4 3	√	3 4 2 1	×	
1 3 2 4	√	4 2 3 1	×	
1 3 4 2	√	2 4 3 1	×	
1 4 2 3	√	3 2 4 1	×	
1 4 3 2	√	2 3 4 1	×	
2 1 3 4	√	4 3 1 2	×	
2 1 4 3	√	3 4 1 2	×	
2 3 1 4	×	4 1 3 2	×	
2 4 1 3	×	3 1 4 2	×	
3 1 2 4	×	4 2 1 3	×	
3 2 1 4	×	4 1 2 3	×	

1 2 3 4	×	4 3 2 1	√
1 2 4 3	×	3 4 2 1	√
1 3 2 4	×	4 2 3 1	√
1 3 4 2	×	2 4 3 1	√
1 4 2 3	×	3 2 4 1	√
1 4 3 2	×	2 3 4 1	√
2 1 3 4	×	4 3 1 2	√
2 1 4 3	×	3 4 1 2	√
2 3 1 4	×	4 1 3 2	×
2 4 1 3	×	3 1 4 2	×
3 1 2 4	×	4 2 1 3	×
3 2 1 4	×	4 1 2 3	×

(c) $Cex(B)$ (d) $Cex(\overline{B})$

FIGURE 14.9: A partial order and its inverse is shown in the top row. The next row shows all 24 permutations in $Prm(\cdot)$ for each. The elements of $Cex(\cdot)$ are marked with √ and the rest are marked with ×. Also, each permutation in the boxed array is the inverse of the one to its left in the unboxed array. Notice that there are some permutations that belong to neither $Cex(B)$ nor $Cex(\overline{B})$.

It is instructive to study the two examples shown in Figures 14.8 and 14.9. We leave the proof of the following lemma as Exercise 190 for the reader.

LEMMA 14.3
Let B be a nondegenerate partial order.

1. *If $q \in Cex(B)$, then $q \notin Cex(\overline{B})$.*

2. *The converse of the last statement is not true, i.e., there may exist $q \in Prm(B)$ such that*
$$q \notin Cex(B) \text{ and } q \notin Cex(\overline{B}).$$

3. *If $q \in Cex(B)$, then $\overline{q} \in Cex(\overline{B})$.*

4.
$$|Cex(B)| = |Cex(\overline{B})|,$$
$$|Cex(B)| \leq \frac{|Prm(B)|}{2}.$$

LEMMA 14.4
(Symmetric lemma) *For a nondegenerate partial order B, if the DAG of B is isomorphic to the DAG of $\overline{B}$, then*

$$|Cex(B)| = |Cex(\overline{B})| = \frac{n!}{2}.$$

The proof is left as an exercise for the reader (Exercise 191).

14.4.1   Computing $Cex(B)$

Given a partially ordered set B, how hard is it to count the number of complete extensions? There are a few special cases where there is a simple algorithm, but in general it is hard. Specifically, it is #P-complete [BW91a, BW91b], i.e., it is as hard as counting the number of satisfying assignments of an instance of the satisfiability problem (SAT). However, it can be approximated by using the polynomial time algorithm for computing the the volume of either the order polytope or the chain polytope of B [DFK] and the fact that the number of complete extensions of an n-element partial order B is equal to $n!$ times the volume of either of the polytopes.

A discussion on this method is beyond the scope of this book. Other researchers have assumed restrictions, e.g., Mannila and Meek [MM00] restrict their partial orders to have an MSVP (minimal vertex series-parallel) DAG to get a handle on the problem. However, we give here an exposition on a much simpler scheme to estimate the lower and upper bounds of the size of $Cex(B)$.

Consider the DAG

$$G(V, E)$$

of a nondegenerate partial order B. Let the depth of a node $v \in V$, *depth(v)*, be the largest distance from the start symbol S, where the distance is in terms of the number of nodes in the path from S to v (excluding S).

q' is defined to be a *subsequence* of q if

1. $\Pi(q') \subseteq \Pi(q)$ and

2. if for any pair $\sigma_1, \sigma_2 \in \Pi(q')$, σ_1 precedes σ_2 in q', then σ_1 must precede σ_2 in q.

For example, given

$$q = a\,b\,c\,d\,e,$$
$$q_1 = a\,c\,e,$$
$$q_2 = c\,a,$$

q_1 is a subsequence of q but q_2 is not.

Let q_1 and q_2 be such that

$$\Pi(q_1) \cap \Pi(q_2) = \emptyset.$$

Then

$$q = q_1 \oplus q_2,$$

is defined as follows:

1. $\Pi(q) = \Pi(q_1) \cup \Pi(q_2)$ and

2. q_1 and q_2 are subsequences of q.

Columns of the grid.   For each v, assign depth

$$col(v) = depth(v).$$

For each distinct depth i,

$$C_i = \{v \mid col(v) = i\}.$$

The *depth(v)* of each v can be computed in linear time using a breadth first traversal (BFS) of the DAG (see Chapter 2).

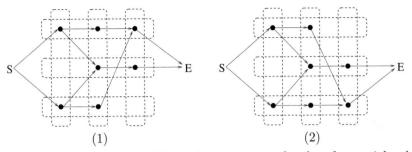

FIGURE 14.10: Two possible grid assignments of nodes of a partial order DAG. The C's are the same but the R's differ in the two assignments.

Rows of the grid. Let

$$v_1 v_2 v_3 \ldots v_l$$

be the vertices along a path on the DAG such that

$$col(v_1) < col(v_2) < \ldots < col(v_l).$$

Then

$$row(v_1) = row(v_2) = \ldots = row(v_l).$$

A depth first traversal (DFS) of the DAG (see Chapter 2) can be used to compute $row(v)$ for each v satisfying these constraints.
Let

$$R_i = \{v \mid row(v) = i\}.$$

Let the number of nonempty C's be c and let the number of nonempty R's be r. We use the following convention:

$$n_i = |R_i|, \text{ for } 1 \leq i \leq r.$$

At the end of the process $row(v)$ and $col(v)$ have been computed for each v. It is possible to obtain different values of $row(v)$ satisfying the condition. But that does not matter. We are looking for a small number of R sets with as large a size ($|R|$) as possible. However, this is only a heuristic to simply put the vertices on a virtual grid (i, j). See Figure 14.10 for a concrete example.
Let

$$col(B) = \{q = q_1 q_2 \ldots q_c \mid \Pi(q_i) = C_i, \text{ for } 1 \leq i \leq c\}.$$

The following observation is crucial to the scheme:

$$\boxed{\text{If } q \in col(B), \text{ then } q \in B.}$$

Note that $col(B)$ does miss a few extensions of B, since the vertices of each column are in strict proximity. Thus

$$col(B) \subseteq Cex(B). \tag{14.1}$$

Also, the size of $col(B)$ is computed exactly as follows:

$$|col(B)| = \prod_{i=1}^{c} |C_i|!$$

Let

$$row(B) = \left\{ q_1 \oplus q_2 \oplus \ldots \oplus q_c \;\middle|\; \begin{array}{l} q_i = v_1 v_2 \ldots v_{n_i}, \\ \text{for } 1 \leq i \leq c \end{array} \right\}.$$

Again, the following observation is crucial to the scheme:

> If $q \in B$, then $q \in row(B)$.

Note that each $q \in B$ must also belong to $row(B)$, since no order of the elements is violated. However, some $q \in row(B)$, may violate the order, since there are some edges that go across the R rows, which is not captured by the $row(B)$ definition. Thus

$$Cex(B) \subseteq row(B). \tag{14.2}$$

Also, the size of $col(B)$ is computed exactly as follows (see Exercise 187 for details of this computation):

$$|row(B)| = \binom{n_1 + n_2}{n_2} \binom{n_1 + n_2 + n_3}{n_3} \ldots \binom{n_1 + n_2 + .. + n_r}{n_r}.$$

In conclusion,

$$col(B) \subseteq Cex(B) \subseteq row(B).$$

For a nondegenerate partial order B,

$$|col(B)| \leq |Cex(B)| \leq \min\left(|row(B)|, \frac{|V|!}{2}\right),$$

where V is the set of vertices in the DAG of B. Thus the sizes of $row(B)$ and $col(B)$ can be used as coarse lower and upper bounds of $|Cex(B)|$. It is possible to refine the bounds by trying out different assignments of $row(v)$ (note that for a v, $col(v)$ is unique).

Back to probability computation. We pose the following question:

> *What is $pr(B)$, the probability of a permutation of all elements of F, being compatible with a given partial order B defined on F?*

The total number of permutations of the elements of F is

$$|F|!$$

Then, the probability is given as

$$pr(B) = \frac{|Cex(B)|}{|F|!}$$

An alternative view. Let B be a partial order on F. We label the nodes of the partial order by integers

$$1, 2, \ldots, |F|$$

in any order. Let q be a random permutation of integers integers

$$1, 2, \ldots, |F|.$$

See Section (5.2.3) for a definition of random permutation. Then the probability, $pr(B)$, of the occurrence of the event

$$q \in Cex(B)$$

is given by

$$pr(B) = \frac{|Cex(B)|}{|F|!}.$$

14.5   Redescriptions

We have already developed the vocabulary to appreciate a very interesting idea called *redescriptions*, introduced by Naren Ramakrishnan and coauthors in [RKM$^+$04]. This can be simply understood as follows. For a given incidence matrix I, if distinct expressions

$$e_1 \neq e_2$$

are such that

$$O(e_1) = O(e_2),$$

then e_1 is called a *redescription* of e_2 and vice-versa. In other words, given a data set I, e_1 and e_2 provide alternative description for some sample set (or set of rows in I) denoted by $O(e_1)$. Usually if

$$\Pi(e_1) \cap \Pi(e_2) = \emptyset,$$

then the implications are even stronger since the alternative description or explanation is over a different set of features (or columns).

A redescription is hence a shift-of-vocabulary; the goal of redescription mining is to find segments of the input that afford dual definitions and to find those definitions. For example, redescription may suggest alternative pathways of signal transduction that might target the same set of genes. However, the underlying premise is that input sequences can indeed be characterized in at least two ways using some definition (say boolean expression or partial orders or partial orders of expressions). Thus for instance in *cis*-regulatory regions, existence of redescriptions signify that possibly these must be under concerted combinatorial control (and likely to lie upstream of functionally related genes).

This is only a brief introduction to an exciting idea and the reader is directed to [RKM+04, PR05] for further reading on this topic.

14.6   Application: Partial Order of Expressions

We discuss here a possible application of the detection of partial order on expressions. Such a complex mechanism is termed *combinatorial control* in the following discussion [RP07].

Combinatorial control [RSW04]—the use of a small number of transcription factors in different combinations to realize a range of gene expression patterns—is a key mechanism of transcription initiation in eukaryotes. Many important processes, such as development [LD05], stress response, and neuronal activity, universally rely on combinatorial control to accomplish a diversity of cellular functions.

How do a given set of transcription factors determine which genes to activate or repress? Since genomic DNA is tightly packaged with proteins inside the nucleus, initiation of transcription is an elaborate process involving chromatin remodeling around the gene of interest, recruitment of transcription activators, promoter-specific binding and assembly of the transcription apparatus, culminating in formation of protein-DNA complexes. Furthermore, for a given regulatory module, varying degrees of occupancy by one or more transcription factors, specificity of binding, co-operativity among activators and repressors, and the resulting stability of the assembled complex all contribute to the richness of gene expression observed in the cell.

Understanding how multiple signals are combinatorially integrated in a given regulatory module is thus a complex problem and can be broken down into meaningful subproblems [LW03]. First, what are the binding sites (*cis*-regulatory regions) influencing gene expression? This question has been studied by elucidating structures of regulatory protein-DNA complexes, promoter prediction algorithms [PBCB99], as well as by comparative genomic sequence analysis [KPE+03].

Second, which transcription factors occupy these sites? Genome-wide location analysis techniques such as ChIP-chip [SSDZ05] are relevant here.

Third, how can transcription factor occupancy be related to gene expression? Many researchers have exploited the computational nature of this problem and proposed predictive models ranging from simple boolean gates [BGH03] to complex circuits [ID05]. Others have adopted a descriptive approach, and identified combinations of promoter motifs that characterize or explain observed co-expression of genes [SS05, PSC01].

Finally, what is the cellular context or range of environmental conditions that actually trigger transcription? The condition-specific use of transcription factors is studied in, e.g., Segal et al. [SSR$^+$03] and Harbison et al. [Hea04].

An emerging trend is to model the entire sequence and structural context in which transcription occurs, in particular capturing the physical layout and arrangement of *cis*-regulatory regions and how variations in their ordering and spacing affect the specificity of the set of genes transcribed. Some transcription factors influencing combinatorial control are not sensitive to spacing between binding sites whereas others can be influenced by changes of even one nucleotide in the spacer region. At a higher level [Chi05], overlapping sites might prevent simultaneous binding whereas far away sites may require DNA looping to initiate transcription.

Furthermore, since the transcription machinery is a multi-protein complex with structural constraints, accessibility to binding sites is crucial and even a slight permutation of the *cis*-regulatory regions can prevent binding. A striking example is given in [Chi05], of yeast sulfur metabolism, where one permutation of binding sites (*Cbf1*, *Met31*) induces expression of the *HIS3* reporter gene, whereas the reverse permutation (*Met31*, *Cbf1*) does not.

Hence, although *cis*-regulatory regions are very short ($\approx$ 5-15 base pairs), they can co-occur in symbiotic, compensatory, or antagonistic combinations. Hence, characterizing permutation and spacing constraints underlying a family of transcription factors can possibly help in understanding how genes are selectively targeted for expression in a given cell state.

14.7 Summary

This chapter discusses a very general, almost descriptive characterization of sequences in terms of permutations and partial orders on (boolean) expressions on motifs. This attempts to answer complex questions: For instance, given a set of sequences, can the order constraints, say in their upstream regions, be characterized? Also, can the concerted clusters of sequences (genes) be identified that exhibit distinctive order constraints?

14.8   Exercises

Exercise 175 (Boolean expression) *Consider the following incidence matrix:*

$$I = \begin{bmatrix} m_1 & m_2 & m_3 & m_4 & m_5 \\ 1 & 0 & 1 & 1 & 0 \\ 1 & 1 & 0 & 1 & 0 \\ 0 & 0 & 1 & 0 & 0 \\ 1 & 1 & 0 & 1 & 0 \end{bmatrix}$$

1. *Construct boolean expressions, e_1, e_2 and e_3 on the four motifs where each satisfies the equation below.*

$$O(e_1) = \{1, 3\},$$
$$O(e_2) = \{1, 3, 4\},$$
$$O(e_1) = \{1, 2, 3, 4\}.$$

2. *Show that for any subset, Z, of $\{1, 2, 3, 4\}$, there exists e such that*

$$O(e) = Z.$$

3. *Enumerate all the distinct expressions on the five variables*

$$m_1, m_2, \ldots, m_5$$

defined bu I.

Hint: 2. For each row i construct an expression e_i and then e is constructed from this collection of e_i's.

Exercise 176 (Boolean variables)

1. *For boolean variables m_1 and m_2 show that*

$$\overline{m_1 + m_2} = \overline{m}_1 \overline{m}_2,$$
$$\overline{m_1 m_2} = \overline{m}_1 + \overline{m}_2.$$

2. *Show that any expression e can be written as a CNF.*

3. *Show that any expression e can be written as a DNF.*

Hint: 1. This is called the De Morgan's laws. Use definition of negation of expressions to prove this. 2. & 3. Use De Morgan's laws.

Exercise 177 (Monotone expression) *Let e be a monotone expression.*

1. *Show that if the value of any variable in $\Pi(e)$ is changed from 0 to 1, then the value of e only 'increases', i.e.,*

   (a) *either from 0 to 1 or remains at 0 or remains at 1,*

   (b) *but never from from 1 to 0.*

2. *Similarly, show that if the value of any variable in $\Pi(e)$ is changed from 1 to 0, then the value of e only 'decreases', i.e.,*

   (a) *either from 1 to 0 or remains at 0 or remains at 1,*

   (b) *but never from from 0 to 1.*

This is called the monotone *behavior of e.*

Hint: 1. & 2. A monotone expression has only conjunctions and disjunctions. Consider conjunctions and then disjunctions.
Note that the notion of monotone boolean expression is analogous to that of a *monotonically increasing* function defined on $\mathbb{R}$.

Exercise 178 (Flip lemma) *For a given incidence matrix I, let a polymorphic $\mathcal{S}$ be defined as follows:*

$$\mathcal{S}[1] = \{i \mid i \text{ is a row in } I\},$$
$$\mathcal{S}[2] = \{f \mid f \text{ is a column in } I\}.$$

Next, given a polymorphic $\mathcal{S}$, let $\overline{\mathcal{S}}$ be defined as follows:

$$\overline{\mathcal{S}}[1] = \{i \mid i \notin \mathcal{S}[1]\},$$
$$\overline{\mathcal{S}}[2] = \{\overline{f} \mid f \in \mathcal{S}[2]\}.$$

If

$$e_1 = m_1 m_2 \ldots m_l$$

and the pair

$$\mathcal{S}[1] = \Pi(e_1) \text{ and } \mathcal{S}[2] = O(e_1)$$

is a minimal (maximal resp.) bicluster in I then the pair

$$\overline{\mathcal{S}}[1] = \Pi(e_2) \text{ and } \overline{\mathcal{S}}[2] = O(e_2)$$

is a minimal (maximal resp.) bicluster in $\overline{I}$ where

$$e_2 = \overline{m}_1 + \overline{m}_2 + \ldots + \overline{m}_l.$$

Hint: See the example in Figure 14.2.

Exercise 179 (Minimal bicluster) *Consider I of Figure 14.2.*

$$
I = \begin{bmatrix}
m_1 & m_2 & m_3 & m_4 & m_5 & m_6 & m_7 \\
0 & 1 & 1 & 1 & 1 & 1 & 1 \\
1 & 0 & 0 & 1 & 0 & 0 & 0 \\
0 & 1 & 1 & 1 & 0 & 1 & 1 \\
1 & 0 & 0 & 1 & 0 & 1 & 1 \\
0 & 1 & 1 & 1 & 1 & 1 & 0 \\
0 & 1 & 0 & 0 & 0 & 0 & 1 \\
0 & 1 & 0 & 1 & 1 & 1 & 0 \\
1 & 0 & 0 & 0 & 0 & 0 & 1
\end{bmatrix}.
$$

1. *Let $V = \{m_2, m_4, m_6\}$. Then, discuss why each of the following biclusters is not minimal.*

   (a) *V and $O_1 = \{1, 3, 5\}$,*

   (b) *V and $O_2 = \{3, 5, 7\}$,*

   (c) *V and $O_3 = \{1, 3, 7\}$,*

   (d) *V and $O_4 = \{1, 5, 7\}$,*

2. *Why is the bicluster V, $O_5 = \{3, 7\}$ minimal?*

Hint: 1. For each O_i, is the corresponding V set the same? 2. Fixing O_5, what is the corresponding set of variables? Is it the same as V?

Exercise 180 (Set intersections to biclusters) *Discuss how the problem of detecting biclusters from a discrete matrix M, given a quorum K, can be solved using the maximal set intersection problem of Chapter 13.*

Exercise 181 *For a partial order B defined on $|F| > 1$, show that*

$$|Prm(F)|$$

is an even number.

Hint: Note that $|Prm(F)| = |F|!$
Yet another argument is by noticing that for $|F| > 1$, a permutation $q \neq \bar{q}$.

Exercise 182 (On transitive reduction) *Show that the following two statements are equivalent.*
Let B be a partial order defined on F.

1. Let $G(F, E)$ be the graph representation (DAG) of B. i.e., if

$$m_1 \preceq m_2$$

   in B, then there is a path from m_1 to m_2 in E. If E is the smallest set of edges satisfying this condition then $G(F, E)$ is called a transitive reduction of B.

2. If $G(F, E)$ is the transitive reduction of B, then for any pair,

$$m_1, m_2 \in F,$$

   if there is a directed path of length larger than 1 from m_1 to m_2 in $G(F, E)$, then edge

$$(m_1 m_2) \notin E.$$

Hint: Use proof by contradiction.

Exercise 183 (Existence and uniqueness of partial order DAG) *Prove that the solution to Problem 27 always exists and the constructed DAG is unique.*

Hint: For any pair $m_1, m_2 \in F$, an edge is introduced between them if m_1 precedes m_2 in all the input sequences in. Next construct the unique transitive reduction.

Exercise 184 (Edge in partial order DAG) *Let*

$$G(F, E)$$

be the solution to Problem 27. Show that

$$(m_1 m_2) \in E,$$

if and only if for each q_i, the following two conditions hold:

1.

$$\left\{ m \in F \;\middle|\; \begin{array}{l} m_1 \text{ precedes } m \text{ and} \\ m \text{ precedes } m_2, \text{ in each } q_i \end{array} \right\} = \emptyset.$$

   and

2. m_1 *precedes* m_2.

Hint: Use proof by contradiction.

Exercise 185 (Partial order algorithm) *Consider the algorithm described in Section (14.3.2) to construct the partial reduction of a DAG given a set of sequences.*

1. *Identify the edge sets E' and E'' at each step in the example shown in Figure 14.3.*

2. *Prove that the algorithm is correct.*

3. *What is the time complexity of the algorithm?*

Hint: 2. Use Theorem (14.1).

Exercise 186 (Complete extensions) *Enumerate the sets $Cex(B)$ and $Cex(\overline{B})$ for the partial orders shown below.*

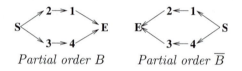

Partial order B Partial order $\overline{B}$

Exercise 187 (Oranges & apples problem) *In the following, q_i's are permutations such that*

$$\Pi(q_i) \cap \Pi(q_j) = \emptyset, \ \text{for } i \neq j,$$

and $n_i = |q_i|$, for each i.

1. *Let*

$$S_2 = \{q \mid q = q_1 \oplus q_2\}.$$

   Show that

$$|S_2| = \binom{n_1 + n_2}{n_1} = \binom{n_1 + n_2}{n_2}.$$

2. *Let*

$$S_r = \{q \mid q = q_1 \oplus q_2 \oplus \ldots \oplus q_r\}.$$

   Show that

$$|S_r| = \binom{n_1 + n_2}{n_2}\binom{n_1 + n_2 + n_3}{n_3} \ldots \binom{n_1 + n_2 + .. + n_r}{n_r}.$$

Hint: 1. This is the problem of arranging n_1 oranges and n_2 apples along a line. Then out of the $n_1 + n_2$ slots, in how many ways can n_1 positions be picked for the oranges? The rest of the slots will be filled by the apples. 2. Now the orange is further categorized as a *pumelo* or a *tangerine* or a *mandarin* and so on. The arrangement of these varieties in the orange slots were determined in the previous step.

Exercise 188 (Excess)

1. *Discuss how (recursive) PQ structures may be used for reducing excess in a partial order DAG.*

2. *What are the issues involved in using multiple DAGs to represent some m permutations.*

Hint: 1. Use the following PQ tree for the example of Figure 14.4.

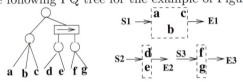

2. How many distinct DAGs? m ? How does excess reduce with the increase in number of DAGs? What criterion to use?

Exercise 189 (Unique degenerate) *Show that the only nonempty partial order that is degenerate $(Cex(B) = Cex(\overline{B}))$ is of the following form:*

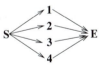

Exercise 190 (Complete extensions) *Let B be a nondegenerate partial order defined on F. Then show the following.*

1. *If $q \in Cex(B)$, then $q \notin Cex(\overline{B})$.*

2. *The converse of the last statement is not true, i.e., there may exist $q \in Prm(F)$ such that $q \notin Cex(B)$ and $q \notin Cex(\overline{B})$.*

3. *If $q \in Cex(B)$, then $\overline{q} \in Cex(\overline{B})$.*

4.

$$|Cex(B)| = |Cex(\overline{B})|,$$

$$|Cex(B)| \leq \frac{|Prm(F)|}{2}.$$

Hint: 1. & 3. Since B is nondegenerate there is at least one ordering of a pair of elements v_1 and v_2 in the DAG. If $q \in Cex(B)$, then this ordering is honored in q, thus cannot belong to $Cex(\overline{B})$. 2. Pick an example from Figure 14.9 to demonstrate this fact. 4. Follows from 3.

Exercise 191 (Isomorphic partial order)

1. If the DAG of a partial order B is isomorphic to the DAG of a partial order B', then show that

$$|Cex(B)| = |Cex(B')|.$$

2. For a nondegenerate partial order B, if the DAG of B is isomorphic to the DAG of $\overline{B}$, then show that

$$|Cex(B)| = |Cex(\overline{B})| = \frac{n!}{2}.$$

Hint: 1. The nodes can be simply relabeled to make the DAGs identical. 2. Follows from 1.

Exercise 192 (Complete extensions) *Let B be a nondegenerate partial order.*

1. Under what conditions does the following hold?

$$|Cex(B)| = \prod_{v \neq E} outDeg(v, B)!.$$

2. Is the following true? Why?

$$|Cex(B)| \leq \min\left(\prod_{v \neq E} outDeg(v, B)!, \; \prod_{v \neq S} inDeg(v, B)!, \; \frac{n!}{2}\right).$$

References

[AALS03] A. Amir, A. Apostolico, G. M. Landau, and G. Satta. Efficient text fingerprinting via parikh mapping. *Journal of Discrete Algorithms*, 1(5-6):409 – 421, 2003.

[ACP03] M. Alexandersson, S. Cawley, and L. Pachter. SLAM—cross-species gene finding and alignment with a generalized pair hidden Markov model. 13:496–502, 2003.

[ACP05] A. Apostolico, M. Comin, and L. Parida. Conservative extraction of over-represented extensible motifs. *ISMB (Supplement of Bioinformatics)*, 21:9–18, 2005.

[ACP07] Alberto Apostolico, Matteo Comin, and Laxmi Parida. Varun: Discovering extensible motifs under saturation constraints. *Preprint*, 2007.

[AGK+04] W. Ao, J. Gaudet, W.J. Kent, S. Muttumu, and S.E Mango. Environmentally induced foregut remodeling by pha-4/foxa and daf-12/nhr. *Science*, 305(5691):1743–1746, 2004.

[AIL+88] A. Apostolico, C. Iliopoulos, G. M. Landau, B. Schieber, and U. Vishkin. Parallel construction of a suffix tree with applications. *Algorithmica*, 3:347–365, 1988.

[AMS97] T.S. Anantharaman, B. Mishra, and D.C. Schwartz. Genomics via optical mapping II: Ordered restriction maps. *Journal of Computational Biology*, 4(2):91–118, 1997.

[AP04] A. Apostolico and L. Parida. Incremental paradigms for motif discovery. *Journal of Computational Biology*, 11(4):15–25, 2004.

[BÖ4] S. Böcker. Sequencing from compomers: Using mass spectrometry for DNA de novo sequencing of 200+ nt. *Journal of Computational Biology*, 11(6):1110–1134, 2004.

[BBE+85] A. Blumer, J. Blumer, A. Ehrenfeucht, D. Haussler, M.T. Chen, and J. Seiferas. The smallest automaton recognizing the subwords of a text. *Theoretical Computer Science*, pages 31–55, 1985.

[BCL+01] John W.S. Brown, Gillian P. Clark, David J. Leader, Craig G. Simpson, and Todd Lowe. Multiple snoRNA gene clusters from *arabidopsis. RNA*, 7:1817–1832, 2001.

[BE94] T. L. Bailey and C. Elkan. Fitting a mixture model by expectation maximization to discover motifs in biopolymers. In *Proceedings of the Second International Conference on Intelligent Systems for Molecular Biology*, pages 28–36. AAAI Press, 1994.

[BGH03] N.E. Buchler, U. Gerland, and T. Hwa. On Schemes of Combinatorial Transcription Logic. *PNAS*, 100(9):5136–5141, 2003.

[BHM87] S. K. Bryan, M. E. Hagensee, and R. E. Moses. DNA polymerase III requirement for repair of DNA damage caused by methyl methanesulfonate and hydrogen peroxide. In *Journal of Bacteriology*, volume 16, pages 4608–4613. ACM Press, 1987.

[BL76] K. Booth and G. Leukar. Testing for the consecutive ones property, interval graphs, and graph planarity using PQ-tree algorithms. *Journal of Computer and System Sciences*, 13:335–379, 1976.

[BLZ93] B. Balasubramanian, C. V. Lowry, and R. S. Zitomer. The Rox1 repressor of the saccharomyces cerevisiae hypoxic genes is a specific DNA-binding protein with a high-mobility-group motif. *Mol Cell Biol*, 13(10):6071–6178, 1993.

[BMRS02] M. -P. Bal, F. Mignosi, A. Restivo, and M. Sciortino. Forbidden words in symbolic dynamics. *Advances in Applied Mathematics*, 25:163–193, 2002.

[BMRY04] K.H. Burns, M.M. Matzuk, A. Roy, and W. Yan. Tektin3 encodes an evolutionarily conserved putative testicular micro tubules-related protein expressed preferentially in male germ cells. In *Molecular Reproduction and Development*, volume 67, pages 295–302. ACM Press, 2004.

[Bsh95] N.H. Bshouty. Exact Learning Boolean Functions via the Monotone Theory. *Information and Computation*, Vol. 123(1):146–153, 1995.

[BT02] Jeremy Buhler and Martin Tompa. Finding motifs using random projections. In *Journal of Computational Biology*, volume 9(2), pages 225—242, 2002.

[BW91a] G. Brightwell and P. Winkler. Counting linear extensions. In *Order*, pages 225–242, 1991.

[BW91b] G. Brightwell and P. Winkler. Counting linear extensions is #P-complete. In *Proc. 23rd ACM Symposium on the Theory of Computing (STOC)*, pages 175–181, 1991.

[Chi05] D. Y.-H. Chiang. *Computational and Experimental Analyses of Promoter Architecture in Yeasts*. PhD thesis, University of California, Berkeley, Spring 2005.

[CJI⁺98] W. Cai, J. Jing, B. Irvine, L. Ohler, E. Rose, H. Shizua, U. J. Kim, M. Simon, T. Anantharaman, B. Mishra, and D. C. Schwartz. High-resolution restriction maps of bacterial artificial chromosomes constructed by optical mapping. *Proc. Natl. Acad. Sci. USA*, 95:3390–3395, 1998.

[CLR90] T. H. Cormen, C. E. Leiserson, and R. L. Rivest. *Introduction to Algorithms*. The MIT Press, Cambridge, Massachusetts, 1990.

[CMHK05] Joseph F. Contrera, Philip MacLaughlin, Lowell H. Hall, and Lemont B. Kier. QSAR modeling of carcinogenic risk using discriminant analysis and topological molecular descriptors. *Current Drug Discovery Technologies*, Vol. 2(2):55–67, 2005.

[CP04] A. Chattaraj and L. Parida. An inexact suffix tree based algorithm for extensible pattern discovery. *Theoretical Computer Science*, (1):3–14, 2004.

[CP07] Matteo Comin and Laxmi Parida. Subtle motif discovery for detection of DNA regulatory sites. *Asia Pacific Bioinformatics Conference*, pages 95–104, 2007.

[CR04] F. Coulon and M. Raffinot. Fast algorithms for identifying maximal common connected sets of interval graphs. *Proceedings of Algorithms and Computational Methods for Biochemical and Evolutionary Networks (CompBioNets)*, 2004.

[DBNW03] O. Dror, H. Benyamini, R. Nussinov, and H. J. Wolfson. Multiple structural alignment by secondary structures: Algorithm and applications. *Protein Science*, 12(11):2492–2507, 2003.

[DFK] M. Dyer, A. Frieze, and R. Kannan. A random polynomial-time algorithm for approximating the volume of convex bodies. In *Journal of the ACM (JACM)*, number 1.

[DH73] R.O. Duda and P. E. Hart. *Pattern Classification and Scene Analysis*. John Wiley & Sons, Menlo Park, California, 1973.

[Did03] G. Didier. Common intervals of two sequences. In *Proc. of the Third Wrkshp. on Algorithms in Bioinformatics*, volume 2812 of *Lecture Notes in Bioinformatics*, pages 17–24. Springer-Verlag, 2003.

[DS03] Dannie Durand and David Sankoff. Tests for gene clustering. *Journal of Computational Biology*, 10(3-4):453–482, 2003.

506 *References*

[DSHB98] T. Dandekar, B. Snel, M. Huynen, and P. Bork. Conservation of gene order: a fingerprint of proteins that physically interact. *Trends Biochem. Sci.*, 23:324–328, 1998.

[ELP03] Revital Eres, Gad M. Landau, and Laxmi Parida. A combinatorial approach to automatic discovery of cluster-patterns. In *Proc. of WABI*, September 15-20, 2003.

[EP02] Eleazar Eskin and Pavel Pevzner. Finding composite regulatory patterns in DNA sequences. In *Bioinformatics*, volume 18, pages 354–363, 2002.

[Fel68] William Feller. *An Introduction to Probability Theory and its Applications*. Wiley, 1968.

[Fis87] D.H. Fisher. Knowledge Acquisition via Incremental Conceptual Clustering. *Machine Learning*, Vol. 2(2):139–172, 1987.

[FRP+99] Aris Floratos, Isidore Rigoutsos, Laxmi Parida, Gustavo Stolovitzky, and Yuan Gao. Sequence homology detection through large-scale pattern discovery. In *Proceedings of the Annual Conference on Computational Molecular Biology (RE-COMB99)*, pages 209–215. ACM Press, 1999.

[GBM+01] S. Giglio, K. W. Broman, N. Matsumoto, V. Calvari, G. Gimelli, T. Neuman, H. Obashi, L. Voullaire, D. Larizza, R. Giorda, J. L. Weber, D. H. Ledbetter, and O. Zuffardi. Olfactory receptor-gene clusters, genomic-inversion polymorphisms, and common chromosme rearrangements. *Am. J. Hum. Genet.*, 68(4):874–883, 2001.

[GJ79] M.R. Garey and D.S. Johnson. *Computers and Intractability: A Guide to the Theory of NP-Completeness*. W.H. Freeman and Co., San Francisco, 1979.

[GPS03] H. H. Gan, S. Pasquali, and T. Schlick. Exploring the repertoire of RNA secondary motifs using graph theory: implications for RNA design. *Nucleic Acids Research*, 31(11):2926–2943, 2003.

[Hea04] C.T. Harbison et al. Transcriptional Regulatory Code of a Eukaryotic Genome. *Nature*, 431:99–104, Sep 2004.

[HETC00] J. D. Hughes, P. W. Estep, S. Tavazoie, and G. M. Church. Computational identification of cis-regulatory elements associated with groups of functinally related genes in Saccharomyces cerevisiae. *J Molec. Bio.*, 296:1205–1214, 2000.

[HG04] X. He and M.H. Goldwasser. Identifying conserved gene clusters in the presence of orthologous groups. In *Proceedings of the Annual Conference on Computational Molecular Biology (RE-COMBo4)*, pages 272–280. ACM Press, 2004.

[HS99] G. Z. Hertz and G. D. Stormo. Identifying DNA and protein patterns with statistically significant alignments of multiple sequences. *Bioinformatics*, 15:563–577, 1999.

[HSD05] Rose Hoberman, David Sankoff, and Dannie Durand. The statistical significance of max-gap clusters. *Lecture Notes in Computer Science*, 3388/2005, 2005.

[ID05] S. Istrail and E.H. Davidson. Logic functions of the genomic cis-regulatory code. *PNAS*, 102(14):4954–4959, Apr 2005.

[IWM03] A. Inokuchi, T. Washio, and H. Motoda. Complete mining of frequent patterns from graphs: Mining graph data. *Machine Learning*, 50(3):321–354, 2003.

[KK00] D. Kihara and M. Kanehisa. Tandem clusters of membrane proteins in complete genome sequences. *Genome Research*, 10:731–743, 2000.

[KLP96] Z. M. Kedem, G. M. Landau, and K. V. Palem. Parallel suffix-prefix matching algorithm and application. *SIAM Journal of Computing*, 25(5):998–1023, 1996.

[KMR72] R. Karp, R. Miller, and A. Rosenberg. Rapid identification of repeated patterns in strngs, arrays and trees. In *Symposium on Theory of Computing*, volume 4, pages 125–136, 1972.

[KP02a] Keich and Pevzner. Finding motifs in the twilight zone. In *Annual International Conference on Computational Molecular Biology*, pages 195–204, Apr, 2002.

[KP02b] Uri Keich and Pavel Pevzner. Subtle motifs: defining the limits of motif finding algorithms. In *Bioinformatics*, volume 18, pages 1382–1390, 2002.

[KPE+03] M. Kellis, N. Patterson, M. Endrizzi, B. Birren, and E. S. Lander. Sequencing and comparison of yeast species to identify genes and regulatory elements. *Nature*, 423:241–254, May 2003.

[KPL06] Md Enamul Karim, Laxmi Parida, and Arun Lakhotia. Using permutation patterns for content-based phylogeny. In *Pattern Recognition in Bioinformatics*, volume 4146 of *Lecture Notes in Bioinformatics*, pages 115–125. Springer-Verlag, 2006.

[LAB+93] C. E. Lawrence, S. F. Altschul, M. S. Boguski, J. S. Liu, A. F. Neuwald, and J. C. Wootton. Detecting subtle sequence signals: A Gibbs sampling strategy for multiple alignment. *Science*, 262:208–214, Oct, 1993.

[LD05] M. Levine and E.H. Davidson. Gene regulatory networks for development. *PNAS*, 102(4):4936–4942, Apr 2005.

[LKWP05] Arun Lakhotia, Md Enamul Karim, Andrew Walenstein, and Laxmi Parida. Malware phylogeny using maximal π-patterns. In *EICAR*, 2005.

[LMF03] A. V. Lukashin, M.E.Lakashev, and R. Fuchs. Topology of gene expression networks as revealed by data mining and modeling. *Bioinformatics*, 19(15):1909–1916, 2003.

[LMS96] M. Y. Leung, G. M. Marsh, and T. P. Speed. Over and underrepresentation of short DNA words in herpesvirus genomes. *Journal of Computational Biology*, 3:345–360, 1996.

[LPW05] Gad Landau, Laxmi Parida, and Oren Weimann. Using PQ trees for comparative genomics. In *Proc. of the Symp. on Comp. Pattern Matching*, volume 3537 of *Lecture Notes in Computer Science*, pages 128–143. Springer-Verlag, 2005.

[LR90] C. E. Lawrence and A. A. Reilly. An expectaion maximization (EM) algorithm for the identification and characterization of common sites in unaligned biopolymer sequences. *Proteins: Structure, Function and Genetics*, 7:41–51, 1990.

[LR96] J. G. Lawrence and J. R. Roth. Selfish operons: Horizontal transfer may drive the evolution of gene clusters. *Genetics*, 143:1843–1860, 1996.

[LW03] H. Li and W. Wang. Dissecting the transcription networks of a cell using computational genomics. *Current Opinion in Genetics and Development*, 13:611–616, 2003.

[MBV05] Aurlien Mazurie, Samuel Bottani, and Massimo Vergassola. An evolutionary and functional assessment of regulatory network motifs. *Genome Biolgy*, 6(4), 2005.

[MCM81] M. Habib M. Chein and M.C Maurer. Partitive hypergraphs. *Discrete Mathematics*, 37:35–50, 1981.

[Mic80] R.S. Michalski. Knowledge acquisition through conceptual clustering: A theoretical framework and algorithm for partitioning data into conjunctive concepts. *International Journal of Policy Analysis and Information Systems*, Vol. 4:219–243, 1980.

[MM00] H. Mannila and C. Meek. Global Partial Orders from Sequential Data. In *Proc. KDD'00*, pages 161–168, 2000.

[MO04] S.C. Madeira and A.L. Oliveira. Biclustering algorithms for biological data analysis: A survey. *IEEE/ACM TCBB*, 1:24–45, 2004.

[MPN⁺99] E. M. Marcott, M. Pellegrini, H. L. Ng, D. W. Rice, T. O. Yeates, and D. Eisenberg. Detecting protein function and protein-protein interactions. *Science*, 285:751–753, 1999.

[MSOI⁺02] R. Milo, S. Shen-Orr, S. Itzkovitz, N. Kashtan, D. Chklovskii, and U. Alon. Network motifs: Simple building blocks of complex networks. *Science*, 298:824–827, 2002.

[Mur03] T. Murata. Graph mining approaches for the discovery of web communities. *Proceedings of the International Workshop on Mining Graphs, Trees and Sequences*, pages 79–82, 2003.

[OFD⁺99] R. Overbeek, M. Fonstein, M. Dsouza, G. D. Pusch, and N. Maltsev. The use of gene clusters to infer functional coupling. *Proc. Natl. Acad. Sci. USA*, 96(6):2896–2901, 1999.

[OFG00] H. Ogata, W. Fujibuchi, and S. Goto. A heuristic graph comparison algorithm and its application to detect functionally related enzyme clusters. *Nucleic Acids Res*, 28:4021–4028, 2000.

[Par66] R. J. Parikh. On context-free languages. *J. Assoc. Comp. Mach.*, 13:570–581, 1966.

[Par98] L. Parida. A uniform framework for ordered restriction map problems. *Journal of Computational Biology*, 5(4):725–739, 1998.

[Par99] L. Parida. On the approximability of physical map problems using single molecule methods. *Procceedings of Discrete Mathematics and Theoretical Computer Science (DMTCS 99)*, pages 310–328, 1999.

[Par06] L. Parida. A PQ framework for reconstructions of common ancestors & phylogeny. In *RECOMB Satellite Workshop on Comparative Genomics, LNBI*, volume 4205, pages 141–155, 2006.

[Par07a] Laxmi Parida. Discovering topological motifs using a compact notation. *Journal of Computational Biology*, 14(3):46–69, 2007.

[Par07b] Laxmi Parida. Gapped permutation pattern discovery for gene order comparisons. 14(1):46–56, 2007.

[PBCB99] A.G. Pederson, P. Baldi, Y. Chauvin, and S. Brunak. The biology of eukaryotic promoter prediction: A review. *Computers and Chemistry*, 23:191–207, 1999.

[PCGS05] N. Pisanti, M. Crochemore, R. Grossi, and M.-F. Sagot. Bases of motifs for generating repeated patterns with wild cards. *IEEE/ACM Transaction on Computational Biology and Bioinformatics*, 2(1):40–50, 2005.

[PG99] L. Parida and D. Geiger. Mass estimation of DNA molecules & extraction of ordered restriction maps in optical mapping imagery. *Algorithmica*, (2/3):295–310, 1999.

[PR05]   L. Parida and N. Ramakrishnan. Redescription Mining: Structure Theory and Algorithms. In *Proc. AAAI'05*, pages 837–844, July 2005.

[PRF⁺00] L. Parida, I. Rigoutsos, A. Floratos, D. Platt, and Y. Gao. Pattern discovery on character sets and real-valued data: Linear bound on irredundant motifs and an efficient polynomial time algorithm. In *Proceedings of the eleventh ACM-SIAM Symposium on Discrete Algorithms (SODA)*, pages 297–308. ACM Press, 2000.

[PRP03] Alkes Price, Sriram Ramabhadran, and Pavel Pevzner. Finding subtle motifs by branching from sample strings. In *Bioinformatics*, number 1, pages 149–155, 2003.

[Prü18] H. Prüfer. Neuer beweis eines satzes über permutationen. *Arch. Math. Phys*, 27:742–744, 1918.

[PS00]   P. A. Pevzner and S.-H. Sze. Combinatorial approaches to finding subtle signals in DNA sequences. In *Proceedings of the Eighth International Conference on Intelligent Systems for Molecular Biology*, pages 269–278. AAAI Press, 2000.

[PSC01] Y. Pilpel, P. Sudarsanam, and G.M. Church. Identifying regulatory networks by combinatorial analysis of promoter elements. *Nature Genetics*, 29:153–159, Oct 2001.

[PZ05]   L. Parida and R. Zhou. Combinatorial pattern discovery approach for the folding trajectory analysis of a β-hairpin. *PLoS Computational Biology*, 1(1), 2005.

[RBH05] S. Rajasekaran, S. Balla, and C.-H Huang. Exact algorithms for planted motif problems. *Journal of Computational Biology*, 12(8):1117–1128, 2005.

[RKM⁺04] N. Ramakrishnan, D. Kumar, B. Mishra, M. Potts, and R.F. Helm. Turning CARTwheels: An Alternating Algorithm for Mining Redescriptions. In *Proc. KDD'04*, pages 266–275, Aug 2004.

[RP07]   Naren Ramakrishnan and Laxmi Parida. Modeling the combinatorial control of transcription using partial order motifs and their redescriptions. *Preprint*, 2007.

[RSW04] A. Reményi, H.R. Schöler, and M. Wilmanns. Combinatorial control of gene expression. *Nature Structural and Molecular Biology*, 11(4):812–815, Sep 2004.

[Sag98] M. F. Sagot. Spelling approximate repeated or common motifs using a suffix tree. *Latin 98: Theoretical Informatics, Lecture Notes in Computer Science*, 1380:111–127, 1998.

[SBH⁺01] I. Simon, J. Barnett, N. Hannett, C. T. Harbison, N. J. Rinaldi, T. L. Volkert, J. J. Wyrick, J. Zeitlinger, D. K. Gifford, T. S. Jaakkola, and R. A. Young. Serial regulation of transcriptional regulators in the yeast cell cycle. *Cell*, 106:697–708, 2001.

[SCH⁺97] A. Samad, W. W. Cai, X. Hu, B. Irvin, J. Jing, J. Reed, X. Meng, J. Huang, E. Huff, B. Porter, A. Shenker, T. Anantharaman, B. Mishra, V. Clarke, E. Dimalanta, J. Edington, C. Hiort, R. Rabbah, J. Skiadas, and D. Schwartz. Mapping the genome one molecule at a time optical mapping. *Nature*, 378:516–517, 1997.

[SLBH00] B. Snel, G Lehmann, P Bork, and M A Huynen. A web-server to retrieve and display repeatedly occurring neighbourhood of a gene. *Nucleic Acids Research*, 28(18):3443–3444, 2000.

[SMA⁺97] J. L. Siefert, K. A. Martin, F. Abdi, W. R. Widger, and G. E. Fox. Conserved gene clusters in bacterial genomes provide further support for the primacy of RNA. *J. Mol. Evol.*, 45:467–472, 1997.

[SOMMA02] S.S. Shen-Orr, R. Milo, S. Mangan, and U. Alon. Network motifs in the transcriptional regulation network of Escherichia coli. *Nature Genetics*, 31:64–68, 2002.

[SS04] T. Schmidt and J. Stoye. Quadratic time algorithms for finding common intervals in two and more sequences. *CPM*, LNCS 3109:347–358, 2004.

[SS05] E. Segal and R. Sharan. A discriminative model for identifying spatial cis-regulatory modules. *JCB*, 12(6):822–834, 2005.

[SSDZ05] A.D. Smith, P. Sumazin, D. Das, and M.Q. Zhang. Mining chip-chip data for transcription factor and cofactor binding sites. *Bioinformatics*, 21 (Suppl1):403–412, 2005.

[SSR⁺03] E. Segal, M. Shapira, A. Regev, D. Pe'er, D. Botstein, D. Koller, and N. Friedman. Module networks: Identifying regulatory modules and their condition-specific regulators from gene expression data. *Nature Genetics*, 34(2):166–176, June 2003.

[Sto88] G. D. Stormo. Computer methods for analyzing sequence recognition of nucleic acids. *Annual Review of Biophysics and Biophysical Chemistry*, 17:241–263, 1988.

[TCOV97] J. Tamames, G. Casari, C. Ouzounis, and A. Valencia. Conserved clusters of functionally related genes in two bacterial genomes. *J. Mol. Evol.*, 44:66–73, 1997.

[TK98]   K. Tomii and M. Kanehisa. A comparative analysis of ABC transporters in complete microbial genomes. *Genome Res,* 8:1048–1059, 1998.

[TKT+94]   S. Taguchi, S. Kojima, M. Terabe, K. I. Miura, and H. Momose. Comparative studies on primary structures and inhibitory properties of subtilisintrypsin inhibitors from streptomyces. *Eur J. Biochem.,* 220:911–918, 1994.

[TLB+05]   Martin Tompa, Nan Li, Timothy L. Bailey, George M. Church, Bart De Moor, Eleazar Eskin, Alexander V. Favorov, Martin C. Frith, Yutao Fu, W. James Kent, Vsevolod J. Makeev, Andrei A. Mironov, William Stafford Noble1, Giulio Pavesi, Graziano Pesole, Mireille Rgnier, Nicolas Simonis, Saurabh Sinha, Gert Thijs, Jacques van Helden, Mathias Vandenbogaert, Zhiping Weng, Christopher Workman, Chun Ye, and Zhou Zhu. Assessing computational tools for the discovery of transcription factor binding sites. *Nature Biotechnology,* 23:137–144, 2005.

[UY00]   T. Uno and M. Yagiura. Fast algorithms to enumerate all common intervals of two permutations. *Algorithmica,* 26(2):290–309, 2000.

[VCP+95]   A. Volbeda, M. H. Charon, C. Piras, E. C. Hatchikian, M. Frey, and J. C. Fontecilla-Camps. Crystal structure of the nickel-iron hydrogenase from *desulfovibrio gigas. Nature,* 373:580–587, 1995.

[Vit67]   Andrew J. Viterbi. Error bounds for convolutional codes and an asymptotically optimum decoding algorithm. In *IEEE Transactions on Information Theory,* volume 13(2), pages 260–269, 1967.

[VPPP00]   R.E. Valdes-Perez, V. Pericliev, and F. Pereira. Concise, intelligible, and approximate profiling of multiple classes. *International Journal of Human-Computer Studies,* Vol. 53(3):411–436, 2000.

[WAG84]   M.S. Waterman, R. Aratia, and D.J. Galas. Pattern recognition in several sequences: Consensus and alignment. *Bulletin of Mathematical Biology,* 46(4):515–527, 1984.

[Wat95]   M.S. Waterman. *An Introduction to Computational Biology: Maps, Sequences and Genomes.* Chapman Hall, 1995.

[WMIG97]   H. Watanabe, H. Mori, T. Itoh, and T. Gojobori. Genome plasticity as a paradigm of eubacteria evolution. *J. Mol. Evol.,* 44:S57–S64, 1997.

[WOB03] S. Wuchty, Z.N. Oltvai, and A-L Barabasi. Evolutionary conservation of motif constituents in the yeast protein interactin networks. *Nature Genetics*, 35(2):176–179, 2003.

[Zak04] M.J. Zaki. Mining non-redundant association rules. *DMKD*, 9(3):223–248, 2004.

[ZKW+05] Lan V Zhang, Oliver D King, Sharyl L Wong, Debra S Goldberg, Amy HY Tong, Guillaume Lesage, Brenda Andrews, Howard Bussey, Charles Boone, and Frederick P Roth. Motifs, themes and thematic maps of an integrated *saccharomyces cerevisiae* interaction network. *Journal of Biology*, Vol. 4(2), 2005.

[ZPKM07] Ruhong Zhou, Laxmi Parida, Kush Kapila, and Sudhir Mudur. PROTERAN: Animated terrain evolution for visual analysis of patterns in protein folding trajectory. *Bioinformatics*, 23(1):99–106, 2007.

[ZZ99] J. Zhu and M.Q. Zhang. SCPD: A promoter database of the yeast *saccharomyces cerevisiae*. *Bioinformatics*, 15:607–611, 1999.

Index